History of Artificial Cold, Scientific, Technological and Cultural Issues

BOSTON STUDIES IN THE PHILOSOPHY AND HISTORY OF SCIENCE

VOLUME 299

For further volumes:
http://www.springer.com/series/5710

Kostas Gavroglu
Editor

History of Artificial Cold, Scientific, Technological and Cultural Issues

Springer

Editor
Kostas Gavroglu
Department of History and Philosophy of Science
University of Athens
Athens, Greece

ISSN 0068-0346
ISBN 978-94-007-7198-7 ISBN 978-94-007-7199-4 (eBook)
DOI 10.1007/978-94-007-7199-4
Springer Dordrecht Heidelberg New York London

Library of Congress Control Number: 2013951643

Printed on acid-free paper

Springer is part of Springer Science+Business Media (www.springer.com)

Foreword

This volume attempts to bring together recent scholarship on the different aspects of the history of artificial cold. Such a history involves history of science and technology, business history, issues in gender studies and issues in cultural anthropology. The physical phenomena of the very cold, the large scale production and utilization of cold, the frozen food industry and the dramatic changes brought about by the home refrigerator are all part of the saga of artificial cold. There are indications that since 1000 BC the Chinese had developed the techniques for storing and preserving natural ice and since then people in many other localities devised analogous methods. It is, however, during the late nineteenth century, and most importantly the twentieth century, that artificially produced cold revealed deeply intriguing phenomena, it became an indispensable part of our lives, created industries with global activities, induced trade worth billions, and brought about a new commodity that of frozen foodstuffs.

The book includes essays which analyse a number of questions in three subject of cold categories: in the history of low temperature physics, in the large scale production and in the consumption of cold. The main reasons for including articles from all three subjects is to display the many sidedness of the activities related to the production, preservation, transportation, distribution and consumption of cold *and* to underline the parallel "lives" of the developments in the understanding of the phenomena in the very cold and those related to the large scale utilization of cold. An attempt is made to bring to surface some of the relevant historiographical issues for a social and cultural history of artificial cold.

I would like to thank Ms. Antonia Pavli, a doctoral student in history of technology, for her immense help with the editorial work and Ms. Lucy Fleet, associate editor for *Philosophy and History of Science* at Springer Publishers for being so helpful and patient throughout the preparation of this volume.

My own interest in the history of artificial cold owes much to the work we had done with Yorgos Goudaroulis (1945–1996). His early death deprived the community of historians and philosophers of science and technology in Greece of a teacher and a uniquely perceptive scholar. We were, also, close friends, and this present volume is an acknowledgment to his memory.

Department of History and Philosophy of Science
University of Athens, Greece
Kostas Gavroglu

Contents

About the Authors

Sébastien Balibar graduated at the Ecole Normale Superieure (ENS) (Paris, 1976) after undergraduate studies at the Ecole Polytechnique (Paris), He is presently a CNRS Directeur de Recherche at the ENS in Paris and a Member of the Academie des Sciences (France). He received the Fritz London memorial award in 2005. He has lectured in Columbus, Konstanz, Kyoto, and Harvard University. His main scientific contributions are in the fields of quantum fluids and solids, crystal surfaces, wetting, phase transitions, cavitation in simple and complex liquids, supersolidity and plasticity in quantum crystals. He published more than 200 scientific articles and gave about 100 invited communications in international conferences. He wrote two books: *La Pomme et l'atome* Odile (Jacob, 2005) [*The atom and the Apple*, Princeton University Press 2008) and *Je casse de l'eau* (Le Pommier, 2008).

Christiana Christopoulou received her D.Phil. in 2008 from the Department of History and Philosophy of Science of the University of Athens. She has taught History and Philosophy of Science at the Hellenic Open University and currently is a member of the Greek Society for the Study and Dissemination of History of Science and Technology in Greece ("EMDIET)". She also holds the position of Managing Director at the Greek publishing house Ekdotike Athenon S.A. She is a co-editor of *Science and Technology, Historical and Historiographical Studies* (Ekdotike Athenon, 2013). She is currently working on the experimental history of cold in the 17th century.

Guillaume de Syon teaches European history and the history of technology at Albright College in Reading, PA, USA, and is a Visiting Scholar in History at Franklin & Marshall College. He is a past contributing editor to the *Collected Papers of Albert Einstein* (volumes 8A and 8B, Princeton). His books include *Zeppelin! Germany and the Airship 1900–1939* (Johns Hopkins,) and *Science and Technology in Modern European Life* (Greenwood.) In 2011, he was a visiting research fellow at the Graduate Institute of International Studies in Geneva, Switzerland. His current research investigates the public understanding of modern technology as reflected in media, advertising and popular culture.

Hans-Liudger Dienel is Professor at the Technische Universität Berlin in the Chair of Industrial Relations, Technology and Society. He is, also, heading the Center for Technology and Society of TU Berlin (www.ztg.tu-berlin.de). Trained as historian, sociologist and mechanical engineer (thermodynamics), Dienel wrote a couple of books on university-industry relationship in the nineteenth and twentieth century in the field of refrigeration, cryogenics, thermodynamics and chemical engineering, including a history of the Linde Company. In recent years, Dienel's research focus has shifted to the history of transport and mobility. Currently, he is president of the International Association for the History of Transport, Traffic and Mobility (www.t2m.org). Dienel is editorial board member of a number of journals (*Forum Qualitative Social Research, Innovation European Journal of Social Science Research, Journal of Transport History*).

Terje Finstad is a postdoctoral researcher at the Norwegian University of Science and Technology (NTNU), Department of Interdisciplinary Studies of Culture. His research interests include science and technology studies, the history of food and environmental history. He wrote his Ph.D. thesis on the building of a frozen food chain in Norway 1920–1970. Publications include: "Cool alliances. Freezers, frozen fish and the shaping of industry-retail relations in Norway, 1950–1960", in Ralph Jessen og Lydia Nembach-Langer (eds.), *Transformations of Retailing in Europe after 1945* (Ashgate, 2012); "Familiarizing Food. Frozen Food Chains and Consumer Trust, Norway 1940–1970", in *Food and Foodways* (forthcoming 2013).

Kostas Gavroglu is a Professor in History of Science at the Department of History and Philosophy of Science at the University of Athens, Greece. His main research interests are the history of physics and chemistry since the mid-nineteenth century, and especially, the formation of new disciplinary fields. His latest book is with Ana Simoes *Neither Physics nor Chemistry, The History of Quantum Chemistry* (MIT Press, 1911). He has, also, been active with the group STEP (Science and Technology in the European Periphery).

Christian Joas is an Assistant Professor in the History of Science at Ludwig Maximilians University Munich. After his Ph.D. in Theoretical Physics at Freie University Berlin, he was a member of the "Project on the History and Foundations of Quantum Physics" in Berlin, first as a Postdoctoral Research Fellow at the Max Planck Institute for the History of Science in 2007–2009, and then as a Research Scholar at the Fritz Haber Institute of the Max Planck Society in 2009–2012. His research focuses on the history of modern physics. His current interests are the history of quantum mechanics as well as the history of processes of knowledge transfer (i) between the emerging subdisciplines of physics, such as solid-state, nuclear and particle physics, and (ii) between physics, chemistry, and biology.

Johannes Knolle studied Physics and Philosophy at Technical University Dresden and as a Fulbright Scholar at Boston University. Before starting his Ph.D. on high temperature superconductivity in theoretical condensed-matter physics at the Max Planck Institute for the Physics of Complex Systems, Dresden, he was a Visiting

Scholar at the Max Planck Institute for the History of Science, Berlin, where he worked on the history of superconductivity. He is mostly interested in knowledge transfer between disciplines, as well as in the interplay of technological developments and basic scientific discoveries.

Faidra Papanelopoulou is a Lecturer in History of Physics at the Department of History and Philosophy of Science of the University of Athens, Greece. She received her D.Phil. in 2004 from the Modern History Faculty of the University of Oxford, and held a post-doctoral position at the Centre A. Koyré (EHESS), and a Marie Curie post-doctoral fellowship at the Fondation Maison des Sciences de l'Homme in Paris. She also teaches History and Philosophy of Science at the Hellenic Open University. She is currently working on the history of artificial cold and low temperature research in the late-nineteenth and twentieth centuries.

Jonathan Rees is Professor of History at Colorado State University - Pueblo. He is the author of *Industrialization and the Transformation of American Life: A Brief Introduction* (M.E. Sharpe, 2012) and *Refrigeration Nation: A History of Ice Appliances and Enterprise in America* (The Johns Hopkins University Press, 2013). His essays have appeared in *Inside Higher Education, Slate* and *The Atlantic*.

Simón Reif-Acherman was born in Palmira, Colombia, the 14th of January, 1958. He studied Chemical Engineering at the Universidad del Valle, where he graduated in 1980. Since this date he belongs to the Staff of the School of Chemical Engineering of the same university. Besides the teaching of courses on material and energy balances in process industries, he had devoted from 10 years ago to studies in history of science and technology. He has published a dozen of papers on international scientific journals on subjects that include early refrigeration techniques, history of theories and scientific instruments, and biographical profiles, among others.

J.S. Rowlinson is a chemist who was educated at Oxford and held posts at Manchester University and Imperial College of Science and Technology in London before being elected Dr Lee's Professor of Chemistry at Oxford in 1973. He held concurrently posts in Cornell University. He retired in 1993 and now works in the history of physics and chemistry in the nineteenth and twentieth centuries. He is a Fellow of the Royal Society and of the Royal Academy of Engineering. He received the Edelstein Award from the American Chemical Society in 2008 for his contributions to the history of chemistry.

Ulrike Thoms has studied history, German and communication science. Her main interest in the history of consumption (with a special focus on food and pharmaceuticals), the history of medicine, the history of the body and the history of marketing. Her work has combined different approaches, thereby integrating perspectives from the history of knowledge and science as well as from business history. She advanced this interdisciplinary approach in her recent project in the history of pharmaceutical marketing. Most recent publications are: "Standardizing and Marketing Drugs in the 20th Century." Special issue of *History and Technology*, edited together with

Jean-Paul Gaudilliére, "Consuming Bodies. The Commodification and Technification of Slenderness in the Twentieth Century" in *Globalising Beauty. Aesthetics in the 20th Century*, ed. Hartmut Berghoff and Thomas Kühne, 41-59, New York: Palgrave Macmillan 2012.

Dirk van Delft (1951) studied physics at Leiden University. From 1992 to 2006 he was senior science editor at the daily newspaper *NRC Handelsblad*. Since 2006 he is Director of Museum Boerhaave, the Dutch national museum for the history of science and medicine. He holds a 1 day a week position as special professor in the Heritage of the Sciences, located at Leiden Observatory. His Ph.D. work involved a biography of Heike Kamerlingh Onnes (Leiden University, 2005). His research is focussed on the Leiden cryogenic laboratory 1880–1960, the international temperature scale, and the Leiden school of instrument makers.

Georges Waysand is honorary director of the Laboratoire Souterrain Bas Bruit (Low Noise Underground Lab) of Rustrel-Pays d'Apt near Avignon (France) which he created from a decommisionned lauching control room of nuclear missiles. He has spent most of his scientific career as an experimenter in superconductivity (materials research and superconducting particle detectors) in the Orsay Group in Superconductivity and later in the Groupe de Physique des Solides, Universities of Paris 6 and 7. In parallel, he has always been involved in science and society activities. He is presently interested in extreme low noise magnetic measurements of Earth-Ionosphere interactions with superconducting magnetometers (SQUID) and is in charge of the " Society in Science" working group of Euroscience. With Jean Matricon, he is co-author of " La guerre du froid" Cold Wars a History of Superconductivity (Editions du Seuil Paris 1994, Rutgers University Press 2003).

Chapter 1
Historiographical Issues in the History of Cold

Kostas Gavroglu

1.1 The Bad, the Good and the Tolerant

"Corpses frozen for future rebirth" ran the title of an article describing the gory business of promising immortality.[1] Ample supply of liquid nitrogen and some technical issues in the freezing process itself, were sufficient to do the trick. The presence of cold in our cultural ethos, has not been exclusively associated with promises of deathlessness. In Dante's *Inferno*, Satan is portrayed as a giant beast, half of his body immersed in ice at the center of Hell. Satan beats his wings, and the cold wind is a continuous reminder that there is not an instant of reprieve for the sinners condemned in the Ninth Circle, where the worst of sinners, those who betrayed their benefactors, are in perpetual suffering. The dominance of cold, however, is not confined, only, to one of the emblematic works of literature of the western world. The Cold War dominated politics for almost half a century (1947–1991) and after it was over what dominated the discussions among historians and political scientists was the attempts to understand what actually happened during those politically scorching years. Cold fusion – the possibility to have endless energy from the light nuclei with almost no energy input –, a fantasy entertained by few, caught the imagination of millions for a short moment when it was announced in 1989. Alas, compared with the eternity of the Ninth Circle and the intensity of the

---

This paper is intended to be an essay attempting to present an overall view of various aspects of low temperature physics and of the large scale production and utilization of artificial cold, aiming to draw some conclusions as to the relevant historiographical issues involved in the history of artificial cold. Hence, the bibliography is not intended to be exhaustive, but includes works where one can find a much more detailed analysis of the topics being discussed as well as many more references to the research of other scholars.

[1] *National Geographic News*, March 18, 2005.

K. Gavroglu (✉)
Department of History and Philosophy of Science, University of Athens, Athens, Greece
e-mail: kgavro@phs.uoa.gr

K. Gavroglu (ed.), *History of Artificial Cold, Scientific, Technological and Cultural Issues*, Boston Studies in the Philosophy and History of Science 299,
DOI 10.1007/978-94-007-7199-4_1,

panic inflicted to the West by the Soviet Union, cold fusion had a minuscule life-time. Materials exhibiting zero electrical resistance would have been at the very heart of another kind of fantasy where, for example, electrical current would be transmitted with almost no losses, trains would travel on these materials with almost no need to supply energy to keep them running. The discovery of superconductivity in 1911 gave an end to these fantasies, since the cost at maintaining the temperatures required in order to make some materials superconducting was unimaginably high. Any practical large scale commercial utilization of such materials appeared to be an impossible dream, until 1986 when such hopes were rekindled by a new generation of materials which were found to exhibit superconducting properties in very high temperatures, still low enough to have a practical use. The genie of a superconductor at room temperatures, however, was out of the bottle.

Artificial cold has brought about dramatic changes in our understanding of a number of physical phenomena and it has deeply affected our everyday lives. From the (re)conceptualization of the notion of a fluid in quantum mechanics to the large scale production of cold to the unparalleled changes in food production and eating habits, the production and harnessing of cold, has probed nature and shaped society. Perhaps, the only situation with "analogous" dramatic changes in the habits and practices concerning food, was the post seventeenth century period of the widespread use of salted fish and meat, easing to a previously unimagined degree the long transatlantic voyages, the military operations and the feeding of the people in the cities which were becoming densely populated.

Though cold has been a rather unappealing state of affairs in our Western culture, artificially produced cold has always been something sought for and, often, a very welcome companion. From the time some poor folk brought ice from the mountains wrapped inside hey and long dried shoots so that the Roman Emperors could enjoy their wine chilled in the summer, to the extensive use of home refrigerators, and from the attempts to achieve the lowering of temperature by adding salt – a technique known to many people of different cultures for many centuries – to the observation of the Bose-Einstein condensates at the end of the twentieth century where an aggregate of atoms behaves as if it is one huge atom, the uses and wonders of low temperatures have been publicly displayed and their mysteries privately pursued. And over the centuries the princes ceased to be the sole beneficiaries of tasting cold. Masses and masses of people have, also, experienced the great varieties of the joys of cold – be it air-conditioned rooms or ice cream which became one of the products symbolizing twentieth century affluence.

The liquefaction of gases, the intriguing phenomena discovered in the very cold temperatures, the developments in thermometer construction, the large scale production and utilization of cold in hospitals, in the preservation of food, in the cooling of buildings, the dramatic effects of the home refrigerator in shaping everyday life and the reflections of cold in language and the arts, form a unified whole of the prevailing dominance of artificial cold in our lives. The history of artificial cold involves techniques of producing it, means of preserving it, ways of transporting and distributing it and rules for consuming it. But what "exactly" is cold? Is it something which can be perceived independent of what it cools? Is what we produce, cold or cold products? These are, indeed, rather interesting questions, but in what follows, there will not be so much emphasis on the character of what "exactly" constitutes cold, a quality which played such a central role in the Aristotelian world view, nor will there be an attempt to understand the ontological status of cold *itself* independent of the ways to produce, preserve, transport, distribute and consume it (for an intriguing discussion of the philosophical implications of the ultracold see Hacking 2006). The aim of what follows is to try and bring out a number of issues related to the social and cultural history of cold, through the ways the life-cycle (production, preservation, transportation, distribution, consumption) of cold is mapped onto the "equivalent" life cycle of the cooling agents and of the products being cooled by these agents, giving rise to a all sorts of social and cultural correlations.

Though research in low temperature laboratories was a relatively low budget activity, its social presence and industrial effects has been from the very first years an increasingly profitable business. The ingenious achievements at reaching low temperatures and the large scale production of liquefied gases or the construction of ever larger cold spaces were processes taking place in tandem. Storage rooms – in warehouses, train wagons, ships and, later, trucks – at varying degrees of cold were extensively used for the preservation and transportation of goods in cities and across continents. Society at large, reluctantly at first and enthusiastically soon afterwards, welcomed the large scale production of cold and the artifacts related to cold, which in turn, created all kinds of needs. Big firms like the *Linde Company* founded in 1879 in Germany, *British Oxygen* founded in 1886 in England and *Air Liquide* founded in 1902 in France, came to symbolize the dominance of the three European superpowers in industrial cold production as well (Dienel 2004).

The First International Congress of Refrigeration in Paris in 1908 and the subsequent establishment of the International Association of Refrigeration in 1909 (which was dissolved right after the First World War and re-established in 1920) have been very closely related to the developments in the refrigeration technologies and the expansion of the international, national and local markets for products whose marketability became possible by the refrigeration technologies. The founding of the Association, heralded one of the very first international bodies where representatives of various countries tried to draw the directions of future scientific pursuits, and agree upon rules which will standardise processes for the production and transportation of frozen food. The commercialization of cold and all the possibilities it offered was the underlying philosophy of the agenda of the meeting. Of course, the internationalism which was so popular in the first decade of the twentieth century was interlaced with all kinds of nationalistic antagonisms, something that left its mark in the deliberations of the Association. The First International Congress was a huge meeting, with more than 5,000 delegates from 42 countries which took place at the Sorbonne, in order to discuss problems concerning developments in the applications of artificial cold in agriculture, public slaughterhouses, and the preservation of food and be informed about the latest scientific developments. The discussions of the meeting were centered around the attainment of low temperatures and their general effects, the refrigerating materials, the application of cold to alimentation, the application of cold to other industries, the application of cold to commerce and transports and issues concerning national legislation.

The founding of the Association was marked by the dominant presence of businessmen, industrialists and politicians and by some of the best known figures in the scientific world, such as Heike Kamerlingh Onnes (1853–1926), Director of the Cryogenic Laboratory of the University of Leiden and Charles E. Guillaume, Director of the International Bureau of Standards, both receiving Nobel Prizes – Kamerlingh Onnes in 1913, Guillaume in 1920. The Association was soon able to define a framework where many individuals, national associations, commercial companies and governments would come to agreements on a number of technical, economic and legislative issues, despite many and strong antagonisms among the national representatives. The Association set up many subcommittees where rules for the proper cooling and transportation of fish, meat, poultry and vegetables were agreed upon and they slowly became the standard processes used by the companies which were in the business of trading cold. Standardization of equipments, of units of measurement, of common nomenclature and, most importantly, of the procedures to be used in the freezing and transportation of foodstuffs, became significant aspects of the deliberations and of the successes of the Association.

The International Association of Refrigeration created a milieu which was able to accommodate the international spirit of science and contain the intense national antagonisms brought about because of the huge prospects for the commercial utilization of cold. Standardization became absolutely indispensable for the phenomenally increasing commercial uses of cold and its products. The Association became the forum where the latest scientific developments would, also, be publicized, thus achieving a rather harmonious coexistence of the scientists together with all those who viewed the Association as a place to help promote the cold industry. In fact, the active involvement in the initial stages of the Association of two persons with unmistaken prestige, that of Kamerlingh Onnes and Guillaume, who convinced the businessmen and industrialists that developments in the science of cold would be of great benefit to their commercial undertakings, was surely of importance in creating such a consensual atmosphere in the Association. The official mandate of the IAR was rather specific: to centralise all information on the production and uses of artificial cold provided by the various national associations and other members of the Association; to encourage the general progress of the industries of cold; to find the best solutions to scientific, technical and industrial questions and the best administrative measures concerning the transportation of perishable produce; to centralise all information useful for the improvement of the existing legislation on transportation and national and international exchanges of refrigerated and frozen produce; to encourage the popularization and the development of the science of cold; to establish relations of solidarity between the different national groups and coordinate the activities of its members (Papanelopoulou 2009; van Delft 2007).

Some like it hot, others claim that making fire heralded the beginnings of human civilization, certain people depict our end through a catastrophe by heat, and sober commentators of our future insist that what will brings us down is neither artificial cold nor artificial heat, but natural cold through the cooling of the sun. Though artificial heat and artificial cold have "parallel" and relatively independent histories, in this article I will attempt to raise a number of historiographical issues for a cultural and social history of artificial cold.

1.1.1 The Physics of the Very Cold

For the master of experimenters, Robert Boyle, cold and experiments with cold were the only subject he found "the most difficult" to work at. Though there had been very systematic experimentation with cold from the time of Robert Boyle and throughout the eighteenth and nineteenth centuries, it was the developments at the end of the nineteenth century and the period up to the end of the 1930s which were decisive for the becoming of both low temperature physics as well as of cold technology. Many people were involved in the study of artificial cold: some liquefied gases, some developed ingenious methods for lowering temperatures, some investigated the properties of materials in low temperatures, some became directors of laboratories specialised in the study of cold, some started industrial firms for the large scale production of cold and some produced and traded frozen food.

"Cold, colder, coldest, Onnes..." was the beginning of an article in *Everybody's Magazine* in September 1915. The article was about the work in low temperature physics at the University of Leiden and was referring to the Director of the Physical Laboratory there, Heike Kamerlingh Onnes. "Mr. Freezer" as some Dutch cartoonists called him, had received the Nobel Prize for physics in 1913 for the liquefaction of helium and he was in charge of the only laboratory that could produce liquid helium and perform measurements of various physical parameters at the lowest of temperatures – barely a couple of degrees above absolute zero. With the liquefaction of the last permanent gas, helium, in 1908, Heike Kamerlingh Onnes' Cryogenic Laboratory at the University of Leiden became the "coldest spot on earth." It hosted many scientists from many different countries and provided them with the opportunity to experiment with this most precious liquid. In 1923, John McLennan from the University of Toronto brought a quantity of helium gas to Kamerlingh Onnes who showed him all the technical details of the liquefier, and, thus, McLennan's laboratory in Toronto became the second place to produce liquid helium (van Delft 2007). In all these activities, the Laboratory had the expert help of graduates from the School of Instrument Makers at the University of Leiden which Kamerlingh, Onnes took the initiative to establish (van Delft 2008; Gavroglu and Goudaroulis 1989, 1991; van Helden 1989).

Almost the same period as the laboratory in Leiden was becoming the low temperature laboratory par excellence, James Dewar was setting up his own laboratory at the Royal Institution, London. Though Dewar managed to have all kinds of fights with his associates, and guarded his place from any external intruders, he did manage to liquefy hydrogen in 1899 and even think he had liquefied helium as well, only to realise that what had happened was the liquefaction of the impurities in the helium gas. The same happened to Kamerlingh Onnes a few weeks before he had actually liquefied helium and who informed Dewar of his "achievement" only to retract it a few days later, saying that he will continue to work with this "most exceptional liquid."[2] Furthermore, Dewar with his aversion to theory but with the amazing dexterity he had developed over the years, was able to perfect and render usable one of the most useful artefacts for conducting low temperature research – the Dewar flask or thermos (Rowlinson 2012).

There are very few instances – if any – in the history of the physical sciences when a newly discovered phenomenon defies even a description in terms of the existing theoretical framework. The discovery of superconductivity in 1911 – of the sudden drop to almost zero of the electrical resistance of some metals at liquid helium temperatures of 4 K – was contradicting the very nature of electrical resistance, the fact that it could not be zero. The observations of the amazingly peculiar properties of liquid helium in the late 1920s and the 1930s, were contradicting the very nature of viscosity. Superconductivity and superfluidity turned out to be pure and rather unique quantum phenomena and brought on a thoroughly new realization about quantum mechanics – that the theory could, indeed, accommodate, macroscopic quantum phenomena.

1.2 Zero Electrical Resistance

In February 1911 Kamerlingh Onnes measured the resistance of platinum, and in April that of pure mercury, at helium temperatures. At 3 °K the resistance of pure mercury fell to 0.0001 times that of solid mercury at 0 °C as extrapolated from the melting point. The temperature at which the resistance first became measurable turned out to be slightly above 4.2 °K. The researchers at the Leiden Cryogenic Laboratory repeated the measurements of the resistance of mercury at helium temperatures during 1912–1913. The work established the sudden drop of resistance and the impotence of impurities (at least in the case of mercury) to hinder disappearance of ordinary resistance. Earlier there were observations of electrical resistance diminishing with falling temperatures, but the abrupt disappearance of the electrical resistance at liquid helium temperatures was a new physical phenomenon, indeed.

Felix Bloch (1905–1983) in 1928 attempted to formulate a quantum mechanical explanation of the phenomenon. He was led to as surprising result: superconductivity could not exist! Bloch used an analogy with ferromagnetism to understand superconductivity. He found that the most stable state of a conductor, in the absence of an external magnetic field, was one with no currents. But since superconductivity was a stable state displaying persistent currents without external fields, it was difficult to see how superconductivity could come about in the first place!

At the beginning of November 1933 there appeared a short letter in *Naturwissenschaften* by Walther Meissner (1882–1974) and Robert Ochsenfeld

---

[2] Kamerlingh Onnes to Dewar, April 7, 1908. James Dewar Papers, The Royal Institution, London.

(1901–1993) which presented strong evidence that, contrary to the expectations of the physicists and their belief of the previous 20 years, a superconductor expelled the magnetic field after the transition to the superconducting state and the magnetic flux became zero. Superconductors were found to be diamagnetic and, hence, superconductivity a reversible phenomenon. Thermodynamics which was earlier used to explain the phenomenon, but with rather arbitrary assumptions, could now be used justifiably.

In 1934 Fritz (1900–1954) and Heinz London (1907–1970) assumed that the diamagnetism was an intrinsic property of an ideal superconductor, and not merely a consequence of perfect conductivity. The Londons, thus, obtained the electrodynamics of a superconductor consistent both with the zero resistance and the Meissner effect. Fritz London, in his discussion of superconductivity in 1936, formulated for the first time the notion of a macroscopic quantum phenomenon. It was believed that ionic masses, being so much larger than the electron mass, could not play an important role in the establishment of the superconductive state. Herbert Frohlich (1905–1991) in 1950 asserted the opposite and using field theoretic methods found that the interaction of the electrons in a metal with the lattice vibrations led to an attraction between the electrons. Experiments verified this assertion. The mass became an important parameter when the motion of the ions was involved, and this, in turn, suggested that superconductivity could be derived from an interaction between the electrons and zero-point vibrations of the lattice. Soon after learning about these results, John Bardeen (1908–1991) was able to show that superconductivity might arise from a new attractive interaction between the electrons and the phonons resulting from lattice vibrations, thus laying the foundations for the electron pair theory. In 1956 Leon Cooper (born in 1930) argued that such an interaction could, in fact, lead to the formation of a bound state of a pair of quasi-particles above the Fermi sea, no matter how weak the interaction. Based on these ideas Bardeen, Cooper and John R. Schrieffer (born in 1931), in 1957 worked out the details of a microscopic account of superconductivity, and shared the Nobel Prize in Physics of 1972 for their theory (Gavroglu 1995; Dahl 1992).

1.3 A Most Peculiar Liquid

If helium managed to confuse everyone until its liquefaction, it guaranteed continuous enchantment to all who experimented with it afterwards. All liquids solidify under their own pressure at low enough temperatures. Helium does not. It can only be solidified under a pressure of 26 atmospheres. The densities and specific heats of all liquids follow a continuous change and increase as the temperature is lowered. Not in the case of helium. These parameters display a maximum at 2.19° absolute and then decrease. The two methods for measuring the viscosity of any liquid – rotating a disk or forcing a liquid through very small capillaries – give identical results. Not so for liquid helium below 2° absolute. When the viscosity of liquid helium is measured with these two methods, the first gives a value that is a million times larger than the second. Finally, all liquids are rather obedient: They can be deposited in open containers, they can be kept in containers with extremely small holes through which they cannot flow, and shining light on them does not make them jump. Liquid helium does not tolerate any such constraints. It goes over open containers, leaks through the smallest capillaries, and is like a fountain when light is shined on it. Liquid helium, below 2.19°, had the characteristics of a superfluid!

No wonder when Lawrence Bragg learned of the properties of liquid helium, he described the situation as resembling Alice in Wonderland!

By arguing, in 1938, that the transition to the superfluid state may be due to a kind of Bose-Einstein condensation mechanism, Fritz London elaborated the possibility of "order" in momentum space and articulated the notion of *macroscopic quantum phenomena.* The peculiar mathematical property derived by Bose and independently by Einstein, turned out to provide an explanation for superfluidity. This approach at the time did not enjoy the success it does today. It hardly had a prediction – only one, in fact – that the isotope of helium, helium-3, for reasons of statistics could not show superfluidity something which in the 1950s was found to be false. Yet, this approach provided us with one of the most radical contributions to the conceptual edifice of quantum mechanics after the early 1930s (Gavroglu 1995).

The second approach was developed in 1941 by Lev Landau (1908–1968) who was working with Petr Kapitza (1894–1984) in Moscow and who already had made significant contributions to quantum theory. Landau developed a theory of quantized hydrodynamics in which the transition to the superfluid state was explained in terms of two different kinds of quanta – rotons and phonons. Despite problems related to the ontological status of rotons, Landau defended the notion of "elementary excitations" and he considered rotons as a special type of such excitations. He firmly asserted that for every slightly excited microscopic system "a conception can be introduced of 'elementary excitations' which describe the 'collective' motion of particles" (Landau 1949: 884). This necessary assumption, according to Landau, was the basis of his microscopic theory, and he thought that every other consideration of the motion of individual atoms in the system of strongly interacting particles was contradicting the first principles of quantum mechanics.

A third approach was proposed by the Hungarian-born physicist Laszlo Tisza (1907–2009) who had been in Kharkov with Landau from 1935 to 1937, and after returning to Hungary was much interested in pursuing the thermodynamic ideas that he had acquired in Landau's group. Thinking that the two measurements of the viscosity of liquid helium were absolutely irreconcilable, Tisza was led to the two-fluid idea (Tisza 1938). Tisza's first step was to examine the concept of viscosity in liquids and gases in view of the discrepancy between the two viscosity measurements, and he concluded that this did not reflect a kinetic coefficient of an unusual value, but the breakdown of the viscosity concept itself. In Tisza's model, helium-II is regarded as a mixture of two (completely interpenetrating) components, the normal and the superfluid. These two components or "fluids" were distinguished by different hydrodynamical behaviors, in addition to the difference in their heat contents.

The phenomenal development of low-temperature physics in the Soviet Union is justifiably tied to the career of Petr Kapitza. An "accidental observation" gave

Kapitza's work an impetus in a totally new direction. Kapitza found that the pressure pulsations transmitted from the helium pipeline in his laboratory into the helium in his capillary caused substantial changes in its thermal conductivity. To study this phenomenon Kapitza devised new experiments, and he was, thus, was able to establish the mechanism of the movement of liquid helium in the capillary tube as a result of a heat current. Kapitza predicted that the thermal conductivity of helium would be strictly normal in the absence of surface phenomena (Kapitza 1965, 1981). Subsequent experiments showed that the entropy of liquid helium flowing through narrow tubes was indeed zero. This had been predicted by both Tisza and London, but Kapitza thought that their schemata could not provide a "rigid theoretical basis" (Kapitza 1965: 638) for his observations.

Landau's formalism led to two different equations for the propagation of sound and hence to two velocities of sound. One was related to the usual velocity of compression, while the other depended strongly on temperature. This was the same phenomenon as Tisza's thermal or temperature waves. Landau named it "second sound." The first attempt to generate and detect second-sound waves was made with acoustic apparatus by A.I. Shalnikov and N.D. Sokolov. They failed, which was interpreted by London whose theory had no such prediction, to mean that Landau's theory was "born refuted" (Brush 1983: 184). Eventually, V. Peshkov was able to demonstrate the existence of standing thermal waves in 1944 for the first time.

By 1956 Richard Feynman (1918–1988) was able present a theory synthesizing the views of London and Landau. Considering all previous theories as phenomenological, he showed that despite the strong forces of interaction between helium atoms which could have undermined the ideal gas approximation by London, did indeed allow the Bose-Einstein condensation. He also showed that some of Landau's assumptions which were rather empirical could be justified quantum mechanically and that the rotons were a kind of quantum mechanical analog of a microscopic vortex ring.

1.4 "Alice in Wonderland"

The process of delineating the disciplinary boundaries of low temperature physics during the end of the nineteenth century and the first two decades of the twentieth, has been concurrent with a rather unique situation in the history of physics. The production of very cold agents (liquid gases) which was a precondition to perform experiments in the very cold, *itself* became an autonomous pursuit. Having sufficient supply of liquid gases became necessary in order to measure the properties of various materials, and this, in turn, gave rise to "two levels of experimentation" – creating cold agents, studying their properties and using them to study the properties of other materials.

The liquefaction of helium became synonymous with the attainment of the lowest temperatures. But such a passive role was not to helium's liking. Slowly it turned into a rather unreliable active agent. Though helium gave to each what was asked of it, it led London and Landau (and to a certain extent Tisza) to different paths, in fact, to antagonistic approaches. Reading its behaviour through their disparate methodological choices and ontological commitments, each thought that helium had fooled the other. And each made strong claims of having harnessed what fooled the other. For some time it appeared that helium had managed to elude them both with its Janus-like behaviour. But helium was eventually trapped by the rules of the very game it set out to play, since later syntheses, especially that of Feynman, provided a reading of Landau's reductionistic approach and London's anti-reductionistic approach not as two incommensurable modes, but rather as two complementary schemata.

Low temperature research has been rather emblematic of creating different experimental and theoretical cultures. Dewar tried every trick through his "brute force" approach, studying properties of materials in liquid oxygen and hydrogen temperatures. Theoretical considerations left him indifferent. This was definitely not the case with Kamerlingh Onnes for whom the theoretical framework developed by Johannes Diderik van der Waals (1837–1923) was decisive in the strategy for liquefying helium. Fritz London elaborated greatly the notion of macroscopic quantum phenomena and accommodated both phenomena of superconductivity as well as superfluidity within this new concept of quantum fluids. Tisza proceeded to a phenomenological explanation by the coexistence of a classical and a non-classical liquid. Landau quantized the liquid and investigated the character of the elementary excitations of this quantization process. Frochlich questioned the expectation that the ionic mass would not be affecting the superconducting properties of various elements and Bardeen, Cooper and Schrieffer proposed the existence of electron pairs. In the establishment of low temperature physics almost all the observed phenomena, the proposed theoretical entities and the implied interpretations, encapsulated the very notion of "unexpectedness."

Each mode of experimental and theoretical practices claimed methodological novelties and novel predictions. Laboratory traditions came into being. Independent of whether some of the "experimenters" did not contribute to theoretical developments and some of the "theoreticians" did not take part in the actual planning and realization of experiments, one remains with the distinct impression that all these traditions and practices aimed at the further conceptual elaboration of quantum theory: macroscopic quantum phenomena, rotons as elementary excitations of liquids, the possibility of actually observing Bose-Einstein condensation which for a long time was considered as simply a mathematical oddity, clarifying the actual nature of fluids as well as the most effective ways to set up a theoretical framework for dealing with liquids and electron pairing. The developments in low temperature physics together with the quantum field theoretic renormalization procedures contributed to the radical reconceptualizations of a number of fundamental issues in quantum theory.

If we are to understand what for a long time appeared to be a problematic symbiosis of liquid helium and quantum mechanics, we will have to understand a peculiarly reciprocal relationship. Liquid helium was as much a probe for quantum mechanics as quantum mechanics was a probe for liquid helium. Quantum mechanics, by providing the framework for the theoretical explanations for liquid helium, itself went through a metamorphosis. And in the course of this metamorphosis, liquid helium forced quantum mechanics to negate one of its basic methodological and historical tenets. Quantum mechanics was developed as a theory of microscopic phenomena and was, at least partly, justified by its classical limits. Liquid helium (and superconductivity) could be understood only when it became possible to talk of macroscopic quantum phenomena. As its properties that appeared to defy hydrodynamic behavior were being accommodated in a redefined notion of a fluid, quantum mechanics could no longer be the framework of only microscopic phenomena but had to allow for the accommodation of macroscopic quantum phenomena as well.

1.5 Large Scale Uses of Cold

There was a counterpart to the magical microcosm of the ultracold. It was the world of the large scale utilization of cold. It was the world of the slaughterhouses, of the breweries, of the mortuaries, of the hotels, of the hospitals, of the ships, of the trains, of the trucks. It was the world of the industries preparing frozen food and of the companies selling them. But above all, it was the world of the house kitchen which became the reference point for all kinds of changes in our eating habits, in our views on the freshness of products, in our shopping mentalities. Since the last years of the nineteenth century, it was becoming progressively apparent that there was much money in the business of consumption of cold. Every decade brought in new innovations which, in turn, gave rise to new products or greatly improved uses of instruments and artefacts, creating new needs, changing attitudes and habits – be it in commercial exchanges, industrial production, public behaviours or private practices (e.g., see Freidberg 2009; Hard 1994; Petrick 2006).

In less than 100 years after William Cullen, in the mid-eighteenth century, was able to make snow by vapourizing water at very low pressure, Jacob Perkins noted the significance of ethyl ether in ice making. The vapour compression refrigeration was developed by Oliver Evans at the beginning of the nineteenth century, but he never constructed an actual refrigerator. Some years later, Jacob Perkins modifying Evans' innovation built a machine that is widely considered as the first refrigerator – he was, surely, the first to file for a legal patent for refrigeration by using the vapour compression method. By 1865 there was a cold storage warehouse in New York and a few years later cold air machines were constructed in ships which were able to transport meat, first, across the Atlantic and, then, from Australia to London (Thevenot 1979).

A person who was at ease with academia and at the same time a particularly resourceful inventor and entrepreneur, was Carl von Linde (1842–1934). He was appointed a lecturer at the Technical University of Munich, then left the post to establish his company and later on, in 1872, returned to academia as a professor of mechanical engineering. He used thermodynamics in the development of a number of refrigeration cycles. In 1875 Linde was able to construct a compression refrigerating machine using methyl ether. His innovative ammonia compressors, in a few years became the standard for the production of cold in many countries. The development of the vapour compression process gave a strong boost to the large scale production of cold, leading, in effect, to Linde's ammonia machines (1876) and the various carbonic acid machines. By the end of the nineteenth century the Linde Company had sold about 2,600 such machines: 1,406 were used in breweries, 403 for cooling land stores for meat and provisions, 204 for cooling ships' holds for transportation of meat and food, 220 for ice-making, 73 in dairies for butter making, 64 in chemical factories, 15 in sugar refining, 15 in candle making, the rest for various purposes. By the end of the first decade of the twentieth century, Linde was able to develop techniques for obtaining pure oxygen and nitrogen.

Though Linde is cited as being the first to develop a compact refrigerator (and he had, in fact, received a German patent for that in 1877), the main improvements in refrigerators for home and domestic use were made by some American engineers. Fred W. Wolf placed the refrigerating machines on top of ice boxes. The design was greatly improved by Nathaniel B. Wales and Alfred Mellowes both American engineers. In 1918, William C. Durant bought Mellowes' small company and established the Frigidaire Company which started to mass produce refrigerators and together with Kelvinator – which was established in Detroit in 1914 – had, by the early 1920s, more than three quarters of the market for electric refrigerators. Other companies with a large share in the market for home refrigerators were Electrolux, founded in Sweden in 1918 and General Electric which had been established much earlier but expanded into the manufacture of electric refrigerators in the 1920s. Just before the Wall Street crash of 1929, two developments would have dramatic repercussions for the production of cold. Thomas Midgley made use of CFCs (Freons) and Clarence Birdseye developed the method of fast freezing for foodstuffs which it was found that it preserved quality. By the end of the twentieth century the market for frozen food was increasing at a rate of about 10 % per year with approximately 25 % of refrigerated food being frozen food. In the 1980s environmental considerations and questions of efficiency of the machinery involved in freezing led to the redesign of much of the technology involved (Evans 2008; Rees 2013).

Clarence Birdseye (1886–1956) developed the technique of fast freezing after he decided to leave his post at the U.S. Agricultural Department and accompany the medical missionary Sir Wilfred Grenfell to Labrador, located in the northeastern part of today's Canada. He realised that the fish the Inuit caught froze almost immediately and when cooked had a superior taste than the frozen fish he had tasted before. Temperatures at Labrador were about −40 °C, much lower than the temperatures used to freeze seafood in places like New York with techniques involving higher temperatures which, however, formed ice crystals and damaged the tissues – hence the bad taste. The company he established in 1923 went bankrupt the next year for lack of consumer interest, but further developments in the freezing techniques initiated by Birdseye himself and a number of patents he was granted, established most of the standard processes to be used in the frozen food industry. These improvements as well as a different approach to packaging, made the next company he founded, General Seafood Corporation one of the most successful enterprises in the trade.

Though Marjorie Merriweather Post's (1887–1973) fame owed much to her lavish lifestyle and philanthropic works, she was, perhaps, the key figure in making Birdseye's company one of the largest in the world. Having inherited a thriving cereal company, Postum, after her father's death, the General Seafood Company (including the patents Birdseye held) was bought by Postum and Goldman-Sachs Trading Corporation in 1929. Then Merriweather Post founded the General Food Company – and became the richest woman in the USA. Birds Eye Frozen Food Company was an offshoot of General Food Company and she conducted a very aggressive marketing for frozen food, having at the same time the services of Birdseye himself who was a consultant in the firm. Frozen food was here to stay and Post made sure that the message passed onto the housewives in as convincing a manner as possible.

But it was the almost messianic insistence of a chemist, Mary Engle Pennington (1892–1952), who in her various roles in government agencies and private companies, wrote and lectured indefatigably about the virtues of ice boxes. Pennington though she completed all the requirements for a B.Sc. Degree in chemistry at the University of Pennsylvania, was given only a certificate of proficiency, since the university did not grant degrees to women. She was, however, able to obtain a doctorate from the same University in 1895 and did research in physiological chemistry. She worked at the Philadelphia Bureau of Health, playing an important role in the improvement of ways milk products were processed. In 1906 Congress had voted the Pure Food and Drug Act and she was offered the position of director of the Food Research Laboratory of the U.S. Department of Agriculture, a laboratory specifically established in order to enforce the new law. She stayed in this post for 12 years and in 1919 she moved to a private firm American Balsa, specialized in the manufacture of insulations for refrigerators. In 1922 she started her own consulting firm, the Household Refrigeration Bureau, where her educational activities around the beneficial effects of the use of ice boxes were supported by the National Association of Ice Industries – of those industries which produced ice and delivered it at homes. "Scientific housekeeping" (Robinson 1997: 254) became the order of the day and housewives came to slowly adopt and appreciate the company of ice boxes. In 1900 there were less than a 1,000 manufacturing plants in the United States. In 10 years the number had jumped to 5,000! For a long time, ice boxes were preferred over electrical refrigerators, since they were cheaper by almost 20 times (in 1922 an electric refrigeration cost about 700 US dollars when a Model-T Ford cost about 450 dollars), but it is to the ice boxes that the industry owes the break in mentality and habits: ice boxes preserved foodstuff and created the conditions for healthier life to the members of the family and, especially, for the children.

Pennington was, in fact, one of the members of the United States delegation to the First Congress of Refrigeration in 1908. The pamphlets she wrote were widely distributed among home economics teachers, home demonstration agents, welfare agencies and women's clubs. She started a training school for home service workers. She even mounted an exhibit on the history of refrigeration at the biennial convention of the Federated Women's Clubs of America (Robinson 1997).

1.6 Some Concluding Remarks

During the twentieth century the domestic kitchen became a rather dynamic site full of social and political correlations mediated and managed by women and the "kitchen became the stage where social actors performed a domesticity that was articulated in explicitly technical terms" (Oldenziel and Zachmann 2009: 8). Though, the consumption of cold was an integral part of these correlations from the very beginning, it only after Second World War that the home electric refrigerator asserted its presence to become an indispensable artefact in domestic kitchens. It is interesting that in countries like the Netherlands and Finland the percentage of homes having an electric refrigerator in the early 1950s was around 5 % and by the early 1970s the figure had jumped to over 70 %! And this particular technology, more than any other artefact in the kitchen – electric or gas cooker, dishwasher, coffee machines and other similar appliances – was catalytic in changing habits, mentalities and the micromanagement of family economics. The home refrigerator forced people to reconceptualize the central notion of healthy nutrition, that of "freshness" and refrigeration brought about immense changes in the ways we have been thinking about a host of everyday issues. The ice boxes and, then, the electric home refrigerator brought about the possibility of self-sufficiency for meals for a prolonged time, home freezers elongated that time to unimaginably long periods. What was fresh was, also, dramatically redefined as was the character of what was nutritive. Eating habits were reconceptualised and a host of new economic activities started to take effect producing "new" kinds of foodstuffs more amenable to being preserved in cold, investing in industries for transporting them, devising new ways of cooking and, of course, bringing into markets exotic foods which could be preserved in cold, thus, contributing towards the gradual globalization of eating trends. In fact, artificial cold created new tastes. Furthermore, there were changes in the organization of family life, in the care of the young, in home architecture. Kitchens became even more pronounced gendered spaces and so did nutrition. The domestication of cold, redefined a sizable portion of the domestic activities and promoted the image of the housewife with lots of free time in her hands. The domestication of cold, especially through the electric refrigerator, became a central tenet of the ideology of the emancipated woman through the new household appliances.

Birdseye, Pennington and Marjorie Merriweather Post were three different types of individuals displaying the complementary characteristics of American commercial success stories: empiricism and experimentation, single-mindedness in educating the population about the benefits of a new commodity and propagandizing its many uses, sui-generis personalities and the articulation of entrepreneurial strategies. Interestingly, it was two American women, whose impact on the consumption of cold was rather decisive. One carved for herself a role, then, strongly sought by many women: that of the educator, reformer, defender of family values, a person ready to make life easier for the housewife. The other, turned a new commercial enterprize into a success story, in a world dominated by men.

The history of artificial cold is a history in cultural re-articulations, theoretical re-conceptualizations and social re-alignments. It is, thus, a topic for history of science and technology, for science policy studies, for business history, for sociology, for cultural anthropology, for cultural studies, for consumption studies, for home economics and for gender studies. From the phenomena of the ultra cold to the activities utilizing what could be done with the moderately cool, artificial cold, especially during the twentieth century, defined the new discipline of low temperature physics, played a dominant role in the discourse of home economics and formed new ways of life. The phenomena incomprehensible to those immersed in classical physics had a parallel life with the industrial and commercial pursuits and the management of a new commodity. The region of the very cold provided a rather novel frame of reference that facilitated the understanding of a reality which could be probed only through quantum mechanics and, during the unravelling of the new phenomena and the attempts to understand them, the explanatory possibilities of quantum mechanics were expanded. But a vast range of possibilities were, also, provided by the large scale production of cold as well. Its use in the preparation and treatment of foodstuffs, has, perhaps, been a development whose significance may have been assessed in terms of the quantities involved, but has not been properly studied in the ways of the deep changes it brought in our everydayness. We had to re-appropriate notions and practices that were thought to have an unchanging reference. Freshness, nutritive values and commodities were some of them. A number of hygienic rules had to be re-defined. New tastes – not necessarily "good" or "bad", but new nevertheless – were acquired. Ever larger segments of the population had to "harmonize" their habits with a new kind of commodity: the frozen version of what they had been buying, cooking and eating before. The character of a frozen product and its relation to its non-frozen state, is not a theoretical problem, but a very practical one, since it refers to our own perception and the way a frozen foodstuff has been appropriated by society at large. Such an appropriation has rendered the frozen products as different products with respect to their unfrozen states, and slowly formed a new culture of everydayness: one started to shop differently, planned meals differently and, perhaps most significantly, cooked differently. The management of cold re-configured the management of time and changes in the management of time always bring about deep changes in our (sub)cultures. There is a historical precedent: salted meat and fish and their role in the post-seventeenth century sea trips, including the discoveries and invasions of the rest of the world by the Europeans.

Popularization and quasi-educational processes in the legitimation of the many uses of cold and especially those for the preservation and consumption of foodstuff, became an indispensable aspect of the technological framework for the utilization of artificial cold. Of course much of this "education" went hand in hand with the propagation of commercial interests or the creation of better conditions for the blooming of new industrial enterprises. It was not self-evident that some foodstuffs should be preserved in cold in order for them to be safe for consumption or to retain their nutritional value. Educating people in order to adopt these new practices, meant persuading people to buy and use the new technologies of cold and, hence, find ways to accommodate the presence of cold in their everyday dealings.

As is the case in any other history, "heros" and, most significantly, "heroines" abound in the history of artificial cold as well. What is intriguing, however, is the diverging paths of these people in our cultural perception of both the phenomena of the very cold and the many facets of industrial cold. The macroscopic quantum phenomena brought about dramatic reconceptualizations in quantum mechanics, and the large scale production, preservation, transportation, distribution and consumption of cold changed ways of life, with the introduction of new practices and the formation of new perceptions of what a commodity is. If historically natural cold was something to be avoided, western societies found a very agreeable companion in artificial cold.

References

Brush, S.G. 1983. *Statistical physics and the atomic theory of matter from Boyle and Newton to Landau and Onsager*. Princeton: Princeton University Press.

Dahl, P.F. 1992. *Superconductivity: Its historical roots and development from mercury to the ceramic oxides*. New York: American Institute of Physics.

Dienel, H.-L. 2004. *Linde. History of a technology corporation, 1879–2004*. London: Palgrave Macmillan.

Evans, J. 2008. *Frozen food science and technology*. Oxford: Blackwell Publishers.

Freidberg, S. 2009. *Fresh. A perishable history*. Cambridge, MA: Belknap Press.

Gavroglu, K. 1995. *Fritz London, a scientific biography*. Cambridge: Cambridge University Press.

Gavroglu, K., and Y. Goudarpoulis. 1989. *Methodological aspects in the development of low temperature physics 1881–1956: Concepts out of context(s)*. Dordrecht: Kluwer Academic Publishers.

Gavroglu, K., and Y. Goudaroulis (eds.). 1991. *Through measurement to knowledge, the selected papers of Heike Kamerlingh Onnes 1853–1926*. Dordrecht: Kluwer Academic Publishers.

Hacking, I. 2006. *Another new world is being constructed now: The ultracold*. Preprint 316. Berlin: Max Planck Institute for the History of Science.

Hard, M. 1994. *Machines are a frozen spirit, the scientification of refrigeration and brewing in the 19th century – A Weberian interpretations*. Frankfurt am Main: Campus Verlag.

Kapitza, P. 1965. Heat transfer and superfluidity of helium-II [1941]. In *Collected papers of P.L. Kapitza*, vol. 2, ed. D. ter Haar. Oxford: Pergamon Press.

Kapitza, P. 1981 Problems of liquid helium. A report at the General Assembly of the USSR Academy of Sciences, 28 December 1940. Reprinted in P.L. Kapitza, *Experiment, theory, practice. Articles and addresses*. Dordrecht: Reidel.

Landau, L.D. 1949. On the theory of superfluidity. *Physical Review* 75: 880–891.

Oldenziel, R., and K. Zachmann. 2009. *Cold war kitchen, Americanization, technology and European users*. Cambridge, MA: MIT Press.

Papanelopoulou, F. 2009. The International Association of Refrigeration through the correspondence of Heike Kamerlingh Onnes and Charles-Edouard Guillaume, 1908–1914. *Annals of Science* 66: 345–370.

Petrick, G.M. 2006. *The arbiters of taste: producers, consumers, and the industrialization of taste in America, 1900–1960*. Doctoral dissertation, University of Delaware.

Rees, J. 2013. *Refrigeration nation, a history of ice, appliances, and enterprise in America*. Baltimore: Johns Hopkins University Press.

Robinson, L.M. 1997. Safeguarded by your refrigerator: Mary Engle Pennington's Struggle with the National Association of Ice Industries. In *Rethinking home economics: Women and the history of a profession*, ed. Sarah Stage and Virginia B. Vincenti, 253–270. Ithaca: Cornell University Press.

Rowlinson, J.S. 2012. *James Dewar 1842–1923, a ruthless chemist*. London: Ashgate.

Thevenot, R. 1979. *A history of refrigeration throughout the world.* Trans. from French J.C. Fidler. Paris: International Institute of Refrigeration.

Tisza, L. 1938. Transport phenomena in helium-II. *Nature* 141: 913.

van Delft, D. 2007. Facilitating Leiden's cold: The International Association of Refrigeration and the Internationalisation of Heike Kamerlingh Onnes's Cryogenic Laboratory. *Centaurus* 49: 227–245.

van Delft, D. 2008. *Freezing physics. Heike Kamerlingh Onnes and the quest for cold.* Chicago: Chicago University Press.

van Helden, A.C. 1989. *The coldest spot on earth. Kamerlingh Onnes and low temperature research, 1882–1923*. Leiden: Museum Boerhaave.

Part I
Investigating the Very Cold

Chapter 2
Early Modern History of Cold: Robert Boyle and the Emergence of a New Experimental Field in Seventeenth Century Experimental Philosophy

Christiana Christopoulou

2.1 Introduction

During the seventeenth century the emerging experimental activities focused, among other categories of knowledge, on the examination of the quality of cold. The interest in cold was closely connected to the new corpuscular theories on matter and the criticism towards Aristotelian and scholastic theory of matter. Three major series of experiments on cold were led by Robert Boyle (1627–1691), the Irish experimental philosopher, and similar experiments took place in two important institutions of that period: the Royal Society of London and the Florentine Accademia Del Cimento.[1] In this paper I focus on Boyle's experiments on cold in order to show his contribution in the development of a new experimental field in natural philosophy.[2] I will firstly describe the historical context surrounding Boyle's experimental practice and focus on presenting the various theoretical schemata related to the quality of cold. I will then set Boyle's experiments on cold in the context of his other experimental inquiries. In the main part of this paper, I will focus on the analysis of the basic categories of experiments on cold conducted by Boyle and their significance for the reconceptualization of the concept of cold. I will try to

[1] During the period between its formation in 1660 and 1694, the members of the Royal Society conducted an important experimental activity on cold. The society's experimental practice on cold is recorded in the society's minutes as well as in a large number of its members' essays and letters. The Accademia Del Cimento was founded by Prince Leopold de Medici in the Florentine court in 1657 and was active until 1667. The experimental practice of its members, as well as that referring to cold is presented in the published treatise of the academy *Saggi* (Accademia del Cimento 1964).

[2] This paper is based on part of my PhD thesis (Christopoulou 2008). There I focus on Boyle's experimental practice on cold and present a comparative study between his experimental examination of cold and that of the Royal Society of London and the Florentine Accademia Del Cimento.

C. Christopoulou (✉)
Greek Society for the Study and Promotion
of History of Science and Technology, Athens, Greece
e-mail: christiana_christopoulou@hotmail.com

K. Gavroglu (ed.), *History of Artificial Cold, Scientific, Technological and Cultural Issues*, Boston Studies in the Philosophy and History of Science 299,
DOI 10.1007/978-94-007-7199-4_2, © Springer Science+Business Media Dordrecht 2014

underline the questions Boyle sought to answer, the new information given by his experiments and its relation to the inquiry into the nature of cold and the concept of cold he tried to put forward.

2.2 The Seventeenth Century Historical Context on Cold

Boyle's experimental examination of cold as presented in *Cold* forms part of his attempts, along with other experimental projects conducted during the period 1650–1666, to formulate a corpuscular theory of matter.[3] During this period the strong criticism of the still predominant Aristotelian and scholastic theories of matter led to the emergence of various new corpuscular theories (Henry 2002; Clucas 1997). These theories took corpuscles to be the constitutive parts of matter and qualities were explained through their basic characteristics: size, shape, volume, and motion. Explaining phenomena through the interaction between particles and their characteristics underlined the mechanical nature of the theories. In this part of the paper, I will introduce those particular theories and concepts that constitute this context and explain Boyle's interest in forming an experimental program on cold. Through this narrative I will also present the existing theories on the nature of cold that played part in discussions of the nature of this quality.

In his *Meteorology,* Aristotle explained the formation of temperature in a body through the participation of the element that carried the corresponding quality (water carried cold, and fire carried heat as each of the qualities were correspondingly predominant in each element). The concept of "antiperistasis", which Aristotle used in his kinematics, explained, in the case of heat and cold, the change in temperature without the alteration of the consistence of a body. According to the phenomenon of "antiperistasis" one quality was intensified when surrounded by its opposite quality (Aristotle 1952). Heat and cold were considered active qualities, in opposition to dryness and wetness that were considered passive. Heat could conjoin homogenous bodies and separate heterogeneous ones. Cold combined both kinds of bodies, which helped explain for example coagulation and freezing phenomena (Ross 1993).

During the twelfth and thirteenth century, scholastic philosophers dwelled on some problematic notions of aristotelianism introducing some modifications which later formed part of the seventeenth century criticism against the Aristotelian theory of matter. During the sixteenth and seventeenth centuries the experimental practices of chemical, alchemical and iatrochemical traditions led to the need for a new theory of matter and qualities that intensified criticism towards Aristotelian principles (Clericuzio 1990; Newman 1996). This growing critique of Aristotelian and scholastic principles concerning matter along with the reconsideration of the

[3] I will use '*Cold*' to refer to his basic treatise on cold *New Experiments and Observations Touching Cold or An Experimental History of Cold, Begun* (Boyle 1665a).

ancient Greek atomic theories of matter of Democritus, Leukippus and Epicure, influenced in multiple ways the formation of new corpuscular theories; most importantly these circumstances begged for further empirical testing and observation on particular qualities in order to unveil their nature.

Central to the emergence of corpuscular explanations and to the growing experimentation on qualities during the seventeenth century, stands Bacons' new method of acquiring knowledge and his reformulation of the concept of "form" (Perez-Ramos 1996). Bacon refuted the Aristotelian notion of "form" pertaining both to the nature of a body and its particular characteristics, and argued that "form" should refer to the particular qualities of bodies (heat, cold) that combine to make its nature. He used an operative notion of science, the idea that by getting to know a quality or a phenomenon we learn to reproduce it, to show that particular forms, particular qualities make up the nature of a body, since their knowledge can lead us to reproducing its nature. The unique and final notion of form was considered by Bacon to consist of the knowledge of an unobservable procedure in matter which caused the qualities. These Baconian concepts led theories of matter a further step away from Aristotelian and scholastic concepts of matter and qualities. They also formulated an instructive tool for the examination of the nature of qualities for experimentalists of this period.

The Baconian ideas along with the theories of matter of three key natural philosophers of the seventeenth century, Descartes, Gassendi, and Hobbes, played an important role in the formation of Boyle's experimental practice and experimental interest on qualities. In *Cold* they were presented as the theoretical background against which Boyle would expose his experimental results on the phenomena of cold. According to Gassendi, qualities were brought about by atoms of certain figure, size and motion; the position and interaction of atoms were also related to the emergence of qualities. Heat, on the one hand, was produced by small, round and light atoms that moved fast. On the other hand, heavy atoms in shape of a tetrahedron, with sharp edges and moving slowly produced cold (Kargon 1966; Hall Boas 1952; Charleton 1966). Thus for Gassendi the nature of cold was positive, as frigorific atoms were being added to bodies and produced the quality. In *Cold* Boyle criticised this position and put forward a privative concept of cold which was based on the deprivation of heat. Hobbes shared with Gassendi the primary principles of Epicurian atomic theory as well as the mechanical explanation of nature but filtered these into a rather different matter theory. Matter was composed of atoms of three different types of hardness; the space between atoms was filled with "ether" thus creating a plenum. Hobbes considered motion as the primary characteristic of atoms and explained the production of qualities through their motion and their interaction with ether. Cold was caused by the motion of violent winds which pressed the atoms together and caused shrinking and freezing. Heat was caused by the motion of ether trapped in between the atoms of a body causing in effect its dilatation (Hobbes 1656; Kargon 1966). Descartes' theory, although averting the use of experiment, proposed a strictly mechanical production of qualities through matter and its three geometrical qualities: shape, size and motion (Descartes 1644; Kargon 1966; Hall-Boas 1952; Shapin 1996; Garber 1992). Since the quality with which matter was identified was extension, the other qualities, shape, size and motion were considered secondary characteristics of matter (Garber 1992). Matter was composed of three kinds of corpuscles that differed in size and motion, allowing for each kind to fill the space between the group of smaller corpuscles thus creating a plenum and leaving no void. Motion, endowed to the universe by God, played the principal role in the production of qualities. Here motion is connected in such a way to heat and cold as to present a mechanical explanation of qualities based on the accelerative motion of corpuscles (heat) and its deprivation (cold) (Hall-Boas 1952; Kargon 1966).

2.3 The Context of Boyle's Experiments on Cold

Boyle's experimental research on cold started off after 1650 and was conducted in various periods until the publication of his main treatise involving phenomena of cold in 1665. Evidence from Boyle's early lists of treatises and essays of the period 1645–1655, indicates that the quality of cold was among his first interests in natural philosophy, such as atomism, mechanics and chemistry, and that Boyle was working on an early essay about it.[4] Boyle presented his extensive experimental practice on cold in the treatise *New Experiments and Observations Touching Cold or An Experimental History of Cold, Begun* which was published in 1665 (Boyle 1665a). During this period (1660–1666), he published the treatises *Spring of the Air* (1660), *The Sceptical Chemist* (1661b), *Certain Physiological Essays* (1661a), *Colours* (1664), and *Origin of Forms and Qualities* (1666–1667), which are indicative of his experimental agenda during that period on qualities of the air, chemical qualities and consistency of bodies, the examination of particular sensible qualities as were fluidity, firmness, heat, cold, and theories of matter.

The emergence of Boyle's experimental program on cold was strongly related to his interest to examine theories of matter experimentally and formulate his own theory of matter and qualities. During the 15 year period of experimentation on cold, Boyle conducted chemical experiments that led to the refutation of the Aristotelian and paracelcian theories on the constitution of matter in the *Sceptical Chymist.* He wrote essays on chemical experiments, as for example the *Essay on nitre* (Boyle 1661a), that were used as a basis for the first presentation of his ideas on corpuscular philosophy in *The Origin of Forms and Qualities.*[5] In parallel, he

[4] Boyle made lists of works he was currently working or designed for the future. According to Hunter and Davies who publish these lists in *The Works of Robert Boyle,* these lists underlined his interests at the particular time they were written. Hunter and Davies 1999–2000; Boyle 1649/1650, 1654. For Boyle's unpublished works of the period 1655–1665 see: "Of the Study of the Book of Nature" in Boyle 1649–1654a, "Essay of the Holy Scriptures" in Boyle 1649–1654b, "Of the Atomicall Philosophy" in Boyle 1649–1654c. For his published treatises see "The Second Essay, of Un-succeeding Experiments" in *Certain Physiological Essays* in Boyle 1661a, pp. 57–82.

[5] The full title of *Essay on nitre* is 'A Physico-Chymical Essay, containing An Experiment with some Considerations touching the differing Parts and Redintegration of Salt-Petre'.

designed and conducted extensive experimental examinations on sensible qualities, as were cold, heat, fluidity, firmness, colors and others. Cold and heat were discussed in the essay "Fluidity and Firmness", as causes of loss of firmness and fluidity and were, therefore, important in the examination of the nature of these qualities. There Boyle examined the change of state of various fluids into solids, explaining partly the phenomenon by the change in the motion of the particles consisting them.[6] In one particular example he explained the difference between the state of matter of water and ice by the motion of the particles in the first and the rest of the particles in the second which he related to the change in temperature of the bodies.[7]

Boyle's experiments and treatises on the nature of chemical and sensible qualities provided the indispensable basic knowledge in order to formulate his corpuscular theory presented in his work *The Origin of Forms and Qualities* which was published in 1666. He considered his research on cold central to the formulation of his theory, as shown by his intention to publish the treatise as an introduction to *Cold* (Boyle 1666–1667:288). In the following section, I will try to show that Boyle's experimental practice on cold, although pertaining to his general interest in forming a new corpuscular theory of matter, grew to be independent of this context, and opened a new experimental field in natural philosophy.

2.4 Boyle's Experimental Practice on Cold

Boyle's experimental practice on cold is here reconstructed mainly through the published treatise *Cold.* His presentations of the treatise and various essays on *Cold* at the meetings of the Royal Society of London, of which he was one of its founding members, have been used as additional sources. Boyle's communication of his experimental practice on cold at the Royal Society meetings consists mainly of the presentation of the main parts of *Cold* before and after its publication in April 1665 as well as of distinct essays presented in meetings and published in the *Philosophical Transactions,* the society's journal.[8] His later essays on cold, "of the Positive or Privative Nature of cold" (1673) and "of the Mechanical Origin of Heat and Cold"

[6] Fluidity and Firmness (Boyle 1661a:139, 173–174, 187).

[7] Boyle 1661a:130.

[8] The chapter on thermoscopes "New Thermometrical Experiments and Thoughts" was presented on the 3rd of February 1663/1664 and the chapters of the section "The Experimental History of Cold" were presented at the meeting of 11th January 1664, in Journal Book Original, vol. 2. For the publication of Boyle's essay "A new Experiment, shewing, How a considerable degree of Cold may be suddenly produced without the help of Snow, Ice, Haile Wind, or Niter, and that at any time of the year" see the Register Book of the Royal Society on the 11th of July 1666 and *The Philosophical Transactions* on 18 July 1666.

(1675–1676) where he discussed the question of the nature of cold and its mechanical explanation are not included in this analysis.[9]

In *Cold* Boyle gave a thorough and well structured examination of cold. Firstly the reader was introduced to the main experimental chapters of the book, by an extended discussion and comparison of the methods used to depict degrees of cold.[10] The main part of the treatise consisted of 21 chapters ("titles") where Boyle categorised the experiments according to the phenomena examined.[11] The examination of cold was completed with two additional chapters; the first focused on the examination and refutation of Hobbes' theory on the nature of cold and the second on the examination of the Aristotelian explanation of the variation of the intensity of cold with the phenomenon of "antiperistasis". The treatise also encompassed, as a separate chapter, an experimental essay on cold written by Boyle's fellow member of the Royal Society, Christopher Merret (1614–1695) (Boyle 1665a:519–542).[12] Boyle had added Merret's experimental essay partly because he considered some of his experiments important to his examinations and partly following his own requirement for a complete experimental examination on cold.

As already mentioned, in the beginning of the seventeenth century, discussion of the quality of cold was conducted as part of the criticism towards Aristotelian and scholastic theories on cold and the theoretical contemplation of the emergence of qualities within the framework set by new corpuscular theories of matter. Knowledge on the production of cold lacked, in large part, the practise of experiments with the exception of some philosophers, like Gassendi, who used experimental methods to verify his theoretical propositions on the quality. Boyle's practise of an organised and systematic experimental examination of the quality created a new context in which the discussion on cold could be reformulated. This context was characterised by the use of experiment and the examination of a series of new questions touching on various phenomena and manifestations of cold.

In his experimental examination of cold, Boyle appropriated two basic features of Bacon's method of knowledge. The first concerns his concept of "form" discussed above. The second concerns Bacon's practice of accumulating observations on nature and formulating "natural histories". Michael Hunter has argued that Boyle adopted the Baconian method of creating categories of experiments and phenomena to organise his experimental practice (Hunter 2007). In *Cold* Boyle's focus on collecting empirical data, categorising the phenomena and examining a big number of

---

[9] "of the positive and privative nature of cold" consists one of the four essays on cold in *Saltness of the Sea*, (Boyle 1673:341–367). Also see (Boyle 1675–1676:313–523) for the discussion on the mechanical nature of cold. These works focus on the debate on the nature of cold and are based on the experimental results of *Cold*. They are not discussed here because this paper focuses on the presentation of Boyle's experimental practice on cold as a whole and not only on the question related to its nature.

[10] See Annex A for the contents of the treatise *Cold*.

[11] The phenomena examined were used as titles to the chapters, Boyle 1665a:226–227.

[12] Merret's experimental essay comprised of a written report he had presented on the 30th of December 1663 at the Royal Society meeting along with some additions he made after that.

phenomena of cold is one of the characteristics of his appropriation of the baconian concept of "history". His attempt to appropriate the Baconian method did not only involve the examination of many categories of phenomena of cold but also the experimental testing of the existing theories on its nature. The basic theories Boyle tried to experimentally refute were the scholastic theory, the epicurean atomic tradition which was expressed mainly by Gassendi, Hobbes' theory on cold and the Cartesian theory.

For a better understanding of Boyle's vast experimental study of cold I have categorised the experiments in five thematic groups. The first category involves experiments on the degrees of cold and the methods of sensing them. The second category concerns the experiments used to examine the effects of cold on liquids and solid bodies. The third category involves the examination of the phenomena of ice; Boyle considered ice to be the manifest phenomenon of cold. The fourth category concerns experiments on the ability of bodies to cause or intensify the quality of cold. Finally, the fifth category involves experiments on the distance, direction and mediums through which cold can be diffused. I begin my presentation of Boyle's experimental practice on cold by the first category.

2.4.1 Can We Measure Cold?

Boyle depicted degrees of cold for four purposes: the examination of the occurrence of phenomena of cold in cold conditions, the examination of the effect of cold on liquids and other bodies, the inquiry into the ability of water, air, and other liquids and bodies to cause phenomena of cold and diffuse cold, the reproduction of the phenomena of cold in cold conditions and the examination of their characteristics. The necessity to record these variations in the degrees of cold led Boyle to evaluate the instruments and methods already in use during this period. In the first category mentioned above I include Boyle's discussion and use of three methods: open air and liquid in glass thermoscopes, human senses (mainly touch), qualitative methods (freezing of various liquids, variations in the density of water).[13]

In the context of the Aristotelian and scholastic tradition, philosophers acknowledged the variation in the intensity of heat and cold but did not consider measuring the degrees of the two qualities. Their belief in a qualitative concept of qualities did not allow them to consider them as part of the same natural process (Barnett 1956). The idea of measuring the temperature of a body was developed in the context of medicinal practices from the beginning of the second century A.D. and onwards. During the sixteenth century physicians were familiar with the concept of a scale and used the human body as a primary method of sensing temperature.[14] During the

[13] The term "thermoscope" is here used for the instrument that indicated the variations in temperature without a standard scale.

[14] Temperature as a concept in physics was adopted in the eighteenth century.

same period the selection of air instead of the human senses as a thermometric substance played a significant part in thermometric practices. The pneumatic experiments of the day worked as a context where the effect of heat and cold on air was examined and led to the construction of experimental instruments, the open air thermoscopes, that were used for the depiction of the degrees of cold.[15] These instruments were used firstly, to indicate the effects of heat and cold on air but not to measure the two qualities.[16]

In the beginning of the seventeenth century there was a growing interest in the use of the air thermoscope in the field of natural philosophy as an instrument that indicated changes of degrees of heat and cold; the scale of the instrument was a matter of improvisation.[17] Some examples consist of Galileo's experiments on the effect of heat on air, Sanctorius' examinations of the temperature of the human body, Van Helmont's work in chemistry as well as the work of Mersenne, Athanasius Kircher and others. In the first half of the seventeenth century scholastic philosophers were also examining the variations in the intensity of the two qualities through the use of the open air thermoscope. Later in the century its use became common in practical activities such as the recording of weather conditions.[18]

In his *Novum Organum* (1620), Bacon set forth a new context for the examination of the degrees of heat and cold which also played a formative role in Boyle's examinations of cold. According to Bacon, the indication and comparison of the degrees of qualities was one of the three methods of categorization of their phenomena in the process of acquiring knowledge about the nature of a quality (Bacon 1960). Phenomena about a quality were first categorized in three tables, "Table of Essence and Presence", "Table of Deviation, or of Absence in Proximity", "Table of Degrees or Table of Comparison" (Bacon 1960:129–137); these tables were later used as a basis for Bacon's inductive method. In the case of heat, Bacon showed that the comparison of the degrees of heat were indispensable for the explanation of its own nature. Boyle followed Bacon in the categorization of phenomena in the baconian tables; he used various methods to indicate the variations of the degrees of cold which became an indispensable parameter in performing his experiments. In Boyle's experimental practice the need to define degrees of cold played an important part in the experimental setup and evaluation of experimental results, the categorization of

[15] The constructions of Philo of Byzantium (second century B.C.) and Hero of Alexandria (62 A.D.) played an important role in the instruments prepared by Robert Fludd (1574–1651), Della Porta (–1615) and Santorio (1561–1636) see Middleton 1964:4–5. Also see the pneumatic instruments prepared by Cornelius Drebbel (1572–1633) in Taylor 1942:154–156, and Middleton 1964:20.

[16] A common air thermoscope consisted of a tube which ended in a round bulb filled up to a point with water, which was inversed into a viol of water in a manner that allowed the entrapment of air in the upper part of the tube and the bulb. The entrapped air worked as a thermometric fluid (See Annex B, Fig. 2a).

[17] Scholastic philosophers used a "philosophical" scale made of 8° of heat and 8° of cold and based on the Aristotelian theory of the four elements. See Taylor 1942:132 and Middleton 1966:4.

[18] See Robert Fludd's *Utriusque Cosmi Historia* (1617) and Bates 1654.

experimental phenomena, the examination of the degrees of cold and the methods of defining them.

According to his study of heat and cold as separate subject-matters, in different experimental practices and in different treatises, as well as his reference to "degrees of cold", it can be argued that Boyle did not use a concept of temperature where degrees of the two qualities could belong to the same scale and explained by the same natural process. Nevertheless, his experimental refutation of many Aristotelian propositions and his strong criticism of Gassendi's theory of the nature of cold, show that Boyle rejected the qualitative nature the two theories gave to heat and cold. In *Cold* Boyle frequently referred to a hypothesis on the nature of cold according to which cold consists of a privation of heat, namely a privation of the motion of the corpuscles of bodies which regulated their temperature. Therefore in reading his experimental examination on cold the reader has to bear in mind that Boyle's treatment of the nature of heat and cold stood at this crucial crossroad of the Aristotelian qualitative concept of cold and a mechanical explanation of cold and heat as parts of the same natural process.

Boyle's discussion on the methods of depicting degrees of cold was important for the developments in thermometry of his time. Firstly, Boyle criticised the information provided by the human senses on the variation of the degrees of cold because the information they gave was always relative to the degree of heat or cold predominant in the human body. Boyle also argued that the human senses could not sense the small variations in coldness. He explained the relative recording of temperature through the interaction of the corpuscles of the cold body, which moved slowly, with the corpuscles of the human body which moved faster. Boyle's criticism against the use of the senses, opened the discussion for the use of open air thermoscopes and phenomena of cold as more reliable methods for depicting degrees of cold (Boyle 1665a:229–240).

Boyle's inquiry for a suitable thermometric substance begun with his strong criticism against the use of air in various constructions and versions of the open air thermoscope. Here he underlined two issues which were central in the process of formulating a scale of degrees of cold. The first concerned the refutation of air as a reliable thermometric substance and the second highlighted the role played by the explanation of the function of the instrument in the formulation of a scale.

The explanation of the function of the open air thermoscope was an issue under discussion in the seventeenth century. Boyle criticised three existing explanations, that of the scholastic philosophers, Hobbes' as well as Gassendi's and gave experimental evidence of his own position.[19] Using his experimental conclusions on the spring of the air in his pneumatic experiments in *Spring of the Air* and *Defence* (1662), Boyle showed in *Cold* that the open air thermoscope was exposed to the alterations of the spring of the air either when the thermoscope was kept in the same place or when it was moved to a place of different atmospheric pressure. Its indications of degrees of cold by the contraction of air, could

[19] Boyle 1665a:253–262.

be taken in consideration only when the pressure of the external air was shown by the barometer to be kept at the same level at all times.[20]

Although the contact of the water with the external air's spring and weight was proven to be the air thermoscope's principal weakness, Boyle experimentally showed that it was indispensable for its function.[21] According to Boyle, when the air in the open air thermoscope became colder, its spring weakened without, however, having as a direct effect the contraction of the air; this was caused by the action of the pressure of the external air on the water because it was stronger than that of the internal air which had suffered a decrease of its spring.[22] Boyle argued that unlike the other three explanations, his could explain more phenomena of cold and could indicate degrees of cold that were undetectable if the function of the instrument was explained otherwise. For instance using his conclusions on the spring of air, he could argue that the power to contract air was inversely proportionate to the space occupied *by air and therefore a double degree of cold would contract air to ¼ of its initial space (Boyle* 1665a*:243). According to Boyle the explanation of the function of the instrument and the way cold affected the thermometric substance where inextricable parts of the formation of a scale of degrees of cold (Boyle* 1665a*:384).* Boyle's conclusions on the problematic use of the open air thermoscope created an ideal context for his effort to validate the use of the sealed liquid thermoscope in the experimental examination of cold.[23] In the first half of the seventeenth century, the Duke of Florence, Ferdinand Medici the 2nd (1610–1670) had developed two versions of the sealed liquid thermoscope as part of the experimental activities he had been exercising in his court (Barnett 1956:281; Middleton 1664:28–32).[24] The first, a copy of which Boyle received in 1662 from a traveller in Florence, Robert Southwell (1635–1702), consisted of a round vessel glued to a tube, which was filled up to a point with distilled wine and then hermetically sealed.[25] The distilled wine served as the thermometric

[20] In Fig. 2a of Part B of the Annex, the air enclosed in section 1 of the tube contracts with cold. In section 2 I note the part of the water that comes in contact with the external air. If the pressure of the external air changes, it affects this part of the water which in consequence affects the water in the tube and the air enclosed in section 1. Therefore contraction of the air in section 1 may occur by the variation of the weight of the external air in section 2.

[21] See section 2 of Fig. 2a in Part B of the Annex. For a better understanding of Boyle's use of the terms "spring" and "weight" of the air see Shapin and Schaffer 1985: 53–55.

[22] According to Gassendi's explanation the coldness of the air was the direct cause of the contraction of the air included in the air thermoscope.

[23] Also there were other construction flaws such as the fact that it could not be immersed in water, problems with length, size of the tubes which Boyle had tried to solve by constructing an air thermoscope which enclosed the air in its lower part (see Fig. 2b in Part B of the Annex). A tube was inserted in a round viol and the space between the tube and the viol was sealed with wax. Then water was inserted in the viol through the tube until some point in the viol thus trapping air in the lower part of the viol (section 1 of Fig. 2b). This allowed for the instrument to be inserted in liquids.

[24] The two thermoscopes probably at use in the Florentine court since 1654, where later developed by the experimental group, *Accademia del Cimento* (Accademia Del Cimento 1964:1–12).

[25] The second consisted of a tube filled with water and some glass balls filled with air. Here heat and cold was thought to affect the density of water which resulted in the ascension or descending of the glass balls.

substance.[26] Boyle's contribution in its construction was confined in the improvements made by his laboratory assistant, Robert Hooke, who had used rectified distilled wine tinted with cochinelle (Boyle 1665a:246).

Boyle's work on the standardisation of the construction of the sealed thermoscope and the validation of its use, indicates three important issues for understanding his attempts to create a scale of degrees of cold. The first concerned the explanation of the way the instrument worked, the second concerned the experimental examinations for a suitable liquid and the third concerned the attempt to standardise the construction of the sealed thermoscope and to define a standard degree on its scale.

In order to validate the use of the sealed liquid thermoscope, Boyle stressed the importance of its obvious practical advantages, as was for instance, the protection of its readings from the variation of the weight of the external air, its safe transportation due to the sealing of the liquid, and its ability to be immersed in liquids. Moreover he refuted the scholastic philosophers' objections to its function on the basis of their view on the impossibility of a vacuum. Boyle experimentally showed that heat and cold could affect the liquid by dilating or contracting it although it was enclosed in the sealed tube. He additionally indicated that heat and cold affected some liquids more than air. Boyle used his knowledge from *Spring of the Air*, to show how air was affected by the dilatation and contraction of the liquid in the tube due to the stronger or weaker resistance it met. Thus he offered an alternative explanation for the ascent of the water in the tube to the Aristotelian theory on water averting a vacuum. Boyle's explanation of the function of the liquid thermoscope was a first step towards an examination of a scale in degrees of cold.

Having shown that a liquid could be affected by cold when sealed in a tube, Boyle focused on the examination of various liquids as to their behaviour towards cold and their ability to freeze. According to Boyle, the correct thermometric liquid should not freeze easily and should be able to depict small variations in degrees of cold. Boyle's general conclusion that liquids that contain aqueous parts froze easily in the climate of England in comparison to "subtle" and "spirituous" liquids that would freeze in lower degrees of cold, had made him use distilled wine rather than water in the sealed thermoscope (Boyle 1665a:273–278). At the same time, his observations that some "spirituous liquids" froze in degrees of cold that were milder than that which could freeze water and that distilled liquids, chemical oils and saline spirits all froze in different degrees of cold showed that his selection of a thermometric fluid was far from being conclusive.

It becomes evident that Boyle's examination of cold was opening more questions on the nature of cold and provided evidence that refuted the existing theories. Boyle had realised that until a theory explaining the nature of cold was adopted, attempts for a standard instrument and scale would remain inconclusive (Boyle 1665a:245). At the same time he acknowledged that there should be a standard of cold by which

[26] Figure 2c of Part B of the Annex.

a scale could be constructed as was a standard for measuring weight and length.[27] Moreover his experimental practice on cold greatly depended on devising standard means of depicting degrees of cold. In this context Boyle had exercised a method in order to construct sealed thermoscopes with comparable indications while being conscious of the fact that his attempt to define a standard in a scale for cold was far from being conclusive (Boyle 1665a:240, 242).

Boyle's key difficulty lay in the definition of the temperature of the air at the moment the thermoscope was sealed. The indication of the liquid at the time of the instruments' sealing would determine what would be considered colder or less cold. Boyle proposed that at the time of the insertion of the liquid in the thermoscope and its sealing, the instrument be immersed in "oyl of aniseed" which had been previously liquefied. As the oil congealed again, he would mark the position of the thermoscope's liquid on the scale, which would serve thereafter as a standard. The particular oil was used because it could congeal both in winter and summer with a mild degree of cold. Boyle proposed that two thermoscopes that had been made from vials and tubes of the same sizes, blown the same way, filled with the same liquid and constructed using this method could have comparable indications. Another standard proposed by Robert Hooke substituted the oil of aniseed with distilled water. This was adopted by the Royal Society and by Boyle in the experiments he added in the second edition of *Cold.*

Boyle tried to define different degrees of cold by using both his thermoscopes and the qualitative phenomena as thermometric methods (Boyle 1665a:278–285); the latter played a crucial role to this end. As an example of a phenomenon used to depict degrees of cold I consider Boyle's method to examine the difference between the cold emitted by ice or snow alone and the cold emitted from the mixture of snow and salt. Boyle filled a glass tube with each of the substances and observed the apparition and freezing of dew on the outside of the glass, the thickness of the ice produced, its duration and the surface that it covered Boyle (1665a:265–272). These observations unveiled the difference in the intensity of cold of the two substances which was taken by Boyle to signify a difference in degrees of cold.

In this context of his thermometrical observations and his qualitative experimental results, Boyle would distinguish between degrees of cold and use them to construct experimental trials for phenomena of cold. Although quantification of degrees of cold was quite scarce in his experimental practice, he claimed to be reaching some important conclusions. Firstly, he experimentally refuted what he considered an opinion commonly held, that the freezing of water was indicative of the highest degree of cold (Boyle 1665a:279); he demonstrated the existence of higher degrees of cold like that of the mixture of snow and salt and tried to quantify this difference, although he considered it small, through the contraction of the liquid of a sealed

[27] "And first, I consider, that we are very much to seek for a Standard or certain Measure of Cold, as we have settled Standards for weight, and magnitude, and time, so that when a man mentions an Aker, or an Ounce, or an Hour, they that hear him, know what he means, and can easily exhibit the same measure", (Boyle 1665a:240).

thermoscope.[28] Secondly, Boyle showed the existence of a variety of degrees of cold higher than that which could freeze water; some of these were even higher than the degree of cold produced by the mixture of snow and salt. Moreover he examined the difference between degrees of cold emitted from ice, snow and air in the climate of England and those emitted by the same phenomena in colder climates which he thought predominant in "cold countries" such as the island of Novia Zemblya, the inner part of Russia, Greenland and Denmark (Boyle 1665a:421). The testimony Boyle had collected on observations in these cold countries had led him to the conclusion that the degree of cold that froze water was not a standard and that the expansion of frozen water varied (Boyle 1665a:309). Boyle's discussion of the differences between the phenomena at different climates was qualitative.

Thirdly, Boyle demonstrated the differences between the degree of cold that froze water and a lower degree of cold. One important instance concerned the difference between a degree of cold that froze liquids and bodies like food in a way that destroyed their texture and a degree that was adequate to cool them and keep their texture intact (Boyle 1665a:294). Boyle also showed that the various phenomena of cold were not only dependent on the degrees of cold but also on their duration. For example, he experimentally demonstrated that the duration of a high degree of cold affected the distance of the effect of cold concerning the depth of the freezing of the ground and the expansion of water when frozen (Boyle 1665a:327).

In his discussion of the comparison of the three methods in *Cold,* the human senses were mainly rejected and the thermoscopes and phenomena of cold were given equal importance in the evaluation of their advantages and disadvantages in depicting the degrees of cold. Although the experimental instances of the use of the open air and sealed thermoscopes were small in number, there is evidence that the use of the thermoscope, especially the sealed one, was used in crucial experiments and had a decisive role in the evaluation of their outcome.[29] Nevertheless, he mostly used his senses to get a feeling of the differences in degrees of cold.

One important example was the case of the discussion of the Aristotelian explanation of the phenomenon of antiperistasis (Boyle 1665a:459–498). In order to explain what happens in terms of heat and cold when bodies get mixed, the scholastic philosophers argued that, when one of the two qualities that reside in a body is surrounded by the opposite quality, it becomes more intense in order not to be overcome by it. One important phenomenon they used to illustrate their theory was the temperature of underground cellars and waters which were warmer in the winter than in summer and colder in summer than in winter. Boyle used as observational evidence against antiperistasis the coldness of the middle layer of the

[28] A degree able to freeze water would contract the liquid by 1/94 of its original bulk, whereas the degree that was produced from the mixture of snow and salt would contract the liquid by 1/47 of its original bulk. In this experiment Boyle used instead of spirit of wine, "oil of turpentine" (Boyle 1665a).

[29] The little use of thermoscopes could be explained by the late arrival of the sealed thermoscope in England. Boyle had performed a big number of the experiments on cold by 1662 when he received it.

atmosphere when the lower and upper is warm, the fall of hail during summer and the heating of quicklime when it is solved in cold water. He also considered that the examination of this phenomenon with yearlong readings of thermoscopes in cellars would give a decisive proof of the refutation of the Aristotelian explanation. Furthermore, he provided many experimental setups with sealed thermoscopes to compare the internal temperature of a cellar with the external air that contradicted the phenomenon. According to these experimental results, he had observed that the warmth of underground waters remained constant when the external air during winter became colder.

The study of Boyle's experiments on the measurement of cold, unveils a rather intriguing dimension of his practices. Though his inquiries and experimental findings did not allow for a standard of cold, Boyle had managed through his examination of the various phenomena and thermometric methods to incorporate and use in his experiments the practice of depicting degrees of cold, thus articulating a new role to instruments like the sealed thermoscope.

2.4.2 *Freezing and Its Effects*

The second category of experiments on cold concerns experiments on the effects of cold on liquid and solid bodies, mainly the freezing effect of cold and experiments that focus on the effect of cold on the texture of bodies. The phenomenon of freezing of liquids was considered to be the principal effect of cold and a qualitative proof of the highest degree of cold as well. Through the examination of freezing, Boyle could test the Aristotelian proposition that cold caused the congregation of heterogeneous and homogeneous bodies. He could also collect information on the relation of cold to fluidity and firmness, since he considered cold to be the primary cause of the loss of the fluidity of a body through the diminution of the motion of its particles. Furthermore he used freezing as a chemical process of analysis and a tool for the categorization of liquids and bodies according to their chemical consistence.

As a basis for the examination of freezing, Boyle gathered information on the ability of various categories of liquids to freeze (Boyle 1665a:273–275 and 275–278); by freezing Boyle meant their solidification into ice. According to the table that depicts Boyle's experimental results, two main categories of liquids that could freeze were those that contained aqueous and phlegmatic parts, like solutions and various types of salts.[30] On the other hand the categories of chemical oils, fermented liquids and strong spirits would not turn into ice. According to Boyle aqueous and phlegmatic parts where indispensable for freezing to occur as the liquids that lacked any of them, could not freeze; in the solution of salts the participation of the solvent was considered a variable for its disposition to freeze. In categories

[30] See the table in Part C of the Annex.

such as natural oils and weak spirits, Boyle could not reach a general conclusion, since results varied among liquids of the same category (Boyle 1665a:276).

Boyle's experiments on the ability of various types of liquids to freeze led him to a new observation. He had shown that water and other liquids that contained aqueous parts expanded in the process of freezing. This observation refuted a basic Aristotelian position according to which water shrunk while being frozen.[31] Boyle used the comparison of the expansion of various liquids as an indicator of the participation of aqueous and phlegmatic parts in them. His experimental results on liquids that did not freeze, verified his conclusion that only liquids that froze expanded. Most of the liquids he had observed not to freeze would suffer a contraction in bulk (for example mercury, distilled wine, common oils, some chemical oils).

The observation of the expansion of water by freezing opened new experimental questions on the phenomenon which focused on its measurement, the observation and measurement of the emergence of a force in the freezing process, and the explanation of the natural procedure that caused expansion. Boyle's experiments on the expansion of water by freezing resulted in the observation of a standard in the expansion of water by a 1/9 of its initial volume. Boyle tried to apply the standard to macroscopic phenomena of expansion of water, as for example were the icebergs in the northern countries and argued that the analogy of the immersed part to the floating one should be 8/9. Although Boyle mentioned some observations of the icebergs of which the immersed part reached this analogy, most of the observations and testimonies Boyle analysed refuted it.

The second question in relation to the expansion of water by freezing concerned the work of a force that he considered to be present in the freezing process. Boyle considered the breaking of glass bottles from the expansion of water by freezing and the dislocation of the leads and corks of the bottles to be demonstrations of this force. He used the durability of various metal weights with which he enhanced the stoppers of the bottles and vessels and of metal vessels which he used as containers, as a method to depict and measure the force of the freezing water (Boyle 1665a:330 335). The force was considered equal to the resistance of a certain quantity of weight.

The third question posed by the experiments on the expansion of water by freezing concerned the examination of the natural process by which this occurred (Boyle 1665a:308–318). Boyle used as given knowledge the position that the expansion and levity of frozen water could be explained through the presence of bubbles in the ice; he was mainly concerned with the examination of their content. According to his experimental results, he argued that a small part of their content could be attributed to the pre-existent air in the water and a large part was either, what he called "a vacuum" or filled by a kind of air like ether.

The experimental questions Boyle examined as a result of his observation on the expansion of water by freezing, can be strongly related to his inquiries on the nature of cold. Through his treatise *Cold,* Boyle showed an interest in refuting Gassendi's

[31] This Aristotelian position had been questioned earlier by scholastic philosophers themselves.

theory of cold as the positive addition of cold particles in bodies and supporting a privative nature of cold, which would explain cold by the lessening of the motion of the particles of bodies. The privative concept of cold could readily explain the shrinking of liquids by a freezing degree of cold, but not the expansion of water by freezing. Boyle's conclusions on the presence of bubbles in ice, as well as his observation on the lack of variation in the weight of water before and after freezing, helped him explain the expansion of water in a way not contradictory to his concept of cold as a privation. The force that appeared when freezing water expanded, remained a phenomenon that Boyle could not reconcile with a hypothesis on the diminution of the motion of the particles of the frozen body.

The second group of experiments of the second category concerned the effect of freezing and of milder degrees of cold on the texture of bodies (Boyle 1665a:288–301). Boyle focused his examination on the preservation of the texture of bodies by a mild degree of cold that could not freeze them and on the preservation and destruction that occurred in bodies in the process of freezing and thawing; his main interest concerned the texture of edible bodies and of the human body. These questions concerned phenomena that were known and used in everyday life activities and therefore affected vital areas of human life.

On the basis of his experiments and testimonies, Boyle verified the opinion commonly held that types of food and human bodies preserved their texture only when they were thawed in cold conditions, like cold water or air, and not when exposed to a degree of heat. Defrosting in water was found a quicker procedure than in cold air. During the thawing process in water, the bodies were usually cased in a layer of ice which was considered part of the reason why they preserved their texture.

Boyle focused on the effects of three temperature conditions on the texture of bodies: the effect of a degree of cold mild enough not to freeze them, the effect of a freezing degree of cold on human bodies and the destructive effect of a freezing degree of cold on various types of food. In the first case, he referred to various practical methods people used to create conditions that preserved the texture of bodies without freezing them. In the second case, Boyle referred to the preservation of human bodies from rotting when frozen and explained it through the pause of the motion and irregular activity of the particles of a body which were responsible for its decay. In the third case, Boyle examined those cases of edible bodies which after their defrosting presented an altered texture and taste; he used this as evidence of the destruction of their texture by freezing. Boyle seems not to have been able to point out experimentally why in some cases the texture of bodies was preserved when defrosted correctly and in others destroyed. He explained the destruction through his recent observations on the freezing of water and aqueous parts of bodies. He supported that the distortion brought about by the expansion through freezing of the juices contained in foodstuff, caused the same effects with natural decay.

Although Boyle used his hypothesis on the nature of cold to explain the preservation and destruction of bodies by cold, he could not derive from his experimental results the cause of the motion or pause of the corpuscles. He took the explanation of the production of ice casing in the process of defrosting in water to play a crucial role to this end. Although he could explain the phenomenon in the case of water, by the freezing of the surrounding water from the coldness of the immersed frozen body, he had conflicting experimental results when he tested the defrosting procedure in various other liquids. Thus he could not refute the corroboration the phenomenon offered to Gassendi's hypothesis of frigorific corpuscles as a cause of the immobility of the corpuscles of bodies.

2.4.3 The Phenomena of Ice

In the third category I classify Boyle's experiments on various characteristics of the phenomenon of ice. Boyle used ice to refer to a solid piece or block of ice which he differentiated from snow and hail. He considered ice a paradigmatic phenomenon of cold as his experimental results on it could be applied to a variety of similar cold phenomena (Boyle 1665a:357–364). Boyle's examination of ice included his own experiments on its cohesion, strength and durability but also observations and testimonies on the characteristics ice exhibited in colder climates (Boyle 1665a:346–357); the latter gave important information on whether these characteristics of ice remained the same at the macroscopic level. As far as the cohesion of its parts is concerned, Boyle focused on measuring the strength and durability of ice as well as the nature of its cohesion. His experiments with weights that were set by a mechanical device on ice, measured the amount of weight a certain piece of ice would bear before breaking. Based on his measurements, he concluded on the strong cohesion of ice which he also thought depended on the degree of cold of the atmospheric air surrounding it. He also verified his laboratory observations through testimonies of travelers according to which in the season of frost, icebergs were stronger than his tests in the laboratory; while during the thawing season, their cohesion became very fragile. Boyle found his conclusion on its cohesion difficult to reconcile with observations of the easy dissolution of ice by salt.

Boyle's examination of the duration of ice and the circumstances under which it can be dissolved is of particular interest (Boyle 1665a:357–364). In this group of experiments Boyle's basic experimental examination comprised of the insertion of equal, in volume and thickness, blocks of ice (ice from water) in air, water and various liquids that were kept at the same temperature conditions. He would then measure the time it took for the pieces of ice to dissolve in each of these media. Boyle would time the melting process with the use of a pendulum. According to his results Boyle concluded that ice would melt or dissolve slower in liquids that were hard to freeze, like oils, than in water and that it dissolved even slower in air.[32] Based on these results, Boyle refuted what was commonly believed that ice melted quicker in oil liquids due to their disposition against freezing and the Aristotelian position that ice melted quicker in "potentially hot" liquids.

---

[32] See Part D of the Annex.

As part of this group of experiments, Boyle also examined the duration of ice in air when the cause of cold conditions and the production of ice had seized its action (Boyle 1665a:360–364). Boyle's inquiry had been triggered by phenomena where snow was conserved in warm and rainy weather and in conditions where the degree of cold was milder than that of freezing water. Some of these phenomena also formed part of everyday domestic practice; for Boyle they were used as an experimental inquiry of particular interest. On the basis of his observational data, Boyle verified the possibility of conservation of ice when the cause of coldness had stopped acting. He further supported that ice would remain in the same condition until an opposite cause of heat would make it dissolve and that it could be conserved in a lesser degree of cold than that needed for water to be frozen. He used this evidence to support his opinion that a phenomenon of cold would continue its presence after the cause of cold had seized its action because as he wrote

> For in most cases, if a certain texture be once produced in a body, it is agreeable to the constancy of nature, that it persevere in that state, till it be forceably put out of it, by some agent capable to overpower it… (Boyle 1665a:360).

He took this conclusion to refute principally those who considered that the phenomena of cold were brought about and conserved only by the continuous action of causes of cold.

2.4.4 Bodies that Cause or Enhance the Action of Cold

In the fourth category of experiments, I have classified two important experimental inquiries that concern the bodies that can cause and enhance cold. The first inquiry involves Boyle's discussion of theoretical schemata related to a "Primum Frigidum" and the second the ability of various substances that could enhance cold emitted by ice or snow. The theories that supported the existence of a "Primum Frigidum" attributed to a single body that was cold by its own nature, the ability to produce cold in all other bodies by its insertion in them (Boyle 1665a:357).

The concept of a "Primum Frigidum" originated from the Aristotelian viewpoint, according to which each of the four primary qualities (heat, cold, dryness, wetness) predominated in each element; cold predominated in water (Boyle 1665a:364). Since the four elements participated in making the mixed bodies, coldness was produced in them by the participation of water. The theories that proposed a "Primum Frigidum" shared in common this notion of qualities in general and cold in particular. In the seventeenth century context of the refutation of the Aristotelian theory of matter and the discussion of new atomistic and corpuscular theories, the inquiry into the way qualities were brought about in bodies was essential and thus the issue of a "Primum Frigidum" became the object of a dispute. In atomic theories like that of Gassendi, the concept of a "Primum Frigidum" was discussed through the qualitative notion given to frigorific corpuscles that were specific in shape and responsible for the cause of cold in other bodies. Gassendi focused his discussion on the

"Primum Frigidum" on refuting the pre-existing propositions and proving his own suggestion, that of niter.[33]

In Boyle's experimental practice on cold, the theories on the "Primum Frigidum" received a much stronger criticism of their fundamental concept. Boyle refuted altogether the existence of a "Primum Frigidum" on the basis of his general position against the existence of a "πρώτο δεικτικό".[34] In continuation of his criticism of the four elements (water, earth, air, fire) of the Aristotelian theory and the three/five elements of the chemical theories (mercury, salt, sulphur, water and phlegm) exposed in the *Sceptical Chemist,* Boyle offered in *Cold* an extended discussion on the issue of a primary cold body. Having already criticised the equivalent concept in heat in his unpublished treatise on heat (Boyle 1650), Boyle offered as an argument his hypothesis on the privative nature of cold. For Boyle cold could be caused by the motion of corpuscles in bodies and could be felt through the comparison between the motion of corpuscles of our senses and the motion of the corpuscles of a cold body. Because he considered his speculations not yet confirmed, he confined himself to the refutation of the predominant theories through the following argumentation (Boyle 1665a:364–381).

He focused his criticism on four "Primum Frigidum" theories: the Aristotelian theory which suggested water, the Stoic philosophers' theory which suggested air, Plutarch's theory which suggested earth, and Gassendi's theory which suggested niter. Boyle argued against each suggestion on the basis of each body's tendency to freeze itself and others and of the comparison of their coldness. In the cases of earth, water and air, Boyle focused on the phenomena of cold exhibited in the sea, on land and in the atmospheric air and not in the surroundings of a laboratory with the exception of his experiments on niter. According to Boyle's conclusion niter, air, water and earth were not considered to be able to cause cold on their own.

In addition to his arguments against the existence of a "Primum Frigidum", Boyle supported that earth, water and air were cold in their natural state. Boyle's hypothesis that cold could be caused by the lessening of the motion of corpuscles of bodies, led him to a relative concept of cold. According to Boyle, the body used as a sensor of cold would always judge temperature in relation to the motion of its particles. Therefore Boyle compared the coldness of the three elements to the human senses which he considered to have a middle temperature in reference to heat and cold (Boyle 1665a:381). Boyle presented this view against Gassendi's position according to which the three elements were neither warm nor cold. For Gassendi, heat and cold were caused by the insertion of calorific and frigorific corpuscles respectively in bodies. In Boyle's view the temperature of a body was determined by the motion of its corpuscles, which meant that it could be hot or cold without an external agent acting on it.

[33] An example of Gassendi's criticism can be found in Charleton 1966:328–339.

[34] Boyle used the Greek phrase to refer to the element that carried a certain quality and by the participation of which other bodies acquired that quality Boyle 1665a: 365.

In the case of air, Boyle went further in his examination of its coldness. Apart from its importance in his inquiries on the use of thermoscopes, Boyle focused on this element as most of the freezing phenomena on earth were brought about by the atmospheric air (Boyle 1665a:381–400). We have already mentioned his experimental results on the contraction of air by cold. Most of his further inquiries concerned the ability of air to cause meteorological phenomena, the seasons and the climatic conditions of the various areas on earth.

Boyle's arguments against the existence of a "Primum Frigidum" allowed him to consider whether bodies of various substances would be able to cause cold or enhance the degrees of cold produced by another cold body (Boyle 1665a:265–272). His comparative experiments of freezing water by the use of snow alone as a source of cold and a mixture of snow and salt, had shown that water froze only when surrounded by the second mixture; Boyle interpreted this result as an indication of the emission of a higher degree of cold from snow and salt than that emitted by snow. He further corroborated this result by using a mixture of snow with other salts like niter, "alum", "sal ammoniack" and sugar.

On the basis of this observation, Boyle designed another experiment to examine the frigorific ability of mixtures of snow or ice with various substances. The mixture was inserted in a glass viol and Boyle observed the freezing of the external vapours of the air by the communication of cold through the glass; the thickness of the sides of the vial would be crafted in a way to allow the communication of cold. The degree of cold emitted by the mixture would be qualitatively judged by the apparition on the external sides of the vial of dew and ice. Boyle would also consider the thickness of the external ice, its duration before melting and the amount of time it took the snow or ice of the internal mixture to melt after its contact with the substance, important indications of the mixture's freezing effect. In some cases he examined the intensity of the cold transmitted by coating the external sides of the vials with substances that did not freeze as easily as water did and observing whether they froze. Moreover some experiments were conducted with closed vials in order to assure that the experimental results would be explained by the communication of coldness through the glass and not the travel of frigorific corpuscles of the internal mixture to the external surface of the glass. Boyle added also the procedure of measuring the weight of the vial before and after the apparition of frost in order to refute the idea that frigorific corpuscles would be communicated through the glass. According to his experimental results, the vials were heavier after the apparition of frost, which meant that cold was caused by the freezing of vapours of the air which had been added to its weight, rather than parts of the mixture that had passed through the glass to the external side.

According to his experimental results, acid spirits (oil of vitriol, spirit of salt, spirit of niter, spirit of sugar, spirit of vinegar), volatile spirits (urinous spirits, sal ammoniack), salts that had not been purified or broken into corpuscles through distillation (sal gem, kitchen sugar, pot-ashes) enhanced the coldness of snow or ice so that the external dew of the viol appeared frozen. Spirit of niter, sal ammoniack and spirit of wine proved, through the appearance of thick ice, to cause intense degrees of cold. Boyle noticed that although spirit of wine was categorized as a vegetable sulphur and was found not to freeze, some of its characteristics suggested its relation to saline liquors which was verified by this experimental result. This last observation underlines one interesting aspect of this line of experiments. Boyle tested various substances which he categorized on the basis of his knowledge of their chemical characteristics, and found that substances of different categories shared in common the ability to freeze.

2.4.5 On the Diffusion of Cold

In the fifth category I have gathered the experiments pertaining to the examination of the distance, direction and mediums through which cold can be diffused.[35] Boyle's inquiries into the diffusion of cold were innovative in relation to the Aristotelian concept of cold and Gassendi's examination. The Aristotelian tradition stressed the importance of the manifestation of cold and heat and their role in the constitution of the four elements and the mixed bodies; its qualitative concept of the two qualities did not allow for a further examination of the way cold and heat would enter and leave bodies. In his treatment of cold, as shown in Charleton's work *Physiologia Epicuro-Gassendo –Charltoniana,* Gassendi focused his analysis of cold on the size of his frigorific corpuscles and the issue of the "primum frigidum".[36] This context left little ground for discussion of the ways of its diffusion. Boyle's manifest refutation of the qualitative nature of cold, as well as his Baconian method to examine many aspects and characteristics of its phenomena, made him focus on the quality and its actions. Looking into the ways of its diffusion allowed him to unveil important information on its nature.

Boyle examined the "sphere of the activity of cold" (Boyle 1665a:335), the distance in which it was diffused, using as mediums air, earth and water. He considered the diffusion of cold to be dependent on the degree of coldness and the volume of the cold body, the medium through which cold was diffused and its state of motion and the thermometric instrument used to depict its diffusion. Baring these factors in mind, Boyle examined the diffusion of cold emitted from pieces of ice through air in order to see whether cold diffused in a similar way to heat and magnetism. Boyle could not manage to sense cold in a distance from ice; coldness was depicted only when the instrument came in contact with the cold body. The phenomena observed in cold countries, like water being frozen when in contact with ice, and cold being felt in a distance from an iceberg, suggested to Boyle that the state of the coldness of the atmospheric air played part in the diffusion of the quality. Boyle's additional experiments on the depth cold would penetrate in the ground, showed that cold could not diffuse itself in those mediums at a big distance.[37]

Boyle's inquiry into the direction of the diffusion of cold stemmed from his effort to examine whether cold would be diffused in an opposite direction than heat. Heat was demonstrated by Boyle as well, to be diffused upwards (Boyle 1665a:285–288).

---

[35] Boyle uses the term "diffusion" Boyle 1665a:335–340.

[36] This work was written by Walter Charleton and served for the dissemination of Gassendi's ideas in restoration England Charleton 1966:296–297.

[37] His experiments concerned mostly the penetration of coldness in the ground than in the water.

According to the Aristotelian tradition, heat had a natural tendency to go upwards in order to find its natural place and cold, as one of the qualities forming the element of earth, had a tendency to follow earth to a downward motion. Boyle's experiments on the direction of the diffusion of cold through air showed that there was neither an upwards nor a downwards tendency. Boyle's investigation of the direction of cold had helped him design an experiment where he could guide the direction of cold in order for water to begin to freeze from the bottom. As an overall conclusion from his experiments on the distance and direction of the diffusion of cold, Boyle had shown that a piece of ice, big or small, could not diffuse cold in a distance or in any direction unless it came in contact with the cause of cold.

Boyle's examination of the mediums through which cold could be diffused, concentrated on those cases where the cause of cold and the body designed to receive it would be in contact. Working with pieces of material of equal thickness, Boyle examined the diffusion of cold through vials made of clay, glass, iron, and pewter, and concluded that glass was a good conductor of cold. Boyle also tested the diffusion of cold in the "vacuum" of his air pump. A mixture of snow and salt was set around the air pump, after pumping the air out, and Boyle observed a tube with water inside it turn into ice. Boyle also attempted to test as mediums, some of the liquids he had tried to freeze, like oil of turpentine, spirit of wine, solution of sea salt, and brine. According to Boyle's general conclusion cold could be diffused easier in the mediums that he had managed to freeze easily.

2.5 Conclusion

Boyle's experimental examination of cold was conducted and published during a period when the Aristotelian and scholastic theory of matter was undergoing strong criticism and the emergence in the seventeenth century new corpuscular theories of matter were producing new explanatory concepts of matter and its qualities. As Steinle has convincingly shown in his study of Ampere, Faraday and Dufay's electromagnetic experiments of the nineteenth century, when the conceptual basis of a theory is questioned, exploratory experiments can lead to the formulation of new concepts and the explanation of new categories of phenomena (Steinle 2002:426). Accordingly, I argue that in Boyle's experimental practice on cold, exploratory experiments formulated a new experimental field in the examination of qualities. Firstly, Boyle's experiments on cold set a new context for the use of methods of depicting degrees of cold. Boyle showed that the examination of the nature of cold went hand in hand with the construction of a method for its measurement and with the definition of a standard of cold and a scale for its measurement. Secondly, his experiments produced new information on the phenomena of cold that most of the times contradicted the accepted views based on Aristotelian propositions. For example, I have argued that the demonstration of the expansion of water by freezing led to new questions on this phenomenon that provided new information on the nature of the quality. Thirdly, Boyle's experiments produced knowledge that helped the reformulation of the concept of cold. For example he unveiled the existence of various degrees of cold, higher and lower than that which froze water which had different qualitative effects in nature. His experimental evidence on the freezing effect on bodies and the ability of some to enhance cold gave important information for the rearticulation of the concept of cold. Fourthly, Boyle provided many experimental results that refuted the existing explanations of the phenomena of cold. At the same time he presented experimental evidence in favour of his own explanatory hypothesis of cold, thus suggesting a new theoretical explanation of the quality on the basis of the diminution of the motion of corpuscles. On the whole, his experiments on cold also showed a rather strong affinity with the Baconian program of which he was such a strong adherent.

References

Archives

Archives of the Royal Society of London.

Classified Papers (Guard Books), vols 4i, 4ii, 5, 20 (1660–1741).

Journal Book Original, vols 1–9 (1660–1696).

Letter Book Original, vols 1–2 (1661–1668), vols 6–7 (1672–1675), vols 8–9 (1677–1684).

Philosophical Transactions vol. 1 (issues 1, 3, 8, 9, 10, 11, 12, 13, 15, 18, 19, 20, 21,) 1665/66, vol. 2 (issues 24, 27, 28, 32) 1666/67, vol. 5 (issue 66) 1670, vol. 7 (issue 90) 1672.

Register Book Original, vols 2–8 (1661–1700).

Bibliography

Accademia Del Cimento. 1964. *Essayes of natural experiments made in the Academie del Cimento*, 1st edn, 1684. New York/London: Johnson Reprint Corporation.

Aristotle. 1952. *Meteorologica*. The Loeb classical library series, vol VII. Trans. H.D.P. Lee. Cambridge, MA/London: Harvard University Press.

Bacon, F. 1960. *The new organon*, ed. Anderson H. Fulton. Indianapolis: The Library of Liberal Arts/Bobbs-Merrill Educational Publishing.

Barnett, M.K. 1956. The development of thermometry and the temperature concept. *Osiris* 12: 269–341.

Bates, J. 1654. *The mysteryes of nature, and art: conteined in foure severall treatises, the first of water workes, the second of fyer workes, the third of drawing, colouring, painting, and engraving, the fourth of divers experiments...partly collected, and partly of the authors peculiar practice, and invention*. London: Bishop, R. for Crook, Andrew.

Boyle, R. 1649/1650. Materialls & Addenda Desing'd towards the structure & completing of treatises already begun or written. January the 25th 1649/1650. In *The works of Robert Boyle (1999–2000)*, vol. 14, ed. M. Hunter, and E. Davies, 329. London: Pickering and Chatto Publishers.

Boyle, R. 1649–1654a. Of the study of the book of nature. In *The works of Robert Boyle (1999–2000)*, vol. 13, ed. M. Hunter, and E. Davies, 145–172. London: Pickering and Chatto Publishers.

Boyle, R. 1649–1654b. Essay of the Holy Scriptures. In *The works of Robert Boyle (1999–2000)*, vol. 14, ed. M. Hunter, and E. Davies, 173–223. London: Pickering and Chatto Publishers.
Boyle, R. 1649–1654c. Essay of the Atomicall Philosophy. In *The works of Robert Boyle (1999–2000)*, vol. 13, ed. M. Hunter, and E. Davies, 227–235. London: Pickering and Chatto Publishers.
Boyle, R. circa 1650. Dialogues concerning flame and heat. In *The works of Robert Boyle (1999–2000)*, vol. 13, ed. M. Hunter, and E. Davies, 259–269. London: Pickering and Chatto Publishers.
Boyle, R. 1654. List of essays and experiments, c. 1654. In *The works of Robert Boyle (1999–2000)*, vol. 14, ed. M. Hunter and E. Davies, 330. London: Pickering and Chatto Publishers.
Boyle, R. 1660. New experiments physico-mechanical, touching the spring of the air and its effects (made for the most part, in a new pneumatical engine). In *The works of Robert Boyle (1999–2000)*, vol. 1, ed. M. Hunter and E. Davies, 141–306. London: Pickering and Chatto Publishers.
Boyle, R. 1661a. Certain physiological essays and other tracts. In *The works of Robert Boyle (1999–2000)*, vol. 2, ed. M. Hunter and E. Davies. London: Pickering and Chatto Publishers.
Boyle, R. 1661b. Sceptical chymist: or Chymico-physical doubts & paradoxes, touching the spagyrist's principles commonly call'd hypostatical, as they are wont to be propos'd and defended by the generality of alchymists. In *The works of Robert Boyle (1999–2000)*, vol. 2, ed. M. Hunter and E. Davies, 205–378. London: Pickering and Chatto Publishers.
Boyle, R. 1662. A defence of the doctrine touching the spring and the weight of the air. In *The works of Robert Boyle (1999–2000)*, vol. 3, ed. M. Hunter and E. Davies, 3–107. London: Pickering and Chatto Publishers.
Boyle, R. 1664. Experiments and considerations touching colours. In *The works of Robert Boyle (1999–2000)*, vol. 4, ed. M. Hunter and E. Davies, 3–201. London: Pickering and Chatto Publishers.
Boyle, R. 1665a. New experiments and observations touching cold, or an experimental history of cold begun. In *The works of Robert Boyle (1999–2000)*, vol. 4, ed. M. Hunter and E. Davies, 203–575. London: Pickering and Chatto Publishers.
Boyle, R. 1665b. The order of my severall treatises. In *The works of Robert Boyle (1999–2000)*, vol. 14, ed. M. Hunter and E. Davies, 331–332. London: Pickering and Chatto Publishers.
Boyle, R. 1665c. Various tracts, as essays &c. In *The works of Robert Boyle (1999–2000)*, vol. 14, ed. M. Hunter and E. Davies, 332. London: Pickering and Chatto Publishers.
Boyle, R. 1666–1667. The origin of forms and qualities. In *The works of Robert Boyle (1999–2000)*, vol. 5, ed. M. Hunter and E. Davies, 281–491. London: Pickering and Chatto Publishers.
Boyle, R. 1673. Saltness of the sea. In *The works of Robert Boyle (1999–2000)*, vol. 7, ed. M. Hunter and E. Davies, 337–451. London: Pickering and Chatto Publishers.
Boyle, R. 1675–1676. Experiments, notes, &c. about the mechanical origine or production of divers particular qualities. In *The works of Robert Boyle (1999–2000)*, vol. 8, ed. M. Hunter and E. Davies, 315–523. London: Pickering and Chatto Publishers.
Boyle, R. 1772. *The works of the Honourable Robert Boyle*, vol. 2, ed. Th. Birch. Hildesheim: Georg Olms.
Boyle, R. 2000. *The works of Robert Boyle (1999–2000)*, 14 vols, ed. M. Hunter, and E. Davies. London: Pickering and Chatto Publishers.
Burckhardt, F. 1867. *Die Erfindung des Thermometers und Seine Gestaltung im XVI Jahrhundert*. Basel: C. Schultze's Universitaets Buchdruckerei.
Charleton, W. 1966 (1st edn, 1654). Heat and Cold *Physiologia-Epicuro-Gassendo-Charltoniana: or a fabrick of science natural, upon the hypothesis of atoms, founded, repaired and augmented by Epicurus, Petrus Gassendus, Walter Charleton*, ed. Robert H. Kargon. New York/London: Johnson Reprint Corporation.
Christopoulou, C. 2008. *Experimental philosophy – 17th century: The case of Robert Boyle and his experiments on cold*. PhD dissertation, Department of Philosophy and History of Science, University of Athens.
Clericuzio, A. 1990. A redefinition of Boyle's chemistry and corpuscular philosophy. *Annals of Science* 47: 561–589.
Clucas, S. 1997. The fate of hylomorphism. 'Matter' and 'Form' in early modern science. *Early Science and Medicine* 2(3): 251–271.

Descartes, R. 1644. *Principles of philosophy*. In *The philosophical writings of Descartes*. Trans. J. Cottingham, R. Stoothoff, D. Murdoch, (1984), vol. 1, 177–291. Cambridge: Cambridge University Press.
Descartes, R. 1984. *The philosophical writings of Descartes*, 2 vols.. Trans. J. Cottingham, R. Stoothoff, and D. Murdoch. Cambridge: Cambridge University Press.
Fludd, R. 1617–1621. *Utriusque cosmi maioris scilicet et minoris metaphysica, physica atque technica historia*, 2 vols. Francofurti: Oppenhemii & Co.
Garber, D. 1992. Descarte's physics. In *The Cambridge companion to Descartes*, ed. J. Cottingham. Cambridge: Cambridge University Press.
Hall-Boas, M. 1952. The establishment of the mechanical philosophy. *Osiris* 10: 412–541.
Henry, J. 2002. *The scientific revolution and the origins of modern science*. Hampshire/New York: Palgrave.
Hobbes, T. 1656. *De Corpore or elements of philosophy, the first section concerning body*. London: Leybourn, R.&W. for Andrew Crooke at the Green Dragon at Pauls Church-yard.
Hunter, M. 1995. How Boyle became a scientist. *History of Science* xxxiii: 61–103.
Hunter, M. 2007. Robert Boyle and the early royal society: A reciprocal exchange in the making of baconian science. *British Journal for the History of Science* 40(1): 1–23.
Hunter, M., and E.B. Davies. 1999–2000. General introduction. In *The works of Robert Boyle (1999–2000)*, ed. M. Hunter and E. Davies, 1, xxi–lxxxviii. London:Pickering and Chatto Publishers
Kargon, H.R. 1966. *Atomism in England from Harriot to Newton*. Oxford: Clarendon Press.
Middleton, K.E.W. 1966. *History of the thermometer and its uses in meteorology*. Baltimore: The Johns Hopkins Press.
Knowles Middleton, W.E. 1964. *The history of the barometer*. Baltimore: The Johns Hopkins Press.
Newman, R.W. 1996. The alchemical sources of Robert Boyle's corpuscular philosophy. *Annals of Science* 53: 567–585.
Perez-Ramos, A. 1996. Bacon's forms and the maker's knowledge tradition. In *The Cambridge companion to Bacon*, ed. M. Peltonen, 99–120. Cambridge: Cambridge University Press.
Ross, W. D. 1993. *Αριστοτέλης*. Trans. Μαριλίζα Καραμήτσου. Αθήνα: Μορφωτικό Ίδρυμα Εθνικής Τραπέζης.
Saggi di naturali esperienze fatte nell'Accademia del Cimento. Firenze: Cocchini, 1666, i.e. 1667.
Shapin, S. 1996. *The scientific revolution*. Chicago/London: The University of Chicago Press.
Shapin, S., and S. Schaffer. 1985. *Leviathan and the air-pump: Hobbes, Boyle and the experimental life*. Princeton: Princeton University Press.
Steinle, F. 2002. Experiments in history and philosophy of science. *Perspectives on Science* 10(4): 408–431.
Taylor, F.S. 1942. The origin of the thermometer. *Annals of Science* 5(2): 129–156.

Annexes

(A) The Table of Contents of the Treatise Cold *(Boyle 1665a)*

- Title Page: New Experiments and Observations touching Cold, or an Experimental History of Cold Begun. To which are added An Examen of Antiperistasis, and An Examen of Mr. Hobs's Doctrine about Cold. By the Honourable Robert Boyle, Fellow of the Royal Society. Whereunto is annexed an Account of Freezing, brought in to the Royal Society, by the Learned Dr. C. Merret, a Fellow of it. Non fingendum, aut excogitandum, sed inveniendum, quid natura fiat, aut ferat, Bacon

- The Publisher to the Ingenious Reader
- The Author's Preface Introductory
- Errata
- The contents of the Experimental History of Cold:

New Thermometrical Experiments and Thoughts

(I) The I. Discourse, Proposing the I. Paradox. Viz. That not only our Senses, but common Weather-glasses, may mis-inform us about Cold.
(II) The II. Discourse Containing Some New Observations about the Deficiencies of Weather-Glasses, together with some Considerations touching the New or Hermetical Thermometers
(III) The III. Discourse, Containing The II Paradox, Viz. Touching the Cause of the Condensation of Air, and Ascent of Water by Cold in Common Weather-glasses.

The Paper that was prefixt (by way of a short Prefatory Address) to the ensuing History of Cold, when being to be brought in, and presented to the *Royal Society*, it was put into the hands of (its most worthy President) the Lord Viscount *Brounker*, was as followeth

The Experimental History of Cold Begun

Title I: Experiments touching bodies capable of freezing others
Title II: Experiments and Observations touching Bodies Disposed to be Frozen- Appendix to the II. Title.
Title III: Experiments touching Bodies Indispose'd to be Frozen
Title IV: Experiments and Observations touching the Degrees of Cold in several Bodies
Title V: Experiments touching the Tendency of Cold Upwards or Downwards
Title VI: Experiments and Observations touching the Preservation and Destruction of (Eggs, Apples and other) Bodies by Cold- An Appendix to the VI Title
Title VII: Experiments touching the Expansion of Water, and Aqueous Liquors, by freezing
Title VIII: Experiments touching the Contraction of Liquors by Cold
Title IX: Experiments in Consort, Touching the Bubbles from which the Levity of Ice is supposed to proceed
Title X: Experiments about the Measure of the Expansion and the Contraction of Liquors by Cold

Title XI: Experiments touching the Expansive Force of Freezing Water
Title XII: Experiments touching a New Way of estimating the Expansive force of Congelation, and of highly compressing Air without Engines
Title XIII: Experiments and Observations touching the Sphere of Activity of Cold
Title XIV: Experiments touching the Differing Mediums, through which Cold may be diffus'd
Title XV: Experiments and Observations touching Ice
Title XVI: Experiments and Observations touching the Duration of Ice and Snow, and the destroying of them by the Air and Several Liquors- Appendix to the XVI Title
Title XVII: Considerations and Experiments touching the Primum Frigidum
Title XVIII: Experiments and Observations touching the Coldness and Temperature of the Air
Title XIX: Of the Strange Effects of Cold
Title XX: Experiments touching the weight of Bodies frozen and unfrozen- Appendix to the XX. Title.

Title XXI: Promiscuous Experiments and Observations concerning Cold- Postscript

- An Examen of Antiperistasis, As it is wont to be Taught and Prov'd-Title Page

An Advertisement

An Examen of Antiperistasis, As it is wont to be Taught and Prov'd
Postscript: A Sceptical Consideration of the Heat of Cellars in Winter, and their Coldness in Summer

- An Examen of Mr Hobs's Doctrine touching Cold-Title Page

An Advertisement

An Examen of Mr Hobs's Doctrine touching Cold

- Philosophical Writings already publish'd by this Author
- An Account of Freezing made in December and January, 1662 by Christopher Merrett- title page

An Advertisement to the Readers of The Following Experiments, by the Author of the foregoing History.
An Account of Freezing made in December and January, 1662

(B) Thermoscopes

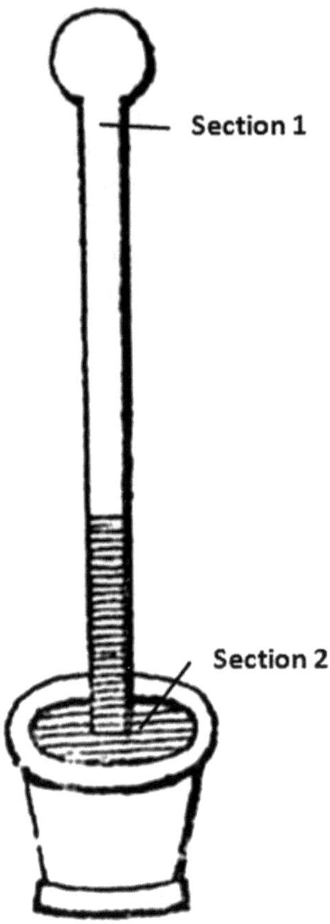

Fig. 2a The open air thermoscope (Burckhardt 1867)

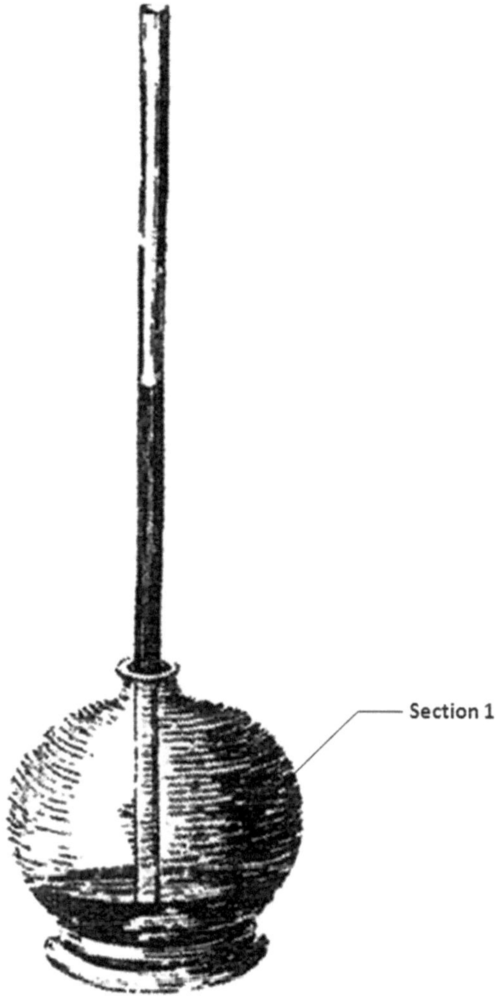

Fig. 2b Boyle's open air thermoscope (Boyle and Birch 1772:482–483)

Fig. 2c Boyle's sealed liquid thermoscope (Boyle and Birch 1772:482–483)

(C) Table of Liquids that Freeze and Liquids that Do Not Freeze

Liquids that freeze	Liquids that do not freeze
Urine	Liquids freed from aqueous or other phlegmatic parts
Beer	Aqua fortis
Ale	Spirit of nitre
Milk	Spirit of salt
Vinegar	Oil of turpentine
French and Rhenish Wine	Chemical oils
Aqueous solutions of:	Spirit of wine
Gum Arabic	Fermented liquids – strong spirits
White sugar	Sack
Solutions of salts:	Lixiviate liquids (not certain)
Alum	Solution of sugar of lead
Nitre	Quicksilver
Salt-petre	Brine
Sea-salt	Mixture of water and sea-salt
Aqueous solution of verdegrease	
Oil of vitriol	
Salts without gross parts:	
Spirit of vinegar	
Spirit of urine	
Aqueous solutions of salt of Pot-ashes	
Oil of tartar	
Liquids that froze in cold countries:	
Soft spirits	
Spirit of wine	

(D) Table of Measurements of the Duration of Ice

Liquid	1st experimental trial (time/minutes)	2nd experimental trial (time/minutes)
Oil of vitriol	5	3
Spirit of wine	12	13
Aqua fortis	12,5	No record
Water	12	26
Oil of turpentine	44	47
Air	64	152
Shallet oil	No record	52

Chapter 3
James Dewar and His Route to the Liquefaction of Hydrogen

J.S. Rowlinson

In the 1840s Kincardine was a small port on the river Forth above Edinburgh where the local inn, the *Unicorn*, was kept by Thomas Dewar and his wife Ann. They had seven sons, of whom six survived infancy, and the last of whom, James, was born in 1842. He was educated first at the local school and then, after the death of both parents, at the nearby Dollar Institution (now Dollar Academy). In 1859 he went to Edinburgh University where he studied under the physicist Guthrie Tait and the chemist Lyon Playfair. As was usual for students from financially modest backgrounds, he took no degree but served as an assistant to Playfair and later to his successor, Crum Brown. His earliest research was in organic chemistry and physiology but he spread himself widely and published steadily, so creating a reputation as an active and coming young man. He applied unsuccessfully for the Regius Chair of Chemistry at Glasgow in 1874, but was then a candidate for the Jacksonian Professorship of Natural Experimental Philosophy at Cambridge where, in 1875, the electors were looking primarily for someone who could teach chemistry to medical students, but where the wide but obsolete conditions of the endowment allowed the occupant considerable freedom to choose his own field of teaching and research. The resources immediately available to the new professor were, however, even less than Dewar had been accustomed to in Edinburgh, so when the Fullerian Professorship of Chemistry at the Royal Institution in London was advertised in March 1877 he applied for that also. He was again successful and held both chairs for the rest of his life. The resources of the Royal Institution, or the RI as it was commonly known, proved over the years to be greater than anything that he could probably have mustered in Cambridge in the last quarter of the nineteenth century.

This chapter is based, with the permission of the publishers, on parts of Chapter 8 of J.S. Rowlinson *James Dewar: a Ruthless Chemist* (Ashgate Publishing, Farnham, Great Britain, 2012)

J.S. Rowlinson (✉)
Department of Chemistry, Physical and Theoretical Chemistry Laboratory,
Oxford University, Oxford, UK
e-mail: john.rowlinson@chem.ox.ac.uk

K. Gavroglu (ed.), *History of Artificial Cold, Scientific, Technological and Cultural Issues*, Boston Studies in the Philosophy and History of Science 299,
DOI 10.1007/978-94-007-7199-4_3, © Springer Science+Business Media Dordrecht 2014

He gave his statutory lectures at Cambridge, and cooperated with George Liveing, the Professor of Chemistry, in a long series of papers on visible and ultraviolet spectroscopy. He was assisted there also by a series of Demonstrators with whom he maintained his interest in organic chemistry but, from 1877, the centre of his research was firmly established in London at the Royal Institution.

Dewar had a strong interest in the history of chemistry and knew of the reputation that Davy and Faraday had acquired at the Royal Institution for research on the liquefaction of gases. Faraday claimed that he had been the first to liquefy six different gases but he never succeeded with the simplest, the so-called 'permanent gases', the elements oxygen, nitrogen and hydrogen. Many gases such as ammonia or chlorine could be liquefied by pressure alone, or by pressure with the modest degree of cooling provided by such refrigerants as 'Thilorier's mixture' – solid carbon dioxide partially dissolved in ethyl ether. Faraday recognised that the liquefaction of the 'permanent gases' would require preliminary cooling to temperatures beyond his reach. It was commonly held that the implication of Boyle's and Charles's laws was that the zero of temperature was at about −273 °C so that there was much thermally unknown territory to be explored. Thomas Andrews in Belfast clarified the position in the 1860s by showing that for each gas there was a 'critical temperature' below which a gas must be cooled before it could be liquefied by pressure. There the matter rested until, a few months after Dewar's appointment to the Royal Institution, the field suddenly sprang to life with a report from a meeting of the Académie des Sciences in Paris on 24 December 1877 that two men had simultaneously and independently liquefied oxygen. They were Louis Cailletet, in the family iron-works at Chatillon-sur-Seine in Burgundy, and Raoul Pictet, a physicist in Geneva. Their methods differed but both depended on compressing a gas to about 300 atm pressure, cooling it with an external refrigerant and then cooling it further by letting a first part of the sample of compressed gas do work in expelling the second part through a narrowly opened valve. This work could come only from the energy of the first part which was thereby cooled and partially liquefied. The liquefaction was visible as a mist either in the gas remaining in the cylinder (Cailletet) or in the jet of gas coming out through the valve (Pictet). So far, so good – the so-called permanent gases *could* be liquefied, but neither man was able to collect the liquid as a static sample of a size large enough for it to be studied further in the laboratory, or for it to be used to cool other bodies down to the temperature of the new liquids.

The next step was taken by Zygmunt Wróblewski and Karol Olszewski in Krakow. They cooled the compressed oxygen in a stout glass tube cooled by liquid ethene. At atmospheric pressure this boils at 169 K which is above the critical temperature of oxygen at 155 K. Their innovation, in 1883, was to cool the ethene further by pumping away its vapour until the pressure was reduced to about 0.04 atm when its temperature is 130 K and so at a temperature at which the oxygen can be liquefied by a modest pressure of about 20 atm. They prepared liquid oxygen in small samples but had still to solve the problem of preserving it from rapid evaporation by the heat conducted from the surroundings. Dewar repeated their process at the Royal Institution and, in 1892, solved the last problem with his invention of the silvered vacuum flask, now known by his name in laboratories throughout the world, or in households by the patented name of 'Thermos flask'. Liquid oxygen, and the physically similar liquid nitrogen, and so also liquid air, were now available, albeit at some trouble and expense, in any laboratory that chose to follow the route mapped out by Cailletet, Pictet, Wróblewski, Olszewski and Dewar. Liquid oxygen was first exhibited in Britain by Dewar in a lecture at the Royal Institution in June 1884, and by 1892 he was making it 'by the pint' (0.6 l) (Dewar 1884, 1886; Brock 2002; Rowlinson 2009).

Hydrogen alone remained as a gas whose 'permanency' was a challenge to the chemists and physicists. Its critical temperature, now known to be at 33 K, was correctly estimated by Wróblewski by a comparison of the behaviour of hydrogen at high pressures with that of the gases that could be liquefied. Such a common family resemblance of different simple gases had been established by J.D. van der Waals in his 'law of corresponding states' in 1880 and was the guiding principle of those seeking to liquefy hydrogen, and later helium, which had been found on Earth in 1895. Dewar was later to suggest that van der Waals's law, or principle, as it is now more correctly described, was the greatest advance in thermodynamics since the time of Carnot, an assessment that overlooks the contributions of Thomson, Clausius and Gibbs (Dewar 1902). But then Dewar was exclusively an experimentalist and so valued highly any generalisation that helped him plan his experiments. Many of his papers and lectures contain such phrases as 'Now what are the facts?', and he often seemed to think of theory as nothing better than speculation.

A temperature of 33 K is too low to be reached by the evaporation of liquid oxygen or nitrogen which solidify at 54 and 63 K respectively. In principle there was no limit to the cooling that could be achieved by making the hydrogen perform external work by, for example, driving a piston against a gas under pressure. This principle was used and well-understood by engineers who built commercial refrigeration plant but difficulties of thermal insulation and of lubrication limited the lowest temperature that could be reached in this way to about 178 K. This temperature was achieved by Ernest Solvay in Belgium in 1885–1887 and the method was patented in several countries. He recalled this work to public notice 10 years later (Solvay 1895) after Linde's success with a radically different method of cooling.

Carl Linde was a German engineer who had learned his thermodynamics from Rudolf Clausius in Zürich and who had appreciated that there was a method of cooling a gas that required no moving parts in the low-temperature section of the machinery and so which avoided the mechanical difficulties of lubrication and minimised the thermal difficulties of insulation. The method depended on expanding a compressed gas in a steady flow through a valve or throttle when the increased mean separation of the molecules in the expanded gas reduced the small attractive potential energy of pairs of molecules at small separations. It is the force of this attractive potential energy (now called the van der Waals force) that is responsible for the condensation of a gas into a liquid. Its origin and molecular magnitude were unknown in the 1880s but its existence and its macroscopic consequences had been established from measurements made in the 1850s by James Joule and William Thomson. They were not concerned with gas liquefaction but needed to know the thermodynamic consequences of such a potential energy in order to validate their calculations of the absolute scale of temperature and to secure the foundations of the two laws of thermodynamics (Chang 2004; Rowlinson 2009). The effect is small, a cooling of about 0.25 K for a fall of pressure of one atmosphere for oxygen and 0.20 K for nitrogen. For hydrogen, for which the molecular attraction is small, and for which the mutual repulsion on molecular collisions is more important, the Joule-Thomson effect, as it is now called, is negative at ambient temperatures; that is a reduction of pressure leads to a rise of temperature, or the gas warms slightly on expansion. It was expected, on the good grounds established by Andrews, that hydrogen could also be cooled by 'Joule-Thomson' expansion if the starting temperature was sufficiently low. The 'inversion point' at which this change of sign occurs is now known to be about 190 K. A so-called perfect gas, which conforms to the Boyle's and Charles's laws, would neither warm nor cool; it has a Joule-Thomson coefficient of zero.

There is little mystery about how Linde came to consider the use of the Joule-Thomson effect for cooling a compressed gas – he had learnt his thermodynamics from its leading exponent, Rudolf Clausius. There is, however, a mystery about how it came to be exploited by British chemists and engineers. By the 1890s Joule was dead and Thomson's many interests now lay outside thermodynamics. But a most unlikely candidate came forward and hit on the cooling of a flowing gas by expansion, without at first any knowledge of the Joule-Thomson effect. William Hampson was an Oxford graduate in classics and philosophy who seems to have never studied physics or engineering (Davies 1989). Nevertheless he conceived in 1894 of a new apparatus that could be used for cooling and so liquefying gases. Its essential component was an 'interchanger' in which a stream of gas at ambient temperature and high pressure is cooled by passing it down a long tube which is at each point in thermal contact with a second tube containing a colder gas flowing in the opposite direction, at a lower pressure but with the same rate of mass flow. The simplest way of ensuring this thermal contact, which was chosen by Linde, is by threading one tube inside the other. In this way the first (or high- pressure) stream is cooled on exit to the temperature of the second (low-pressure) stream at the point at which it enters the combination of tubes; the temperature of the two streams is interchanged. Hampson proposed to take gas at room temperature and pressure, compress it to, say, 200 atm. so warming it by the work done on it, cooling it back to ambient temperature with cold water, and then passing it through an 'interchanger' in which it was cooled further by the contact with the stream of colder gas flowing in the opposite direction. It next went through a partially-open valve at which the pressure fell and, for oxygen and nitrogen, the temperature fell also as a consequence of the Joule-Thomson effect – perhaps by 22 K for oxygen falling from 100 to 10 atm. This cooled and expanded gas now becomes the second stream in the interchanger and so gives up its 'cooling' to the incoming high-pressure gas. The returning gas is re-compressed and starts the cycle again but now already a little cooler than on its first pass. This recycling is continued, with the gas getting cooler in the interchanger on each cycle until it liquefies in part on the low-pressure side of the expansion valve. Hampson conceived such a device but had not the resources to build it, so in

November 1894 he went round to the Royal Institution, where he had had an introduction to Robert Lennox, Dewar's principal technical assistant, to propose collaboration. There is no evidence that Hampson knew then of the Joule-Thomson effect and so we cannot know why he thought that the apparatus would work. From the wording of a patent that he took out the next year (Hampson 1895) it seems that he supposed that the combination of an interchanger (a device already known to refrigeration engineers), expansion through a valve, and the repeated recycling of the gas, constituted a novel and so patentable process. Lennox had learnt some engineering and maybe some thermodynamics from his apparently casual attendance at lectures in Glasgow by James Thomson, the elder brother of William, but it seems that he too had not then heard of the Joule-Thomson effect. Lennox thought about Hampson's proposal and they discussed it more than once. Lennox claimed later to have told him that, on reflection, he thought that it was 'nebulous' and would not work. What seems certain, however, and is critical to our story, is that Lennox told Dewar of Hampson's visit. Dewar was obsessively secretive about his research and within a few days of Hampson's visits he had notified the Managers of the Royal Institution that he had repelled 'endeavours on the part of certain strangers to obtain access to the laboratory.' This report is apparently the only communication of this kind that he ever made to the Managers, and is to be found in their Minutes at the Royal Institution for 3 December 1894. Hampson is not named but there can be no doubt that Lennox had told Dewar of the visit and it is most unlikely that he did not also tell him the name of the visitor and of his proposal for collaboration.

When Hampson got no reply from Lennox to further letters early in 1895 he submitted his application for a patent and started discussions with Brin's Oxygen Company (later British Oxygen Company, or BOC) who were interested in any new way of preparing oxygen more cheaply than by the then usual chemical route via barium peroxide. Meanwhile Dewar and Lennox had not been idle but had decided to follow up Hampson's proposal. It seems that Dewar, although he knew of the Joule-Thomson effect, did not then associate it with this possible new route to the liquefaction of gases. He had delivered a commemorative lecture after Joule's death in 1899 (Dewar 1890) but his remarks there show that he did not attach much importance to the effect and that he had not fully appreciated their results, saying, for example, that Joule and Thomson had found that hydrogen was cooled on expansion at room temperature, whereas they had shown that it was heated.

The first mention of a radically new apparatus is in an undated entry in Dewar's laboratory notebook which, from its position in the book, must have been written in about June 1895. It was called the 'direct liquid air apparatus'. Meanwhile a tight security was maintained; that summer he refused to let either Raoul Pictet, now in Berlin, nor Heike Kamerlingh Onnes from Leiden, visit his laboratory. It was unusual for the monthly meetings of the Managers to discuss anything other than the business side of running the Royal Institution but on 2 December 1895 the new 'direct' apparatus was revealed: 'The Managers after the Meeting went to the Laboratory where Professor Dewar showed the making of liquid air by a simple method.' (Managers' Minutes 2/12/1895) The demonstration was described in the *Times* the next day, when the reporter was more impressed by the simplicity and cheapness of the new process rather by any new principle that had been exploited. Indeed, he said explicitly 'The principle is a familiar one, and has long been practically applied upon a large scale in the laboratory of the Royal Institution' (*Times* 3/12/1895). It is hard to attach any meaning to this sentence, unless the writer had in mind Dewar's invention of the vacuum flask 3 years previously.

Linde, knowing nothing of the discussions and work in London, had built his own apparatus, which he had demonstrated to scientists and engineers in Munich in May 1895. There was an account of his success in a German technical journal in August but this was probably not seen by Dewar or by Hampson since the journal was in neither the British Library nor the library of the Royal Institution. A second account Linde's work was given before a meeting in Aachen and a translation of this report appeared in Britain on 4 October in the *Engineer*, a journal that was taken by the library at the Royal Institution (Schröter 1895). The dependence of the new process on the cooling provided by the Joule-Thomson effect was now made clear, with direct quotation of the quantitative results from Joule and Thomson's papers. The appearance of Schröter's report put Dewar in a tricky position. He was clearly proud of his new 'direct' route to liquid air, which he had not yet published, and he was keen to claim credit for it. Was he to try to acquire as quickly as he could information about this novel use of the Joule-Thomson effect, or was he to claim that the effect was already well-known and it was his ingenious combination of its use with the interchanger that was his claim to novelty? In the event, he privately followed the first line, while in public he followed the second. Within days of the publication of Schröter's paper he had written to Thomson (now Lord Kelvin) to tell him of his success with liquid air (Dewar 1895a):

> Private 29th October 95
>
> Dear Lord Kelvin,
>
> I have just received the results of some experiments which I left going this morning. And I think that they will interest you. For long I have been at the Hydrogen problem and have almost given it up in despair, as beyond my resources. The result has been that I have had to invert the problem and ask myself how can air be liquefied from temperatures far above the critical point.
>
> This I find is easy if I repeat the old experiment you made with Joule. - - - Thus I feel almost certain that if hydrogen in the liquid state can be collected I will do it through and on account of the work of Kelvin and Joule
>
> Ever Yours Truly
> James Dewar

To find out more about the Joule-Thomson effect he went to his usual source of advice on problems in theoretical physics, his brother-in-law, Hamilton Dickson. He was a mathematician and physicist who had trained under William Thomson at Glasgow and was now a Fellow and the mathematical tutor at Peterhouse, Cambridge. He had been giving such advice to Dewar from the time of his marriage to the younger sister of Dewar's wife in about 1879. He now wrote on 11 December 1895 to explain, not very clearly, how the Joule-Thomson coefficient could be calculated, and how the lowering of the temperature could be calculated from Joule and Thomson's experimental results (Dickson 1895). So Dewar was well-prepared when he described and demonstrated his new apparatus before the Chemical Society on 19 December (Dewar 1895b, 1896). He did not, however, mention explicitly his use of the Joule-Thomson effect, nor the work of Linde and Hampson, saying only that 'The experiments of Joule and Thomson and Regnault on the temperature of gas jets issuing under low pressures are well known'. He went further with the extraordinary claim that: 'Apart, therefore, from important mechanical details, and the conduct of the general working, nothing new has been added by any investigator to the principles involved in the construction and use of low-temperature apparatus since the year 1878'. In the discussion of the paper, Bertram Blount, an 'engineering chemist' known for his plain speaking, asked why Dewar had not acknowledged the priority of Linde's work, as reported in Schröter's paper. Dewar could only reply that if his own apparatus had any 'resemblance to the Linde apparatus described by Mr Blount, chemists, for once, may be congratulated that a small laboratory apparatus works in some respects better than a large industrial plant.' He was on safe ground in denying any debt to Linde since there was too little time between the appearance of Schröter's paper in Britain on 4 October and his own demonstration before the Chemical Society on 19 December for the design, building and testing of a new apparatus.

Hampson enlisted the help of Brin's Oxygen Company to build an apparatus that was close in design to that of Dewar, a resemblance that argues again for a common source in Hampson's original drawings of November 1894 – Linde's, although working on the same principles, was of quite different design. Hampson demonstrated his apparatus publicly in March 1896, but it seems that he still did not appreciate the crucial role of the Joule-Thomson effect. A pseudonymous correspondent in *Nature* wrote that there is no evidence that Hampson understood the principle of his apparatus. His patent application and his demonstration contained nothing on the subject, a point to which Linde himself later drew attention ('Zero' 1896; Linde 1899). The almost simultaneous exhibition of this new way of liquefying gases by Dewar and Hampson led to long and acrimonious correspondence about plagiarism and discourtesy in the scientific and technical press, mainly between the two principals but which involved Lennox also, writing under the pseudonym 'Arenal' – his initials were R.N.L. This correspondence led nowhere and added nothing to the science of the field.

The next step was to try to liquefy hydrogen in bulk. This was clearly going to be difficult since its critical temperature had been estimated correctly by Wróblewski at −240 °C (or 33 K). The expansion of hydrogen, pre-cooled by liquid air, in Dewar's new apparatus had led to a jet of a mixture of gaseous and liquid hydrogen but there was no way of collecting the small amount of liquid from the rapidly moving jet. His first idea was to liquefy a mixture of nitrogen and hydrogen and to use this as a first step in cooling pure hydrogen to a temperature at which the Joule-Thomson coefficient was positive. This attempt failed, probably because hydrogen is not sufficiently soluble in liquid nitrogen. He had therefore to embark on building an improved liquid air apparatus with which to pre-cool his compressed hydrogen. There was first a small diversion while he combined forces with Henri Moissan to liquefy fluorine, the only elementary gas other than hydrogen and the recently discovered helium that was still to be liquefied. This problem was soon solved (Moissan and Dewar 1897) and he was then able to turn his attention again to hydrogen. There is no full description of the apparatus he built. None was published and so the apparatus that was subsequently acquired by the Science Museum in London was that designed by Hampson 2 years after Dewar's success.

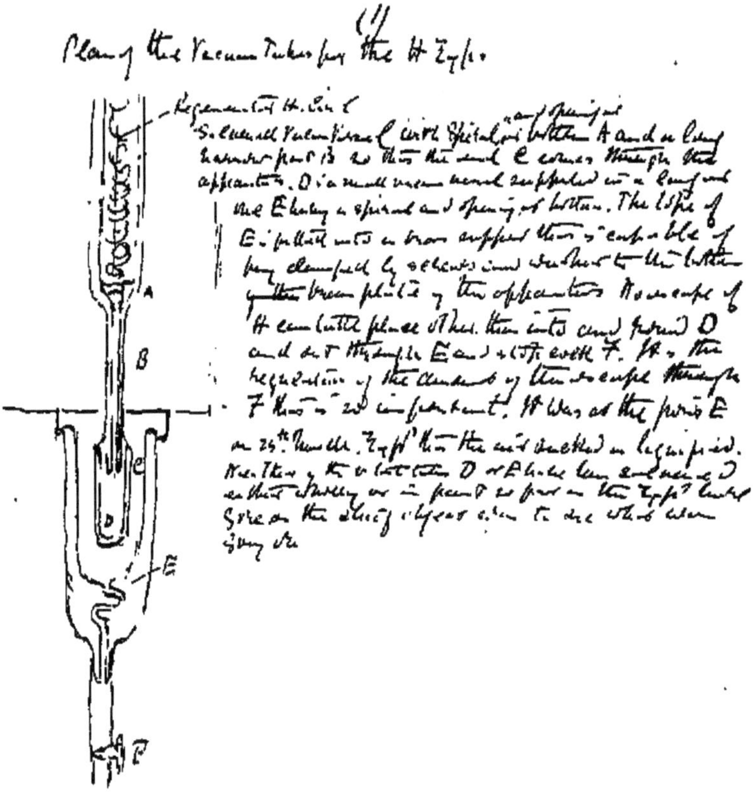

This sketch in Dewar's laboratory notebook shows only the arrangement of the unsilvered vacuum vessels (Gavroglu 1994). These were fastened below the new liquid air plant that allowed the hydrogen to be pre-cooled down to about 57 K by liquid air under a reduced pressure of about 6 in. (15 cm) of mercury. This apparatus took a year to build and test. The sketch is entitled 'Plan of the Vacuum Tubes for the H[ydrogen] Exp.' and the helical tubes at the top are labelled 'Regenerator H coil' (Dewar always used the word 'regenerator' for what is now usually called an 'interchanger'.) The legend alongside the sketch reads, after deciphering Dewar's terrible hand-writing:

> Silvered Vacuum Vessel with spiral, and opening at bottom A and a long narrow part B so that the end C comes through the apparatus. D is a small vacuum vessel supported in a longer one E having a spiral and opening at bottom. The top of E is fitted into a brass support that is capable of being clamped by screws and washer to the bottom brass plate of the apparatus. No escape of H[ydrogen] can take place other than into and round D and out through E and stop cock F. It is the regulation of the [amount?] of the escape through F that

is so important. It was at the point E on 25th March Exp that the air sucked in liquified Neither of the v[acuum] test tubes D or E have been silvered wholly or in part as far as the Exps have gone as the chief object was to see what was going on.

The principal novelty of this apparatus was the helical capillary coil that allowed the liquid hydrogen collected in D and C to pass to the stop cock F. This gave sufficiently elasticity to the whole assembly to allow the glassware to adjust its shape to the changes produced by the severe temperature gradients. He later wrote of this coil:

This device, developed after many unsuccessful attempts at such a vacuum vessel, was found essential for the easy production and collection of liquid hydrogen, and as all the Royal Institution designs for such vessels have been made in Germany, they have been supplied and utilized by other workers unconscious, it may be of where or how they originated (Dewar 1901).

The best published account of his liquefaction of hydrogen is in the Friday Evening Discourse that he gave at the Royal Institution on 20 January 1899, but a fuller picture can be had from the laboratory notebook cited above. Serious experiments started on 25 February 1898:

All apparently went well until after the liquid air was put under exhaustion when attempting to open the H[ydrogen] valve nothing would come. Nothing would remove the obstructions. Thought it was solid air but the stoppage remained hours afterwards so it looks like some solid carried forward at valve. - - - It was found that the obstruction was due to solid particles of solder carried by H[ydrogen].

14th March/98.

Between 25th February and this date had repeated failure in getting Silvered Vacuum. Tube to fit the Regenerator. - - - Started the apparatus as described on the 25th Feb^y - - - . After a few minutes a white solid along with [?] of liquid, say 5 cc, collected in the inner of the two vacuum tubes round which the H[ydrogen] coming from the Regenerator [was] passing.

25th March/98.

Since the previous Experiments we placed an enlarged coil of [?] pipe of 600 cc capacity over the end of the H[ydrogen] coil in the L[iquid] H[ydrogen] chamber before it goes [through] the Regenerator in order to catch liquid air and other impurities in the H[ydrogen]. The whole arrangement seemed to work well. - - -

25th April.

Repeated the above experiment

10th May/98

Started Hydrogen apparatus treating CO_2, C_2H_4 and air circuits as in previous Exp^t. This time the liquid was collected for 2½ hours and the air pump put on for the last half hour.- - - After ½ hour the air supply was stopped and the air pump was allowed to get the pressure reduced to 6 inches of mercury. The H[ydrogen] started running through the valve at 175 atmospheres and about the rate of 15 c. ft a minute [7 litres a second]. Shortly after starting the nozzle plugged but it got free by good luck and almost immediately drops of liquid began to fall from the V[acuum] tube into the [inner] V[acuum tube] and soon accumulated to 20 cc in volume. There was a great escape of H[ydrogen] through the [stopcock] and the outer Vacuum Vessel was covered with ice. The H[ydrogen] was a clear transparent liquid with a well defined meniscus (even better than liquid air) showing no absorption spectrum and as long as the surroundings of the [inner] vessel were cool seemed to evaporate very slowly. I took off the [vacuum] vessel from the apparatus and placed in the liquid H[ydrogen] a narrow quill tube; sealed at the end immersed and open at the other; when immediately solid air appeared in the tube. In the same way the He[lium] tube was placed in the liquid hydrogen and I could see that a liquid was formed. This He[lium] tube gave nothing when placed in liquid air under exhaustion. The H[ydrogen] vapour above the liquid had a very marked appearance and seemed very dense with circulating currents of a greenish blue cloud. - - - The H[ydrogen] gas before use was tested for the presence of air and was found not to exceed 1 %.

By chance Dewar's colleague Lord Rayleigh was present in the Royal Institution when hydrogen was first liquefied and was witness to this success. That day, a Tuesday, Dewar asked the President, Lord Lister, if he might announce his results at the regular meeting of the Royal Society on the Thursday. This he did, but he worked in the laboratory until the last minute. The notebook continued:

12 May

Conducted everything as on the 10th starting with the H gas at 140^{atm} and [?] collected about 60 cc and took it out of the apparatus placing the V[acuum] tube in liquid air so that the outer surface was always [?] -183°. - - - Dipping some cotton wool into the liquid and lifting into the air it ignited with a [?] giving a [?] of flame. The cotton wool saturated with L[iquid] H[ydrogen] between the poles of a magnet was attracted. Believe this due to liquid air being condensed on its surface as solid.

The claim to have liquefied helium was soon tacitly dropped; again an impurity in the hydrogen was probably responsible for the formation of this liquid. The liquefaction of helium was to prove too difficult for the skill of Dewar, for the resources of the Royal Institution, and for the problem of obtaining sufficiently pure helium. It was to be another 10 years before Kamerlingh Onnes in Leiden succeeded with the last of the 'permanent' gases.

In 1899 the solidification of hydrogen followed its liquefaction, when Dewar cooled his liquid hydrogen by evaporation. He estimated the temperature of the melting point by gas thermometry to be about 16 K, a little above the currently accepted value of 14 K. By pumping on the solid he cooled it further to perhaps 10 or 11 K, but this was still a long way from the critical the temperature of helium, now known to be at 5.2 K.

The story of the liquefaction of hydrogen had an unexpected sequel. In 1904 an international exhibition was held in St Louis, Missouri, at which all nations were invited to show what they could do in the arts, science and technology. The centrepiece of the British entry in the Chemistry Section was chosen to be an apparatus for the liquefaction of hydrogen, an operation not yet attempted in the United States. To this end a properly engineered version of Dewar's apparatus of 1898 was designed and built by the firm of Lennox, Benton and Reynolds. This was a small engineering company in the Rodebush Works in Fulham in south-west London that Robert Lennox had established, with Dewar's cognisance, for making his ever increasingly complicated apparatus. The new version, built mainly in brass and other metals, worked on the same principles as the 1898 apparatus but was on a bigger scale. It was assembled and tested at the Royal Institution before being shipped to America. Dewar never intended to go to the exhibition himself but entrusted the demonstration of the apparatus to Joseph Petavel who had worked in the Davy Faraday Research Laboratory at the Royal Institution and had become first known to Dewar when he was the assistant to Ambrose Fleming at University College, London. Dewar and Fleming had cooperated in electric and magnetic experiments at low temperature. Dewar was a difficult man to get on with, and had previously had a row with Petavel, but had now accepted that he was the best man for the job. He ended his career as Director of the National Physical Laboratory. The choice of Petavel did not prevent Dewar from raising last-minute objections to the use of the German-made vacuum tubes with the necessary helical spiral. It is not clear what the difficulty was, but fortunately Robert Hudson, one of the British engineers involved, managed to get a fresh supply directly from the glassblowers in Berlin which reached St Louis just in time for the opening of the exhibition (Hudson 1964). The apparatus proved to be perhaps the greatest hit of the Exhibition. It produced liquid hydrogen twice a week for 2 months without a single hitch. Each public demonstration by Petavel needed between 1 and 3 l of liquid hydrogen, amounts that were then quite unprecedented. Some of this was solidified and this was also exhibited. Around these spectacular demonstrations Petavel gave over 80 lectures to the public at large and to professional physicists, chemists and engineers. At the end of the exhibition the apparatus received a Grand Prize, the highest award, and Dewar and Petavel received respectively a gold and a silver medal. The equipment never returned to Britain but was bought by the US Government, presented to the National Bureau of Standards, and re-erected in Washington. If the liquefaction of hydrogen was the climax of Dewar's career then this superbly engineered apparatus was the finest achievement of his principal assistant, Robert Lennox. But, as so often with Dewar's affairs, the story did not have a happy ending and a few years later, when Dewar blamed Lennox for his part in the failure to liquefy helium at the Royal Institution, Lennox resigned and, in the words of one of his colleagues, 'the Laboratory lost the most outstanding Assistant it had ever known' (Green 1956).

References

Brock, W.H. 2002. Exploring the hyperactic: James Dewar at the Royal Institution. In *The common purposes of life: Science and society at the Royal Institution of Great Britain*, ed. F.A.J.L. James, 169–190. Aldershot: Ashgate.

Chang, H. 2004. *Inventing temperature: Measurement and scientific progress*, 159–219. New York: Oxford University Press.

Davies, M. 1989. William Hampson (1854–1926). *British Journal for the History of Science* 22: 63–73.

Dewar, J. 1884. On the liquefaction of oxygen and the critical volumes of fluids. *Philosophical Magazine* 18: 210–216.

Dewar, J. 1886. Recent researches on meteorites. *Proceedings of the Royal Institution* 21: 541–550.

Dewar, J. 1890. The scientific work of Joule. *Proceedings of the Royal Institution* 13: 1–6.

Dewar, J. 1895a. Letter to Lord Kelvin, 29/10/1895. Cambridge University Library, Cambridge, Add. Ms. 7342, D 87.

Dewar, J. 1895b. The liquefaction of air and research at low temperatures. *Proceedings of the Chemical Society* 11: 221–234.
Dewar, J. 1896. New researches on liquid air. *Proceedings of the Royal Institution* 15: 133–146.
Dewar, J. 1901. Gases at the beginning and end of the century. *Proceedings of the Royal Institution* 16: 730–737.
Dewar, J. 1902. Presidential address to the British Association in Belfast. *Proceedings of the British Association for the Advancement of Science* 72: 1–47.
Dickson, J.D.H. 1895. Letter to James Dewar, 11/12/1895. In RI Archives, D Vb/2.
Gavroglu, K. 1994. On some myths on the liquefaction of hydrogen and helium. *European Journal of Physics* 15: 9–15, from laboratory notebook at the Royal Institution, D Vb/1.
Green, W.J. 1956. Some memories of the Royal Institution and its laboratories 1900–1950. *Proceedings of the Royal Institution* 36: 543–567, reprinted as Appendix 2 in James (2002).
Hampson, W. 1895. Improvements relating to the progressive refrigeration of gases. Patent No. 101325, Provisional application, 23 May 1895, Complete application, 19 Feb 1896.
Hudson, R.S. 1964. *Recollections of a technologist*, 45. London: Pitman.
James, F.A.J.L. 2002. *The common purposes of life': Science and society at the Royal Institution of Great Britain*. Farnham/Great Britain: Ashgate Publishing.
Linde, C. 1899. Zur Geschichte der Maschinen für die Herstellung flüssiger Luft. *Berichte der Deutschen Chemischen Gesellschaft* 32: 925–927.
Moissan, H., and J. Dewar. 1897. On the properties of liquid fluorine. *Proceedings of the Chemical Society* 13: 175–184.
Rowlinson, J.S. 2009. James Joule, William Thomson and the concept of a perfect gas. *Notes and Records of the Royal Society* 64: 43–57.
Schröter, M. 1895. The commercial manufacture of liquid air and oxygen. *Engineer* 80: 325.
Solvay, E. 1895. Sur la production méchanique des températures extrêmes. *Comptes Rendus de l'Académie des Sciences Paris* 121: 1141–1143.
'Zero'. 1896. The Hampson liquid oxygen apparatus. *Engineer* 81: 326.

Chapter 4
The Cryogenic Laboratory of Heike Kamerlingh Onnes: An Early Case of Big Science

Dirk van Delft

"The polar regions of physics appeal to the fighting spirit of scientists in the same way that the extreme North and South Poles appeal to the explorer." (Kamerlingh Onnes 1904: 5) The Leiden-based cold pioneer Heike Kamerlingh Onnes used these words in his founder's day speech in 1904, the year in which he served as rector of the University of Leiden. Polar expeditions fired the imagination. Just 1 year earlier, the Norwegian adventurer Roald Amundsen had succeeded in navigating the north-west passage between the Atlantic and the Pacific Oceans. His heroic contest with British explorer Ronald Scott to be the first to reach the South Pole still had to take place. At the start of 1912, Scott's team found a Norwegian flag planted at the Pole by Amundsen 1 month previously; tragically, Scott's team was not to survive the return journey (Huntford 1985). At the same time, Kamerlingh Onnes was involved in a struggle that itself was not bereft of danger. The goal of his journey was the descent to the absolute zero of temperature, −273 °C.

Heike Kamerlingh Onnes (1853–1926), son of a brick manufacturer from Groningen, was delicately built as a youth, and throughout his life struggled with chronic bronchitis. His first and greatest love was chemistry, and during a *Wanderjahr* in Heidelberg, at the tender age of 18 years, he was as proud as a peacock to be able to spend one semester under the tutelage of the great Robert Bunsen, an experimenter through and through. On the other hand, Kamerlingh Onnes also demonstrated an interest in theory, and in Heidelberg that meant studying under the physicist Gustav Kirchhoff. Having won a *Seminar* prize in Kirchhoff's physical practicals, he was offered an assistantship, and it was in that atmosphere that the physicist Heike Kamerlingh was born.

On 11 November 1882, Kamerlingh Onnes conducted his inauguration speech in Leiden as professor of experimental physics. In that speech, in which he for the first

D. van Delft (✉)
Director Museum Boerhaave, Professor Material Heritage of the Sciences,
Leiden University, Leiden, The Netherlands
e-mail: dirkvandelft@museumboerhaave.nl

K. Gavroglu (ed.), *History of Artificial Cold, Scientific, Technological and Cultural Issues*, Boston Studies in the Philosophy and History of Science 299,
DOI 10.1007/978-94-007-7199-4_4, © Springer Science+Business Media Dordrecht 2014

time used the now celebrated phrase, 'Through measurement to knowledge', he revealed the nature of his research programme, in minute detail. Over the coming years he would undertake to experimentally test the molecular theories (equation of state, law of corresponding states) of his mentor and friend Johannes Diderik van der Waals, and according to the results would improve and refine those theories (Kamerlingh Onnes 1882; Laesecke 2002; Kipnis et al. 1996).

This could only be achieved in a well-equipped research laboratory. Consequently, Kamerlingh Onnes started on the task of rigorously rebuilding the fundamentally sound but not particularly dynamic Physical Cabinet of his predecessor Rijke. Kamerlingh Onnes had not only inherited weak lungs from his father, but also a tremendous thirst for work, and entrepreneurial talent. Just like his father Harm, Heike emerged as a manufacturer, albeit not of roof tiles but of cold.

He was excellently served by his entrepreneurial talent. The establishment of a cryogenic institute of international allure, the likes of which had been seen nowhere in the world in terms of scale and staff, demanded more than the talents of merely a good physicist. Anyone who paid a visit to the Steenschuur in Leiden (opposite the Van der Werf park) and there observed the confusion of pipes and taps, gas bottles, gas holders, pumps, throbbing engines, workshops and the glass-blowing shop, and the instruments and appliances for scientific research, may have imagined himself in a factory – a cold factory, with Kamerlingh Onnes as professor and director, who kept a tight reign on all activities, and who set the course to be followed. He also established a well-oiled organisation with a curator in charge of management, an academic staff (including assistants and PhD students), a head of operations, instrument makers, glass blowers, laboratory assistants, technicians, a mechanic, a housemaster and not to be forgotten a small army of 'blue boys': the apprentices of the instrument makers' school, clad in blue overalls, whose education was fully integrated in the laboratory. Gerrit Jan Flim, master instrument maker and head of operations of the cryogenic laboratory, a man with hands of gold, was himself a 'home-grown' product.

This approach – that on a scale and with a degree of focus never seen before, was a clear predecessor of what later became known as Big Science – could only be successful under a director who demonstrated perseverance, courage, force of will, vision and huge patience; a man who 'ruled' with a hard hand, but who at the same time had the gift of winning people over to his ideas, convincing them and gaining their loyalty; a man who understood the art of incessantly playing the authorities until they eventually gave the 'expensive professor' what he wanted. According to Kamerlingh Onnes, the 'destruction' of what had been established in the Steenschuur so carefully after so many years was a constant threat. Here was a man who was easily capable of establishing contacts, had an excellent eye for those who could advance his aims, who mollycoddled his guests and who was far too tactically-skilled to maintain fights or feuds that could cause him any harm. Kamerlingh Onnes was an inventive, thorough scientist, but the huge success of his cryogenic laboratory was first and foremost due to his organisational talent, social skills and consistent focus on the creation of cold (Delft 2007; Gavroglu and Goudaroulis 1991).

Now back to the Polar regions. Kamerlingh Onnes' scientific programme was based on two spearheads: accuracy and cold. Because his interest was above all focused on *deviations* from the laws of Van der Waals, only precise observations counted, and because single molecules like nitrogen and hydrogen were his preferred mediums, and because those substances only became liquid at extremely low temperatures, Kamerlingh Onnes had no option but to build a cryogenic laboratory (*kryos* is the Greek for cold).

Reaching very low temperatures can only be achieved by making use of a special technique – the liquefaction of gases. In 1845, Michael Faraday, director of the Royal Institution in London, succeeded in liquefying all the then known gases with the exception of just six: hydrogen, oxygen, nitrogen, nitrogen monoxide, methane and carbon monoxide. These 'permanent gases' refused to surrender even under the highest pressures (of several thousand atmospheres). The reason was that the critical temperature of these six was so low that even a mixture of solid carbon dioxide gas (dry ice) and ether (−110 °C) was still too warm, as a coolant (Faraday 1845).

What is the critical temperature? In 1822, the French researcher baron Cagniard de la Tour established a vacuum in a glass tube filled 40 % full with alcohol and melted it closed. As the tube was heated, he saw how the liquid first expanded, at the expense of the volume of the alcohol vapour above it. This continued until a certain point at which the level of the liquid – (the meniscus) the separation between liquid and vapour – actually disappeared. De la Tour concluded that at a certain temperature – the critical temperature – the liquid changed into gas. However, the Frenchman could just as easily have claimed the opposite. Above the critical temperature, the conclusion was, the difference between liquid and vapour simply disappeared; all that was left was gas (Cagniard 1822: 127–132 and 178–182). The value of that critical temperature differs from substance to substance: for water it is 374 °C, for carbon dioxide 31 °C, for oxygen −118 °C and for helium −268 °C, just 5° above absolute zero. Because above the critical temperature only the gaseous phase occurs, any attempt to condense the gas in that situation is doomed to failure. No matter how high the pressure is raised, no droplet of liquid will be created.

Johannes Diderik van der Waals (1837–1923) provided the phenomenon of critical temperature with its theoretical underpinning. This son of a carpenter from Leiden was employed as a teacher, and only started his studies in 1862. In his thesis from 1873, entitled *About the continuity of the gas and liquid state*, Van der Waals came up with an essential alteration to the general gas law. For ideal gases (pointed molecules; no intermolecular forces) the general gas law is as follows: $pV = \mathrm{R}T$. In this formula, p is the gas pressure, V the volume, T the absolute temperature (in kelvin) and R a constant. However, not all gases are ideal. The molecules do occupy space and clearly attract one another.

These ideas were expressed in an altered general gas law. Via study of theory, Van der Waals arrived at the equation $(p + \mathrm{a}/V^2)(V - \mathrm{b}) = \mathrm{RT}$. The constants a and b relate to the own volume of the molecules and their mutual attraction. A graphic reproduction of this *equation of state* in a pV diagram (a graph in which the pressure is shown vertically and the volume horizontally) with lines at constant temperature (*isotherms*) generates practically the same hyperboles as those of an ideal gas, above critical temperature. However, in the vicinity of critical temperature, deviations in the hyperbole form arise, this applies to an even greater extent on and close to isotherms below critical temperature (Levelt-Sengers 2002).

Van der Waals' theoretical isotherms tied in well with the measurement results published in 1869 by the Irish researcher Thomas Andrews. As the volume is reduced, the pressure does not continue to rise permanently but at a certain point becomes 'stuck'. Under even further compression, the vapour at a constant pressure gradually turns into a liquid: condensation. Only once all the vapour is 'used up' does the pressure once again rise, at an accelerated pace: a liquid is very difficult to compress (Andrews 1869).

In 1882, Heike Kamerlingh Onnes, in his inaugural speech, referred to the pumps produced by fellow cold pioneers Louis-Paul Cailletet and Raoul Pictet (the discoverers of liquid oxygen; 1877) as 'essential laboratory tools', and within just 1 year, he had some of these pumps in his possession. A real start could only be made on establishing the cryogenic installations in Leiden during academic year 1885–1886, once the internal reconstruction of the laboratory had been completed.[1] In Châtillon-sur-Seine, Kamerlingh Onnes ordered the mercury compressor described in 1883 by Cailletet. This pump for compressing pure gases was delivered in 1885, but only after years of adjustment did it meet the stringent requirements imposed on it by Kamerlingh Onnes. The remarkable feature of the compressor was that a layer of mercury rested on the piston. By continuously adopting the shape of the cylinder, this prevented a proportion of the inflowing gas from being left behind at the end of each compression stroke. Kamerlingh Onnes was so enthusiastic about this principle that even when the compressor turned out to be "entirely dissatisfactory", he was "more than willing to spend several years" improving Cailletet's design (Kamerlingh Onnes 1894).

The elaboration and testing of rigorous alterations and innumerable minor improvements to the pumps took oceans of time and was not without danger. In April 1889, a newly-installed overpressure valve burst at a pressure of 20 atm. "Glass in every corner of the room, considerable shock," noted Kamerlingh Onnes in his laboratory logbook. "The copper sleeve has been rudely ripped open."[2] A rifle barrel was then used as the basis for a new valve. The Cailletet pump did not function perfectly until 1890. However, Kamerlingh Onnes then had in his possession a "laboratory tool" on which he could rely 100 %. Leiden profited enormously from the Cailletet compressor that in combination with a booster achieved a pressure of 100 atm. It was above all useful in situations where the presence of contaminants in and around the gases to be pumped would be disastrous.

The precise design, fabrication, testing and improvement of a cryogenic installation that produced liquid oxygen caused Kamerlingh Onnes massive headaches.

[1] For the construction of the cryogenic laboratory see Kamerlingh Onnes, H. 1894., 'On the Cyogenic Laboratory at Leiden and on the production of very low temperatutes'. *Communications*, 14; van Helden, A. C. 1989. *The coldest spot on earth*. Leiden: Museum Boerhaave.

[2] Museum Boerhaave, archives of Heike Kamerlingh Onnes, inv. no. 46.

Leiden, too, opted for the step-by-step cooling method, using circuits: a cascade à la Raoul Pictet. At the Société Genevoise pour la Construction des Instruments Physique, where Pictet also had his instruments built, Kamerlingh Onnes ordered two conjugated pumps, in 1885. In this type of pump, two pistons were mounted together, on the same piston rod. One cylinder created a vacuum while the other compressed the gas. The purchased pumps were of trade standard and before Kamerlingh Onnes had raised their performance and reliability to laboratory level, several months had gone by. In 1894, looking back on the job, he wrote, "Eradicating larger and smaller leaks and defects from all the parts; gaskets that seal perfectly; installing suitable pipes; fitting cocks with cork gaskets that do not become stuck by the cold; viewing glasses that constantly show the status of the condensed gas; filter devices that protect the taps; fabricating all these took a huge amount a time. Much of what is currently available in the trade was then not yet known and had to be tediously manufactured. Practice had to be gained in all kinds of unusual activities." In addition, as Kamerlingh Onnes recognised, working with gases such as sulphurous acid, chloromethyl and ethylene engendered explosive hazards and "in an educational institution, hazards are permitted to a far lesser extent".

Kamerlingh Onnes quickly demonstrated his dissatisfaction with the Pictet design. As early as 1888, as revealed by his laboratory logbook, he replaced the toxic sulphurous acid (a substance that also caused internal damage to the circuit) with chloromethyl.[3] To 'spill' as little cold as possible, Kamerlingh Onnes fitted the boiling flask with a regenerator coil: the ethylene from the second circuit travelled via a spiral tube to the chloromethyl boiler (to condense there under 8 atm of pressure). The chloromethyl vapour that was pumped in the opposite direction released part of its cold to the coil, thereby precooling the ethylene. In practice, it took three-quarters of an hour for the boiler in the first circuit to be filled with liquid chloromethane at –70 °C.

As already stated, circuit 2 contained ethylene. This was an excellent coolant, but its use did have some disadvantages. Ethylene with air for example formed an extremely explosive mixture so that a series of safety measures had to be taken: steam heating in the room in question, electric lighting as opposed to gas lamps and the installation of a powerful ventilator fan to extract any escaping ethylene or chloromethyl gas.[4] The purity of the ethylene also caused problems. When Kamerlingh Onnes started his ethylene circuit in 1887, the gas was not available on the market. Kamerlingh Onnes was therefore forced to produce his own ethylene in a time-consuming and costly process.

The construction of the ethylene boiling flask, 22 cm in diameter at the bottom and gradually tapering towards the top, was determined by a range of factors: the optimum balance between the amounts of condensation surface, the metal to be cooled, the ethylene required, the oxygen produced and the heat leakage into the

[3] MB, archives of HKO, inv. no. 46.

[4] HKO to Pieter Zeeman, 22 September 1893, North Holland Archives, Zeeman archives, inv. no. 82.

environment. With a view to thermal insulation, Kamerlingh Onnes had wrapped the appliance in layers of sheep's wool. According to the expert consulted, wool – washed, degreased in warm water, cleaned and "combed in French balls or blanket layers" – delivered the best result. The layers of wool, placed in compartments divided by felt, were separated from one another by cotton and varnished paper – all intended to suppress the convection, conduction and radiation of heat. The result was remarkable: within a few hours, without becoming wet (let alone freezing) on the outside, the boiling flask contained a litre of liquid ethylene, boiling at 2–3 cm of mercury pressure. The device was designed in such a way that the extracted ethylene gas, that precooled the incoming oxygen gas, left the boiling flask at room temperature. All the available cold was used.

In the meantime, the oxygen circuit was also completed. Following in the footsteps of the Polish researcher Zygmunt von Wroblewski, in 1885, Kamerlingh Onnes wanted to tap liquid oxygen from the bottom of the construction and then transfer it to a separate glass vessel (a cryostat) in which it could boil, whether or not under reduced pressure. The vessel also contained thermometers and preparations to be investigated.

After 18 months of tinkering with the ethylene boiling flask and oxygen boiling cabinet, the sealing of innumerable leaks and the distilling of hectolitres of ethylene, the celebrations rang out on 23 June 1892: Kamerlingh Onnes captured his first liquid oxygen in the 'Wroblewski glass'. Pieces of dirt were floating on it and at just 20 ml the volume was modest. "This cold is awe-inspiring," wrote the otherwise so sober Groningen-born researcher in his laboratory logbook. Unfortunately the joy was only short-lived. During its very first use, the oxygen boiling cabinet developed a defect. By December 1893, however, an installation worthy of its name was completed. Around Christmas it produced a quarter of a litre of liquid oxygen, which Kamerlingh Onnes then went on to boil under low pressure. Over a period of several hours, he revealed the blue liquid to a number of scientific friends. In May 1894, Kamerlingh Onnes even took a vessel of oxygen with him to the monthly meeting of the Academy of Sciences in Amsterdam.

The cold experiments in Leiden drew broad international interest. In August 1896, *Nature* was full of praise for Kamerlingh Onnes' cryogenic installations. The article emphasised that Leiden had opted for a totally unique approach. Ideas from Pictet (cascade), Cailletet (use of chloromethane and ethylene) and Wroblewski (system for capturing liquid oxygen) had been integrated in an installation that could produce a half a litre of liquid oxygen in a permanent bath. By focusing sufficient attention on safety and efficiency – the cascade delivered an extremely high yield – in the eyes of *Nature*, Leiden was a "model cryogenic laboratory". "One cannot but admire the perseverance and skill which the development of this system reveals," commented *Nature*. The British magazine was also full of praise for the experiments currently underway in the Leiden laboratory. "In short, the place is rich in apparatus of all kinds, and possesses numeral appliances; so much so, that one would rank it amongst the best provided (and, one may add, most productive) research laboratories."[5]

[5] 1896. 'The physical laboratory at Leiden (Holland)', *Nature* 54: 345–347.

The race to be the first to liquefy hydrogen gas had still not been won in 1894, when Kamerlingh Onnes finally perfected his cascade for liquid oxygen. The race between laboratories in Leiden, London Krakow and Berlin was not some subsequently dreamed-up historical, scientific invention; Kamerlingh Onnes himself said in 1895 that he was "permanently involved in a competition with foreign players".[6] As far back as 1884, the Poles Wroblewski and Olszewski had succeeded in briefly liquefying hydrogen, but there was no possibility of gentle boiling, and in 1895, James Dewar in his Royal Institution, had also not yet achieved success. On the other hand, in 1891 and 1895, Olszewski released rough determinations of the critical pressure and critical temperature of hydrogen, both vitally important elements in deciding whether the gas would allow itself to be defeated.

However, before Kamerlingh Onnes could truly launch his attack on liquid hydrogen, he was faced with a setback from an unexpected corner. In January 1895, a letter from the Municipal Council of Leiden was delivered, in which the question was asked whether the professor director in fact had a permit to store in the gardens of his physics laboratory "substances that could represent a threat to the surrounding area".[7] The area was a sensitive one, because the laboratory was built at the same location where in 1807 an exploding gunpowder ship had caused an enormous disaster. According to a historian a century after a the event, following the explosion "the houses trembled, the walls split and collapsed, and the most beautiful district of Leiden was converted into one gigantic, dismal heap of rubble and an icy grave for any number of its citizens" (Knappert 1906: 22).

Kamerlingh Onnes himself had been the author of his own downfall, by complaining to the police about wild behaviour in the gardens where "the stored gas cylinders ran the risk of exploding, if inexpertly handled". In his response to the Council of Leiden, Kamerlingh Onnes revealed that he was storing in the laboratory a cylinder containing 11.8 kg of chlorine, one with 4.4 kg of ammonia and one with 28.8 kg of sulphurous acid. "All these gases are toxic if released in large quantities into the air." He also revealed that he was storing in the garden demijohns of ether and oil, and methane in a gas container. In the same statement, he went on – quite unasked – to inform the Municipal Council of the substances he used in his laboratory: "drums under high pressure filled with carbonic acid, oxygen, nitrogen, hydrogen, ethylene, and methane in varying amounts of each substance, while the need could certainly arise to also store in that same location other compressed gases in drums."[8]

The result was a dispute with the authorities that dragged on and on. While waiting for a decision from the Council of State, Kamerlingh Onnes was required to halt his cold research. It took 3 years before the Nuisance Law permit was finally awarded. Only on 4 June 1898 was the cold-free period in Leiden brought to an end. By that time, the race to liquefy hydrogen had just come to a conclusion.

---

[6] HKO to *curatoren*, 5 April 1895, Leiden University Library, Archief curatoren.

[7] B&W (Leiden municipality) to HKO, 19 January 1895, archives Huygens Laboratory.

[8] HKO to B&W (Leiden municipality), 26 January 1895, archives MB, inv. no. 74.

On 12 May, in the Royal Society in London, James Dewar had presented his 'Preliminary Note on the Liquefaction of Hydrogen and Helium'. Two days earlier, he explained, in the cellars of the Royal Institution in London, he had succeeded in capturing 20 ml of liquid hydrogen in a double-wall insulated glass vessel. The liquid was clear and colourless. Dewar would have been delighted to produce more, but his installation (a circuit) had become blocked up after just 5 min by air freezing in the pipes. Nonetheless, he had succeeded in inserting two thin tubes into the liquid hydrogen. The first was open at the top, and upon reaching the liquid hydrogen, the air in it froze immediately. The second tube was connected to a globe containing helium gas, which upon immersion immediately started to condense. "All known gases have now been condensed into liquids," announced a proud Dewar. It was however half a triumph: not the helium itself had condensed, but a contaminant present in the gas (Dewar 1898).

In order to arrive at static hydrogen (boiling gently in a dewar flask), the cascade method proved insufficient. With pumped-off liquid hydrogen, a temperature of 54 K (−219 °C) could be achieved; colder was not possible because any oxygen present itself then froze. Because the critical temperature of hydrogen was 33 K, a gap of 20° had to be bridged before liquid hydrogen could even be envisaged. The gap was bridged by the Joule-Thomson effect. In 1852, James Joule and William Thomson (Lord Kelvin) were experimenting with air that penetrated into a tube via a porous bung, behind which it expanded. Thermodynamic calculations revealed that this process resulted in a minor rise or fall in temperature. Below the *inversion temperature,* the result was cooling; above it, heating (Joule and Thomson 1852). The inversion temperature of hydrogen is −193 °C: a temperature which was easily within the range of pumped-off liquid oxygen; against that background, attempts to condense hydrogen gas using the Joule-Thomson effect were not entirely without prospects of success, in advance.

In July 1895, Kamerlingh Onnes wrote his British colleague a letter – the start of a correspondence full of mutual respect, that only ended in 1922 following Dewar's death, "I have many times shown the exquisite blue liquid, that we got after so much labour, to scientific friends, coming even from Russia"; thus Kamerlingh Onnes described liquid oxygen, and; "I am sure that your vacuum glass is the greatest advance in low temperature work since 1883."[9] Dewar in fact underlined this final sentence. In 1896, Dewar was sent an article from Leiden, in which Kamerlingh Onnes wrote that for years he had been considering a method for the liquefaction of hydrogen. A recent patent by Solvay (soda manufacturers in Brussels) that in principle was similar to his own thoughts, had encouraged him to reveal his own ideas. The key to Kamerlingh Onnes' approach was Van der Waals' law of corresponding states. If an installation were able to render oxygen liquid from room temperature, in one process, the obvious conclusion was that the same could be achieved with hydrogen, on condition the starting temperature taken was that of liquid oxygen. Kamerlingh Onnes' original thoughts turned to a self-cooling expansion motor, an

[9] HKO to Dewar, 18 July 1895, archives of the Royal Institution, London.

original idea that eventually stranded due to insolvable mechanical problems. As soon as he heard of the Joule-Thomson method in 1895, he too turned in that direction. In the design of the eventual hydrogen liquefactor, the 1896 analysis played a key role (Kamerlingh Onnes 1896).

During the cold-free period (1895–1898), Kamerlingh Onnes never lost his belief in a successful outcome and behind the scenes the attack on liquid hydrogen was still being prepared. In 1898, the umpteenth addition to the laboratory was intended for a more powerful gas motor (40 hp), a room for 'chemical processes' and a room where the hydrogen liquefactor would be installed. The fact that the liquefactor only functioned as intended by 1906 was due entirely to the severe demands imposed by Kamerlingh Onnes on the installation. Following Dewar's performance in May 1898, there was no sense whatsoever in quickly assembling an installation that produced just a small amount of liquid hydrogen, at best good for a few initial tests. Leiden had lost that particular race. Kamerlingh Onnes' goal now was a hydrogen liquefactor that was capable of producing several litres of liquid hydrogen an hour, continuously, in the most economic manner possible. Drained into dewar vessels and transferred to a measuring cryostat elsewhere in the laboratory, the liquid hydrogen should then result in a liquid bath with temperatures which could be stabilised to the nearest one-hundredth of a degree.

Before this stage was reached, a great deal of work had to be undertaken. Alterations to the additional necessary vacuum and compression pumps, the construction of the liquefactor itself, the perfection of an installation to purify hydrogen available on the market, the construction of a cryostat, the insertion in the cascade of a circuit for liquid air, the fabrication of supporting appliances meant that the workshops of the Physics Laboratory had to work overtime. However, the results were worth it. "The Leiden hydrogen liquefactor for continuous use," as described by Kamerlingh Onnes in 1906, "comprises sufficient unique elements to occupy a position of its own as an independent construction alongside the installations of Travers and Olszewski, none of which would be able to meet the demands of the Leiden measurements" (Kamerlingh Onnes 1906).

In 1905, everything was ready for the first attempt to liquefy hydrogen. The Crossley gas motor, installed in 1900 in fact let the system down badly in November 1904 when a bolt broke off the crankshaft – a construction error that the company from Manchester refused to accept.[10] It took a year before the parties agreed on a repair contract, and the repair was undertaken. By that time, in room Aa, two slow-running, oil-lubricated two-stage compressors from Burckhardt (that brought the gas to 250 atm) and the huge vacuum pump for pumping off the liquid air at 1.5–2 mm of mercury pressure (also produced by Burckhardt; by this time, Kamerlingh Onnes was a regular customer in Basel) had been in place for some considerable time. An initial trial with a small liquefactor, inspired by installations produced by Hampson (liquid air, 1898) and Olszewski (liquid hydrogen, 1902), failed because the liquefactor broke down. However, on 16 June 1905, the

[10] Correspondence on the Crossley engine can be found in the archives of the Huygens Laboratory.

experiment advanced well, and Leiden at last had its own liquid hydrogen. The yield was quarter of a litre, of which some 15 cc could be poured off.

This was not a large volume, and it was quickly decided to build a large hydrogen liquefactor, capable of delivering 3–4 l an hour. The new coil with expansion valve, ending in a dewar flask, was successfully tested for soundness using nitrogen, after which the installation produced its first litre of liquid hydrogen in February 1906. At the Academy meeting in May, Kamerlingh Onnes took with him a flask, prepared on the previous day in Leiden, so as to be able to carry out a number of demonstrations. The new liquefactor exceeded their wildest dreams, and was the best in the world, producing 4 l an hour. Dewar's installation suffered regularly from frozen taps, while Travers and Olszewski's installations (both built in 1902) were unable to produce hydrogen in a continuous process, and in each production run delivered no more than several hundred millilitres (Olszewski 1903). In 1906, 71 l of hydrogen were drained off in Leiden; a year later, the output had risen to 167 l. These developments put Kamerlingh Onnes in an ideal starting position for the final stage: liquid helium.

The first attempt to liquefy helium dates back to 1895. In March of that year, William Ramsay announced that he had succeeded in his chemical laboratory at University College in London to demonstrate the presence of helium on earth. The fact that this noble gas occurred on the sun (Greek: *helios*) had been known since 1868. To determine the critical pressure and the critical temperature of the gas, Ramsay sent a sphere filled with the new gas to Karol Olszewski, whom he knew from his student days in Heidelberg. In August 1894, when Ramsey had first isolated argon, he had also sent a sample of that gas to Krakow. The stubborn Dewar, who had dared in letters to the *Times* to cast doubt on the discovery of argon, was in this case the loser.

At the end of 1895, Olszewski hoped to subdue the helium from London via a single rapid expansion of compressed helium gas at the temperature of liquid oxygen. The tried and tested Cailletet method was doomed to failure. With pumped-off oxygen, –219 °C was the lowest achievable temperature, while the critical temperature of helium (as later emerged) was –268 °. Condensing a gas is only possible at temperatures below the critical value. Initial calculations by the disappointed Olszewski led him to believe that the boiling point of helium was below –264 °C (9 K) (Olszewski 1896).

The struggle for liquid helium was then played out between Dewar, Olszewski and Kamerlingh Onnes. The Briton obtained his helium from a spring in Bath, a city close to Bristol. This gas (mainly nitrogen) contained a 0.5 per mille concentration of helium. By passing the helium through a tube immersed in liquid hydrogen, contaminants were frozen out as far as possible. Dewar estimated the critical temperature at –264 °C (9 K) or lower.

When it emerged that helium could not be subdued in a Cailletet tube, there was no other option than to build an installation with circulating helium and a Joule-Thomson valve, and to precool the entire system with liquid hydrogen. In his Bakerian Lecture of 1901, Dewar calculated that in this way, the liquefaction of helium could be achieved, on condition the critical temperature of the gas was not below 6–8 K, corresponding to a boiling point of 4–5 K (Dewar 1901). In a letter to Leiden, the Scot made it known that he intended to start the circulation as quickly as possible. "It is however a very complicated and risky business as you well know," Dewar wrote. "I have already lost 1 cylinder of helium by the breaking of vacuum vessels during the course of its circulation at liquid air temperatures and I dread any repetition of the disaster." He then paid his Dutch competitor a compliment: "I only wish that I had again the gift of growth so that I might begin my scientific career after a training in your Dutch school of science."[11]

In his reply, Kamerlingh Onnes – with a distinct lack of diplomacy – informed Dewar that he would have avoided the disaster with the lost helium if he had worked with the Leiden mercury compressor for pure gases, the Cailletet pump that Kamerlingh Onnes had adapted at the start of his career with much blood, sweat and tears. Nonetheless, he had hit the nail on the head. Looking back on his loss, Dewar reflected in 1908 that he had been defeated by two problems: obtaining a sufficient quantity of pure helium, and the production of sufficient liquid hydrogen to keep the helium cycle operating. He admitted that both achievements required an industrial approach, and that the state of affairs at the Royal Institution in London was anything but industrial. "In my work I have never been able to do anything unless substantially with my own hands," Dewar wrote in 1904 to Leiden, and: "In pioneering work assistants are a waste."[12] At that moment, Leiden was in fact staffed by a small army of 'blue boys' (apprentice instrument makers) headed by excellent technicians like Gerrit-Jan Flim and the master glass blower Oskar Kesselring, brought to Leiden from Thüringen.

There was however one thing that Kamerlingh Onnes did not have: helium. At the start of June 1905, he contacted the management of the Bath springs, but they referred him to Dewar. Kamerlingh Onnes informed his competitor that he had by now advanced to the stage where he was ready to produce an accurate determination of the helium isotherms. He asked whether the isotherms were not in fact the ideal route to determining a reliable critical temperature, which in turn could provide an answer to the urgent question, namely whether helium could be liquefied at all via the Joule-Thomson cooling. In brief, Kamerlingh Onnes asked Dewar whether he was willing to share his helium gas from Bath with Leiden, in return for a consideration. By way of reassurance, he informed the Scot that it would take him at least 2 years to purify the helium sufficiently to be able to do anything useful with it.[13]

However, Dewar's kindred spirit did not go that far. "We both want the same material in quantity from the same place at the same time and the supply is not sufficient to meet our great demands," the Scot informed him. "It is a mistake to suppose the Bath supply is so great."[14] Kamerlingh Onnes drew his own conclusions and went urgently in search of a new supplier, whom he found in Ramsay.[15]

[11] Dewar to HKO, 7 May 1903, MB, archives of HKO.

[12] Dewar to HKO, 5 January 1904, MB, archives of HKO.

[13] HKO to Dewar, 8 June 1905, archives of the Royal Institution.

[14] Dewar to HKO, 12 July 1905, MB, archives of HKO.

[15] Ramsay to HKO, 19 July 1905, MB, archives of HKO.

In Leiden, the attack on helium was due to take place in E', a small room completed in 1903 at the same time as a room for the electric motors that drove the pumps of the cascade. The determining factors for the possible success of the attack were the inversion temperature of helium (below which the Joule-Thomson process delivers a cooling effect) and the critical temperature (below which helium condenses under pressure). Olszewski estimated the critical temperature at 1.7 K; Dewar suggested 5–6 K. It was up to Kamerlingh Onnes to identify the correct temperature via precision measurements – Leiden's great strength.

The measurement of helium isotherms was started in 1905. It took a year before Kamerlingh Onnes revealed his first results. The inversion temperature of helium was approximately 23 K, such that Joule-Thomson expansion of helium through a valve would indeed deliver a cooling effect if a start was made at the freezing point of hydrogen (14 K). Kamerlingh Onnes estimated the critical temperature of helium at 5.3 K. Strengthened by this number, despite the previous failures of Olszewski, Dewar and Travers (Ramsay's assistant), he decided to make an attempt to liquefy helium via a single expansion in a glass Cailletet tube. Failure he already had; success was a possibility.

On 28 February, Kamerlingh Onnes had 7 l of helium expanded from 100 atm to 1 at −259 °C (the freezing point of hydrogen). To his considerable surprise, the glass tube beneath the expansion tap filled with a flocculated, snowy mass, while the best Leiden had in fact hoped for was a faint mist. The interpretation Kamerlingh Onnes gave to his observations was that under rapid expansion, helium immediately became a solid, much like carbon dioxide snow from a fire extinguisher. The test was repeated 8–9 times, and on each occasion the same phenomenon was revealed – although over time the view was hindered when the tube misted up, on the outside.[16]

The news spread like wildfire through the laboratory: 'the helium is solid!' Delighted, Kamerlingh Onnes telegraphed James Dewar that the race was run. "Converted helium into solid." At the Academy session on 29 February, in other words on the very next day, he announced the spectacular news. "As the helium was relaxed", explained Kamerlingh Onnes, "a thick grey mist formed, from which solid masses were deposited, that were suspended in the gaseous helium like cotton wool and other more coherent masses in a syrupy liquid, and attached to the wall then toppled downwards while evaporating rapidly (in 20 s). No trace of melting could be observed" (Kamerlingh Onnes 1908a).

The victory over helium did not go unnoticed. The *Globe*, the *Daily Telegraph* and *Nature* immediately reported the news about 'solid helium'. Dewar revealed himself a sporting loser in London. The Scot sent congratulations and informed the *Times*. Six weeks later, Dewar sent the paper another letter, this time with a message that the solid helium from Leiden had turned out to be a mistake. Leiden had celebrated too early, and the students derisively referred to 'halvium'. In his keenness to be first, Kamerlingh Onnes had completely ignored his own lessons relating to

[16] Notebook of Gerrit Jan Flim, MB, archive 444.

experimental research. "For some time," Kamerlingh Onnes recapitulated in his revised Academy notice in March, "I was truly convinced, as a result of observations that later gave grounds for doubt, or indeed proved incorrect, that I had seen rapidly evaporating helium in solid state giving off vapour [...]" (Kamerlingh Onnes 1908a). Upon further inspection, the helium, it turned out, was contaminated with hydrogen. During the expansion of the helium, that gas had been frozen out, and evaporated rapidly. When the test was repeated with pure helium, there were no further traces of flocculation.

All this meant the procedure had to be undertaken using a circulation. While Dewar plodded on in London with tight-frozen taps and blocked pipes, things moved forward fast in Leiden. Kamerlingh Onnes had not waited for the results of the helium isotherms, and had already made a start on constructing a helium circulation appliance (liquefactor) with a regenerator coil (along which the helium flowed back following the expansion stage thereby precooling the incoming helium gas) and a dewar flask (in which to capture the liquid helium). The design was a 'translation' of the successful hydrogen liquefactor, in the smallest possible format, all governed by Van der Waals' law of corresponding states. After 3 months of solid work – the blowing of the necessary large double-walled vacuum flasks pushed even Kesselring's arts to their maximum limits – the appliances could be assembled.

On 10 July 1908, a windy and wet summer's day shortly before the holidays, the helium was tested for the second time. At around 5 o'clock in the morning, Kamerlingh Onnes was collected by coach and horses from Huize ter Wetering, his attractive country estate on the Galgewaard, and brought to the Steenschuur. At the Physics Laboratory, he immediately donned his white jacket. Under the direction of head of operations Gerrit Jan Flim, his technicians were already hard at work. On the previous day, they had raised the stock of liquid air to 75 l, and together with the professor director, had checked the appliances for leaks, pumped them to vacuum condition, and filled them with pure gas. Today, the liquefaction of hydrogen was started very early in the morning. While the pumps ran at full capacity in room E', Kamerlingh Onnes and his staff wildly turned taps, disconnected and reconnected gas bottles, and fearfully monitored pressure gauges and thermometers. They did not even take a break to eat. At half past one in the afternoon, 20 l of liquid hydrogen had been drained into dewar flasks (thermos bottles), sufficient to open the attack on helium.

Kamerlingh Onnes gave himself little likelihood of success. The embarrassing failure in February was still fresh in his memory. The new attempt was made with 200 l of helium gas with a further 160 l in reserve – considerably more than in February. The helium gas had been extracted from thorium-bearing monazite sand, a mineral that had been shipped in huge quantities from America via Hamburg to Leiden, through the mediation of Onno Kamerlingh Onnes, director of the Office of Trade and Information in Amsterdam, and Heike's own brother. Above all the purification of the helium gas turned out to be a tiresome task, which had stretched four chemists to their very limits, for a full 4 months. By heating, the gas erupted from the grains of sand in which it was trapped. To remove contaminants, the gas was cooled with liquid air, burned by copper oxide under the addition of oxygen, and passed over carbon in a compressed state at the temperature of liquid hydrogen (–253 °C). All the effort proved worthwhile: on 10 July, the helium was shown to be sufficiently pure.

At the same time, Kamerlingh Onnes was not entirely certain of his helium isotherms; it seemed to him that at a lower temperature, they resulted in a lower critical temperature (and hence less likelihood of liquid helium) than at a higher temperature. He was also terrified that a fractured vacuum glass would nullify so many months' work. Finally, did the Leiden Cailletet pump, with its maximum capacity of 1,400 l of gas per hour, deliver sufficient power? There was also the problem of liquid hydrogen: that had to be produced in the morning, before a start was made on helium. The law of corresponding states informed Kamerlingh Onnes how much liquid hydrogen he needed, and how long the helium test would last. "They remained precisely below the threshold, such that the structuring of the test in the designed manner would be advised against," wrote Kamerlingh Onnes, "but just how close that threshold was only emerged later." In other words, it was a close run thing.

From eight thirty in the morning, while the hydrogen installation reliably delivered its 4 l of hydrogen per hour, the helium system was vacuum pumped, and the final contaminants removed from the helium gas. Liquid air was passed into the outer vacuum glass, followed by liquid hydrogen in the inner glass. At two thirty in the afternoon, a start could be made on cooling the helium. After half an hour, the temperature had already fallen to 180° below zero. At four twenty in the afternoon – his wife Betsy had by this time anxiously popped by to check on progress, and to feed scraps of bread to her husband as he slogged away – the pump was switched on to allow the helium to circulate. An hour later, the gas was at a pressure of 100 atm. The idea was to allow it to expand in a vacuum space – surrounded by glasses containing liquid hydrogen and liquid air – whereby it cooled down slightly, thereby allowing the return flow of helium to precool the incoming flow. The same principle was employed in the liquefaction of hydrogen, but whether it would work with helium would only be revealed in practice. It would be a tight run thing in room E'.

Once again things looked set to fail. The helium thermometer refused almost entirely to fall, and a second thermometer had given up the ghost. To Kamerlingh Onnes' relief, however, the cooling process did begin, and at around six thirty in the evening, as the helium was rapidly expanded, the temperature was already below that of liquid hydrogen. With fluctuating progress, the readout fell to –267 °C, 6° above absolute zero. By this time, Kamerlingh Onnes had connected the final bottle from the store of liquid hydrogen to the apparatus. The helium had flowed around the installation 20 times, and still there was little more to be seen than the occasional swirl.

At –269°, the thermometer suddenly started to indicate a remarkably stable value. This took place at 7.30 p.m. What was going on? An acquainted chemist, who like so many of Kamerlingh Onnes' colleagues had dropped in to room E' to check on progress, pointed out that it appeared as if the thermometer was already in a bath of liquid. He was right. When Kamerlingh Onnes took a closer look at the glass containing helium, through the dewar flasks, the liquid air and the liquid hydrogen, aiming a light at the flask at an angle, from below, he saw the surface of the liquid. Contact wires could clearly be seen penetrating the surface. "Once we had seen the surface," Kamerlingh Onnes would later write in his notice to the Academy, "we never let it out of our sight. It remained clearly visible against the glass wall."

The experiment had resulted in 60 ml of liquid helium, just about a teacup full. Once circulation in the assembly was halted, it started to evaporate slowly. To Kamerlingh Onnes' distress, the large Burckhardt pump from the liquid air circuit that was able to pump down to 2 ml of mercury pressure, could not be connected quickly enough. As a result, the helium could only be placed under a vacuum of not more than 2.3 cm of mercury pressure. That equated to 1.7 K, which was insufficient to freeze the helium; it later emerged that this process in fact required high pressure, and it was Willem Keesom who first solidified helium in 1927. There was no doubt whatsoever that the helium in Leiden was pure; all the taps in the helium circuit continued to function perfectly, and even the final residue of liquid was crystal clear. Kamerlingh Onnes determined the boiling point at 4.5 K (0.3° too high). The density of liquid helium was surprisingly low and it was also noticeable how mobile the liquid was.

At nine forty in the evening, with just a few millilitres of liquid helium left in the system, Kamerlingh Onnes halted the process – enough was enough. "Not only had the equipment been tested to its absolute limit in this experiment and in its preparations," he wrote in his Academy report. "The maximum possible had also been demanded of my assistants. Without their dedication and inspired cooperation, no part of the programme would have been so impeccably achieved as was necessary to ensure the success of this attack on the helium. I above all owe a great debt of thanks to Mr G.J. Flim, who not only stood by in managing the work as head of the cryogenic laboratory but also as head of the workshop, managing the construction of the appliance according to my instructions, and in both respects offering me the most intelligent of assistance" (Kamerlingh Onnes 1908b).

Having telegraphed to Dewar, Kamerlingh Onnes quickly set to writing an Academy report. Methods, the construction of the helium appliance (including schematic diagrams and a photograph of the installation) and helium preparation were extensively described. Dewar received much praise for his discovery of the dewar vessel, Van der Waals for his theory, and a huge debt of gratitude was expressed to the technical staff. What followed was a compelling report of the events of the 10th of July, and a description of the characteristics of the liquid helium – a highpoint in scientific literature. In July, the notice was added and distributed as a loose appendix to the already published report of the June meeting by the Academy, undoubtedly through the intervention of secretary Van der Waals. The fact that a repetition of the test, whereby the capture vessel was illuminated with a light from the side resulted in not a single drop of liquid helium, is further evidence of just how close run an event the 10th of July had been.

The liquefaction of helium was the highpoint in Kamerlingh Onnes' career – at the time he was 55 years old. It was also a milestone in the history of international physics. On the day on which the last permanent gas was conquered a new temperature area unfolded, a barren ground of extreme cold, full of surprises.

Suddenly the road to the discovery of superconductivity was open – a feat achieved in 1911 – alongside other spectacular phenomena (Delft and Kes 2010). For his helium work, Kamerlingh Onnes received the Nobel Prize in 1913. For years, Leiden was the coldest place on earth. Only in 1923, when James McLennan also managed to liquefy helium in Toronto (having first collected drawings from Kamerlingh Onnes), was Leiden's monopoly on liquid helium broken.

With his Big Science approach, his carefully orchestrated research programme, in which, instead of working individually, everyone contributed to a team effort in pursuit of a well-defined goal, he set an example that other laboratories later emulated. "Heike Kamerlingh Onnes drove his people like the wind drove the clouds", said his pupil Pieter Zeeman (winner of the Nobel Prize in 1902, together with Lorentz). The amazing thing is that a frail boy who missed a whole school year through illness when he was 11 (and sat at home reading Plutarch) and who was compelled to go to the Alps every summer for a health cure throughout his life, nonetheless possessed the energy and endurance to set up a large-scale enterprise such as his cryogenic laboratory and to make it such a great success. Helped by his wife, Betsy, Heike was able to focus his limited energy on that one goal. It is true that his programme made no provision for trying things out on impulse, but frivolities of that kind would never have produced liquid helium.

References

Andrews, T. 1869. On the continuity of the gaseous and liquid states of matter. *Philosophical Transactions of the Royal Society of London* 159: 575–590.

Cagniard de la Tour, C. 1822. Exposé de quelques resultats etc. *Annales de Chimie et de Physique* 21: 127–132 and 178–182.

Dewar, J. 1898. Preliminary note on the liquefaction of hydrogen and helium. *Proceedings of the Royal Society of London* 63: 256–258.

Dewar, J. 1901. The nadir of temperature and allied problems. Bakerian lecture. *Proceedings of the Royal Society of London* 68: 360–366.

Faraday, M. 1845. On the liquefaction and solidification of bodies generally existing as gases. *Philosophical Transactions* 135: 155–177.

Gavroglu, K., and Y. Goudaroulis. 1991. *Through measurement to knowledge. The selected papers of Heike Kamerlingh Onnes 1853–1926*. Dordrecht/Boston/London: Kluwer Academic Publishers.

Huntford, R. 1985. *The last place on earth*. London: Pan Books.

Joule, J.P., and W. Thomson. 1852. Porous plug experiment. *Philosophical Magazine* 14: 481–492.

Kamerlingh Onnes, H. 1882. *De beteekenis van het quantitatief onderzoek in de natuurkunde*. Leiden: IJdo.

Kamerlingh Onnes, H. 1894. On the Cryogenic Laboratory at Leiden and on the production of very low temperatutes. *Communications* 14.

Kamerlingh Onnes, H. 1896. Remarks on the liquefaction of hydrogen, on thermodynamical similarity, and on the use of vacuum vessels. *Communications* 23.

Kamerlingh Onnes, H. 1904. *De beteekenis van nauwkeurige metingen bij zeer lage temperaturen*. Leiden: IJdo.

Kamerlingh Onnes, H. 1906. Methods and apparatus used in the cryogenic laboratory at Leiden. X. How to obtain baths of constant and uniform temperature by means of liquid hydrogen. *KNAW Proceedings* 9(1): 156–180. *Communications* 94f.

Kamerlingh Onnes, H. 1908a. Experiments on the condensation of helium by expansion. *KNAW Proceedings* 10(2): 744–747. *Communications* 105.

Kamerlingh Onnes, H. 1908b. The liquefaction of helium, *KNAW Proceedings* 11: 168–185. *Communications* 108.

Kipnis, A.Y., B.E. Yavelov, and J.S. Rowlinson. 1996. *Van der Waals and molecular science*. Oxford: Clarendon Press.

Knappert, L. 1906. *De ramp van Leiden*. Schoonhoven: Van Nooten.

Laesecke, A. 2002. Through measurement to knowledge: The inaugural lecture of Heike Kamerlingh Onnes (1882). *Journal of Research of the National Institute of Standards and Technology* 107: 261–277.

Levelt-Sengers, J. 2002. *How fluids unmix. Discoveries by the school of Van der Waals and Kamerlingh Onnes*. Amsterdam: KNAW.

Olszewski, K. 1896. Ein Versuch das Helium zu Verflüssigen. *Annalen der Physik* 59: 184.

Olszewski, K. 1903. Apparate zur Verflüssigung von Luft und Wasserstoff. *Annalen der Physik* 10: 768–782.

van Delft, D. 2007. *Freezing physics; Heike Kamerlingh Onnes and the quest for cold*. Amsterdam: Edita/KNAW.

van Delft, D., and P. Kes. 2010. The discovery of superconductivity. *Physics Today*, September, 63: 38–43.

Chapter 5
Superconductivity—A Challenge to Modern Physics

Christian Joas and Georges Waysand

The discovery of superconductivity could not have happened without the liquefaction of helium by the Dutch physicist Heike Kamerlingh Onnes in 1908, which allowed physicists to reach temperatures close to absolute zero. Helium liquefaction was the result of Kamerlingh Onnes's lifelong enterprise to apply large-scale industrial means to fundamental research. It delivered the final blow to nineteenth-century conceptions about the existence of non-liquefiable "permanent" gases. Until 1923, his Leiden cryogenic lab would remain the only place in the world where helium could be liquefied (see, e.g., van Delft 2007). For Kamerlingh Onnes, the ability to conduct experiments at low temperatures opened the gates toward investigations into some of the most fundamental issues of contemporary physics:

> Further discoveries in another field have made necessary, in an unexpected connection, the examination of many properties in the neighbourhood of absolute zero. The Nernst heat theorem and, especially, Planck's quantum theory and zero-point energy theory have made measurements at low temperatures of prime interest to physicists. (Kamerlingh Onnes 1913)

Walther Nernst's heat theorem, which would later be reformulated into the Third Law of thermodynamics, and Max Planck 's 1900 quantum theory—as well as his second theory of radiation of 1911 which included the concept of zero-point energy (Kuhn 1978: 246–248; Badino 2012)—were matters of intense debates within the community of physicists. Kamerlingh Onnes hoped to resolve these debates through

C. Joas (✉)
Ludwig Maximilians University, Munich, Germany

Fritz Haber Institute of the Max Planck Society/Max Planck Institute for the History of Science, Berlin, Germany
e-mail: Christian.Joas@lmu.de

G. Waysand
Laboratoire Souterrain Bas Bruit de Rustrel-Pays d'Apt
(UMS 3538 UNS/UAPV/CNRS/AMU/OCA), 84400 Rustrel, France
e-mail: Waysand@orange.fr

K. Gavroglu (ed.), *History of Artificial Cold, Scientific, Technological and Cultural Issues*, Boston Studies in the Philosophy and History of Science 299,
DOI 10.1007/978-94-007-7199-4_5,

carefully-conducted experiments close to absolute zero, very much in line with the famous *dictum* from his 1882 inaugural lecture in Leiden, "Door meeten tot weten" ("By measurement to knowledge") (Laesecke 2002).[1]

The discovery of superconductivity in 1911 is a direct result of this ambitious program, rooted in questions of nineteenth-century chemistry and thermodynamics, yet consciously modeled to contribute to pressing questions of contemporary basic research. Aiming at establishing a standard for thermometry at low temperatures under the supervision of Kamerlingh Onnes, Gilles Holst measured the resistivity of mercury and observed a sudden disappearance of electrical resistivity at T = 4.19 K.

The phenomenon of superconductivity was a genuine surprise to the emerging community of low-temperature scientists and opened a labyrinth of speculations about its nature (Dahl 1992; Matricon and Waysand 2003; Schmalian 2011). In the decades after its discovery, it would attract the attention of many, transforming an initial curiosity into a challenge to modern physics, and into a laboratory for new physical concepts (Joas and Waysand 2011). Along the way, the very conception of what constitutes the phenomenon underwent deep transformations. Both unsuccessful and successful attempts at explaining superconductivity led to novel concepts in physics. Some of these (e.g., macroscopic quantum states, spontaneously broken symmetry) would prove fruitful far beyond the explanation of the phenomenon itself.

In the early twentieth century, the question of electrical and thermal conductivities in metals was a wide-open problem. Attempts to solve it had been made by Eduard Riecke and Paul Drude around the turn of the century (see, e.g., Hoddeson and Baym 1980; Kaiser 2001). Drude had assumed that metals contain a gas of charged particles (and soon thereafter, electrons, i.e., the charge-carrying particles that had been experimentally confirmed in 1897 by Joseph John Thomson). In 1905, Hendrik Antoon Lorentz had extended Drude's model through consequent application of the methods of statistical mechanics. Lorentz's extension, however, while based more firmly on kinetic theory, produced a prediction for the Wiedemann-Franz ratio of the electric to the thermal conductivity of metals that was at odds with empirical results. Within Drude-Lorentz theory, also the sudden loss of resistivity in a superconductor could not be readily explained. The initial reaction of physicists was to interpret superconductivity as a limiting case of normal metallic conduction at low temperatures. Several attempts were undertaken to explain it within the Drude-Lorentz picture, initially without invoking any concepts stemming from quantum theory. In 1915, for instance, Frederick Lindemann proposed a model in which a rigid lattice of electrons passed through the ionic crystal lattice without dissipating energy (Lindemann 1915; see also Sauer 2007).

One of the first theories of metallic conduction that did exploit concepts of quantum theory—more precisely, Bohr's 1913 atomic model—was developed by the German physical chemist Fritz Haber in 1919.[2] Haber (1919a, b) assumed that metals

[1] That Onnes often strayed from the path indicated by his famous *dictum* is demonstrated in (Matricon and Waysand 2003:18ff).

[2] Similar models had already been proposed independently by the Swedish metallurgist Carl Benedicks and the American physicist Percy Bridgman (Benedicks 1916; Bridgman 1917; see also Kaiser 2001).

are made of a regular lattice of metal ions and valence electrons orbiting these ions. For Haber

> [t]he point of view that naturally comes to mind is to conceive of superconductivity as a state in which valence electrons of the metal move in orbits that have common tangents in points of equal velocity. (Haber 1919b:1003, quoted after Sauer 2007:194)

Electrons on a Bohr orbit do not radiate, hence an electron sliding from the orbit of one ion to the same orbit of a neighboring ion remains at the same energy and therefore does not dissipate energy. This could lead to dissipationless transport— i.e., infinite conductivity—and thus might constitute a possible explanation for superconductivity.

In 1922, Albert Einstein laid out his own theory of superconductivity in a contribution to a *Festschrift* for Kamerlingh Onnes (Sauer 2007). Like Haber, Einstein assumed that metals did not contain any free electrons, in stark contrast to Drude-Lorentz theory. While Haber had stressed the valence origin of the ideally-conducting channels, Einstein used the notion of preferred channels of conduction, whatever the mechanism behind, suggesting that superconductivity is associated with spatial ordering. Thinking of superconductivity as a limiting case of metallic conduction, the notion of a superconducting "phase" that is commonplace today was totally absent.[3] Of course, the spatial ordering had to be related to the ionic lattice. Einstein suggested an experiment where two different superconductors were put into contact, expecting that they would exhibit resistive behavior due to different lattice constants. The experimental result was negative, which made the notion of molecular channels and similar models based on spatial ordering and chemical considerations rather dubious.

During the early 1920s, it became increasingly clear that the methods of old quantum theory were insufficient to treat many-body problems. Einstein (1922) said about his own suggestion for a model of superconductivity:

> Owing to our wide-ranging ignorance of the quantum mechanics [Quanten-Mechanik][4] of composite systems, we are far from able to condense this vague idea into a theory. (Einstein 1922:434)

After the advent of quantum mechanics in 1925–1926, earlier attempts at understanding the physics of many-body systems were taken up again, and new concepts were developed in the context of the study of solids. In late 1926, Wolfgang Pauli explained the paramagnetism of metals using the new Fermi statistics. In mid-1927, Arnold Sommerfeld extended Drude-Lorentz theory by taking into account Pauli's exclusion principle. Werner Heisenberg felt that a quantum-mechanical treatment of electrons in metals was still missing and in 1928 suggested it to Felix Bloch as a thesis project. Bloch soon realized that, within wave mechanics, *"a periodic*

[3] At the time, thermal measurements were not sensitive enough to detect the jump in specific heat at the transition that would have suggested a phase transition (see below).

[4] The use of the word quantum mechanics [*Quanten-Mechanik*] in (Einstein 1922), to our knowledge, is one of the earliest—if not the earliest—appearance of this term that would raise to prominence with the advent of quantum mechanics in 1925. It is to be noted, however, that Einstein refers here to a quantum-theoretical many-body mechanics, rather than to a new "quantum" mechanics that is to replace classical mechanics, which is how the term would be later used by Born and Heisenberg.

arrangement is not really an obstacle for waves, but it's only the thermal vibration" (Mehra and Rechenberg 2000:596). As Walter Heitler and Fritz London had done for the hydrogen molecule in the first quantum-mechanical treatment of the chemical bond in 1927 (Heitler and London 1927; see also Gavroglu and Simões 2012), Bloch constructed electron wavefunctions starting from a basis of unperturbed single-atom ground-state orbitals while neglecting electron-electron interactions and was able to explain the large mean free paths of electrons in metals (Bloch 1928; see also Hoddeson et al. 1987).

A second notable development, also heavily influenced by Heitler and London's work, was Heisenberg's 1928 quantum-mechanical treatment of ferromagnetism. Already by 1926, Heisenberg had realized that parallel orientation of spins might be the energetically most favorable configuration. While Weiss's 1907 theory had to postulate the existence of a molecular field, Heisenberg was able to explain it as originating in quantum-mechanical exchange forces. The success of Heisenberg's theory of ferromagnetism and of his own model for metallic conduction prompted Bloch to attempt to explain superconductivity through an analogy with ferromagnetism in the years 1928–1929. Just as ferromagnets below the Curie point support a permanent magnetization (which since the days of Ampère had been described in terms of persistent currents), Bloch thought, superconductors below the critical temperature could assume a state that would support persistent currents. His attempts at finding such a state failed, and Bloch never published his work.[5] His failure led Bloch to an often-mentioned "theorem" that *"the only theorem about superconductivity which can be proved is that any theory of superconductivity is refutable"* (Dahl 1992:151).

In 1933, Paul Ehrenfest introduced the notion of phase transitions of second order after Willem Keesom and J. N. van den Ende had observed a jump in the specific heat of tin (Ehrenfest 1933).[6] Ehrenfest's graduate student A. J. Rutgers immediately applied the notion of phase transition to the case of superconducting tin to derive a relation between the jump in the specific heat and the derivative of the dependence of the critical magnetic field on temperature. It described the experimental data well, which suggested that superconductivity might be due to a novel thermodynamical phase. Cornelius Gorter set out to develop a thermodynamic theory of superconductivity (Gavroglu and Goudaroulis 1985). Its success only came with another astonishing discovery: The Meissner-Ochsenfeld effect—the complete expulsion of the magnetic field from the volume of a very long superconducting tin cylinder—was discovered in 1933 (Meissner and Ochsenfeld 1933): superconductors behaved as diamagnets. This immediately attracted the attention of theorists as it contradicted prior theoretical expectations. Based on both the discovery of Meissner and Ochsenfeld and the idea of a superconducting phase, Gorter and Hendrik Casimir presented a

[5] In 1933, Lev Landau published a similar idea in a paper that foreshadows elements of his later work with Ginzburg (Landau 1933).

[6] For a detailed historical analysis of Ehrenfest's classification of phase transitions, see (Jaeger 1998).

> consistent thermodynamical treatment of the transitions from the supraconductive state to the normal state, supposing B to equal zero in the supraconductive state. (Gorter and Casimir 1934:318)

Assuming perfect diamagnetism to be a general characteristic of the superconducting phase, Gorter and Casimir formulated what would come to be known as the "two-fluid model"— a model that would be very influential well beyond the 1950s.

The Meissner-Ochsenfeld experiment could not be easily reproduced, so there was considerable debate as to its universal validity. Meissner concluded from his experiment that there is a possible analogy with ferromagnetism, thus apparently following the initial models of Bloch and Landau. In 1935, the brothers Fritz and Heinz London adopted a radically different point of view and developed a phenomenological theory of superconductivity (London and London 1935, see also Gavroglu (1995)). Fritz London introduced the notion of a macroscopic quantum state, viewing "*the whole supraconductor as a single big diamagnetic atom*"(London 1935:32). In the absence of a magnetic field, there is strictly no current, hence there is no contradiction with the principle that the lowest energy state is a zero-current state (see, e.g., Gavroglu and Goudaroulis 1989). Fritz London presented the new view in lectures in Paris in April 1935, where he stated that

> We will show that the problem that one has approached in such an unfortunate way is not being posed by the nature of the facts, that the interpretation of the experiments has gone farther than implied by the observed facts; it is for this reason that [hitherto] one has asked of electronic theory to solve a problem that is definitely unsolvable. (London 1939:7–8).[7]

In May 1935, he stated[8]

> The foundation of our macroscopical equations by the theory of electrons in metals has not yet been undertaken. But it is rather attractive to try to sketch the programme which seems to be set by our equations to a future microscopical analysis. [...] But suppose the electrons to be coupled by some form of interaction, in such a way that the lowest state may be separated by a finite interval from the excited ones. Then the disturbing influence of the field on the eigenfunctions can only be considerable if it is of the same order of magnitude as the coupling forces. (London 1935:31)

London would be a major source of inspiration for John Bardeen on his way to a microscopic theory of superconductivity (see Bardeen 1963).

Liquid helium—the prerequisite for the study of superconductors—itself became an object of study in the 1920s because of an observed anomaly around 2.19 K that suggested a phase transition. Like in superconductors, no latent heat was detected at

[7] "*Nous allons montrer que le problème que l'on a attaqué de façon si malencontreuse n'est pas posé par la nature des faits, que l'interprétation des expériences a dépassé les faits observés; et c'est pour cette raison qu'on a posé à la théorie électronique un problème certainement insoluble.*"

[8] A similar passage can be already found earlier (London and London 1935:87), received by the journal in October 1934 ("*But now suppose the electrons to be coupled by some form of interaction. Then the lowest state of the electrons may be separated by a finite distance from the excited ones and the disturbing influence of the field on the eigenfunctions can only be appreciable if it is of the same order of magnitude as the coupling forces.*").

the transition. Since the liquid lost its entropy without solidifying, Franz Simon in 1927 introduced the notion of "liquid degeneracy," reminiscent of the well-known gas degeneracy (Simon 1927:808). Viscosity measurements by Allen and Misener (1938) as well as Kapitza (1938) suggested a new liquid state with extremely low viscosity that Peter Kapitza called superfluid (Kapitza 1938), in analogy to superconductors. London (1938) suggested that liquid helium below the lambda point might be a Bose-Einstein condensate, a peculiar state of a Bose gas whose possibility had been pointed out by Einstein (1925). Laszlo Tisza introduced a two-fluid model for helium that was analogous to the one introduced for superconductors by Gorter and Casimir. The story of superfluidity is told in (Gavroglu and Goudaroulis 1984).

World War II interrupted many strands of development in low-temperature physics: Helium research was almost completely abandoned, and the study of superconductors stalled. The War-oriented research programs, however, strongly affected the postwar development of low-temperature physics. On the one hand, radar technology provided a host of affordable microwave equipment which could be used for precise surface impedance measurements and microwave spectroscopy. Brian Pippard, using this technique, realized that superconductors have to be described using nonlocal electrodynamics and introduced the coherence length as a characteristic length scale of macroscopic dimension with respect to interatomic distances. On the other hand, the laboratories at Oak Ridge and Los Alamos provided the first isotopes of superconducting elements. In 1950, Maxwell and Serin and co-workers discovered independently that for one and the same element, the critical temperature varies with the square root of the mass of the isotope (Maxwell 1950; Reynolds et al. 1950). For Herbert Fröhlich, who had already made theoretical predictions along this line (Fröhlich 1950:856), and for Bardeen, who had been interested in superconductivity from the late 1930s onwards through discussions with Slater, this so-called isotope effect was the proof for the importance of the ionic lattice, and thus of electron-phonon interactions.

After the end of the War until the mid-1950s, Soviet physics was practically isolated from the West (Gorkov 2011). For four years, the 1950 theory by Vitaly Ginzburg and Landau providing a full phenomenological treatment of the transition from the normal to the superconducting state (Ginzburg and Landau 1950) was ignored in the West. In contrast to the London brothers, Ginzburg and Landau had been willing to give up the rigidity of the macroscopic quantum state by allowing for the spatial variation of an abstract quantity which they called the order parameter. The few copies of the journal containing the paper by Ginzburg and Landau that were shipped to the United States ended up in the Hudson river because of a boycott by the Longshoremen Union (Matricon and Waysand 2003:129). In the West, Bardeen concentrated his efforts on electron-phonon interactions in solids by the early 1950s. Like Fröhlich, he observed that the difference in energy between superconducting and normal state was so small that only the electrons in the immediate vicinity of the Fermi surface could play a role in superconductivity. In 1955, Bardeen hired the young quantum field theorist Leon Cooper in order to develop a field-theoretical treatment of the interacting Fermi gas of electrons in metals. Together with Bob Schrieffer, one of Bardeen's PhD students, they held informal meetings about quantum field theory and about metals and superconductors. Bardeen was convinced that the insights of Fritz London would play an important role in an eventual microscopic theory of superconductivity.

Assuming an arbitrarily small, attractive interaction of unspecified origin between the electrons, Cooper was able to show that the Fermi gas becomes unstable with respect to the formation of pairs of electrons, later to be known as Cooper pairs. Electron pairing was able to explain how an energy gap, already predicted by London, might arise. The idea had already occurred around 1955 to Schafroth and Blatt who had proposed a kind of resonant two-electron state leading to localized "molecules" that effectively acted like bosons (see, e.g., Schafroth and Blatt 1955, and references therein). This "quasi-chemical" method, however, had proved inconclusive. Richard Feynman, who had successfully formulated a theory of superfluid ^{4}He, also was working intensively on the problem of superconductivity (Feynman 1957; see also Matricon and Waysand 2003:141–145; Goodstein and Goodstein 2000). Therefore, Bardeen and his co-workers worked frantically on extending Landau's recently-published Fermi liquid theory to include the interaction between phonons and electrons and to connect it to Cooper's idea. Schrieffer eventually wrote down a variational many-body ground-state wavefunction in terms of a variational "gap" parameter that took Cooper pairing into account. When computing the energy gap in the excitation spectrum, he found that it agreed with the variational parameter in the ground state energy. This convinced Bardeen, and his group set out to calculate the thermodynamic, electromagnetic, and transport properties of superconductors within the new theory, finding good agreement with known properties of superconductors, e.g., for the Hebel-Slichter peak in NMR. In 1957, Bardeen et al. (1957) published their new "BCS" theory of superconductivity.[9] The basic idea was that superconductivity arose from the attractive interaction mediated between the electrons through their coupling to the lattice vibrations, and that the superconducting state was a macroscopic, coherent state in which the electrons appeared as coupled pairs of opposite spin and wavevector. They were able to show that the normal-to-superconductor transition was indeed a phase transition of second order, and how persistent currents arose.

While experimentalists in the field were convinced rather quickly of the adequacy of the new microscopic theory, prominent theorists remained skeptical (Hoddeson et al. 1992:558ff).[10] This might explain why Bardeen, Cooper, and Schrieffer only received a Nobel prize in 1972. Questions about the violation of gauge invariance were debated and ultimately resolved by a consistent quantum-field-theoretical formulation of BCS theory by the likes of Anderson, Nambu, Gorkov, and Bogoliubov. It would lead to the quantum-field theoretic formulation of the concept of spontaneously broken symmetry that later would cross-fertilize back into the theory of elementary particles (e.g., in the Anderson-Higgs mechanism which lies at the foundation of the Standard Model of particle physics). Even stronger

[9] For a detailed account of the history of BCS theory, see (Matricon and Waysand 2003:146ff). See also (Hoddeson et al. 1992; Hoddeson and Daitch 2002).

[10] For an example, see (Bloch 1966).

criticisms of BCS theory arose with respect to its predictive power: Neither the phenomenological Ginzburg-Landau theory nor the microscopic BCS theory could explain which materials would become superconductors or how high critical temperatures of superconductors would be. Theorists called for a quantitative theory of superconductivity from first principles that would also include retardation effects into the field-theoretic treatment of the electron-phonon interaction.

In 1959, G. M. Eliashberg extended A. B. Migdal's field-theoretic treatment of the electron-phonon interaction in normal metals to superconductors, using Gorkov's formulation of the Fröhlich Hamiltonian and Bogoliubov's method of transformations. This so-called "strong-coupling" theory was the beginning of a quantitative theory of superconductivity, which was developed by Swihart, Morel and Anderson, Schrieffer, Scalapino, Wilkins, and others. Novel tunneling experiments by Ivar Giaever, which allowed to measure the energy gap in superconductors to high precision, played a key role in this development. They showed systematic deviations from the simple BCS behavior for some superconductors which could be related to the details of the phonon spectrum. In 1965, McMillan and Rowell extracted information about the phonon spectrum in the normal state of a superconducting material from experimental data provided by tunneling experiments in the superconducting state by inverting—numerically—the Eliashberg-BCS gap equation.[11]

The 1986 discovery of high-temperature superconductivity in cuprates by Karl Bednorz and Alex Müller opened up yet another chapter in the history of superconductivity, both with respect to its potential technological applications and with respect to its explanation. Reminiscent of the decades after Kamerlingh Onnes' discovery, a host of new mechanisms and concepts was proposed to explain superconductivity in these novel materials. The only thing upon which theorists seemed to agree was that they disagreed about the nature of superconductivity in these so-called "unconventional" superconductors. As of yet, no final consensus has been reached. Perhaps, the history of early theories of superconductivity told here can serve as a lesson also with respect to the riddle of high T_c, or of superconductivity in the novel iron-based materials: While not all attempts will contribute to an eventual solution, the problem itself serves as a theoretical laboratory in which new concepts are developed that might become useful in the future, be it for the explanation of superconductivity in new materials, or in distinctly different areas of physics.

Acknowledgments Substantial parts of this text are based on a German article published in *Physik Journal* (Joas and Waysand 2011). The authors wish to thank Dieter Hoffmann, Jeremiah James, Stefan Jorda, Johannes Knolle, Jean Matricon, and Alexander Pawlak for helpful comments and suggestions.

[11] See the contribution by Knolle and Joas, Chap. 7 in the present volume, and the references therein.

References

Allen, F., and A.D. Misener. 1938. Flow of liquid helium II. *Nature* 141: 75.

Badino, M. 2012. Productive conservatism: Planck's second theory of radiation. *Annalen der Physik* 524(2): A29–A31.

Bardeen, J. 1963. Developments of concepts in superconductivity. *Physics Today* 16(1): 19–28.

Bardeen, J., L.N. Cooper, and J.R. Schrieffer. 1957. Theory of superconductivity. *Physical Review* 108: 1175–1204.

Benedicks, C.A.F. 1916. Beiträge zur Kenntnis der Elektrizitätsleitung in Metallen und Legierungen. *Jahrbuch der Radioaktivität und Elektronik* 13: 351–395.

Bloch, F. 1928. Über die Quantenmechanik der Elektronen in Kristallgittern. *Zeitschrift für Physik* 52: 555–600.

Bloch, F. 1966. Some remarks on the theory of superconductivity. *Physics Today* 19(5): 27–36.

Bridgman, P.W. 1917. Theoretical considerations on the nature of metallic resistance with especial regard to the pressure effect. *Physical Review* 9: 269–289.

Dahl, P.F. 1992. *Superconductivity. Its historical roots and development from mercury to the ceramic oxides*. New York: American Institute of Physics.

Ehrenfest, P. 1933. Phasenumwandlungen im üblichen und erweiterten Sinn, classifiziert nach den entsprechenden Singularitäten des thermodynamischen Potentiales. *Verhandlingen der Koninklijke Akademie van Wetenschappen (Amsterdam)* 36:153–157. Also published in *Communications from the Physical Laboratory of the University of Leiden, Suppl. No. 75b (1933)*.

Einstein, A. 1922. Theoretische Bemerkungen zur Supraleitung der Metalle. *Gedenkboek Kamerlingh Onnes, Leiden (11 March 1922)*, 429–435.

Einstein, A. 1925. Quantentheorie des einatomigen idealen Gases. Zweite Abhandlung. *Sitzungsberichte der Preußischen Akademie der Wissenschaften zu Berlin*, 3–14.

Feynman, R.P. 1957. Superfluidity and superconductivity. *Reviews of Modern Physics* 29: 205–212.

Fröhlich, H. 1950. Theory of the superconducting state. I. The ground state at the absolute zero of temperature. *Physical Review* 79: 845–856.

Gavroglu, K. 1995. *Fritz London. A scientific biography*. Cambridge: Cambridge University Press.

Gavroglu, K., and Y. Goudaroulis. 1984. Some methodological and historical considerations in low temperature physics: The case of superconductivity 1911–1957. *Annals of Science* 41: 135–149.

Gavroglu, K., and G. Goudaroulis. 1985. From the history of low temperature physics: Prejudicial attitudes that hindered the initial development of superconductivity theory. *Archive for History of Exact Sciences* 32: 377–383.

Gavroglu, K., and G. Goudaroulis. 1989. Quantum mechanics and macroscopic quantum phenomena: The case of superconductivity and superfluidity. *Zeitschrift für allgemeine Wissenschaftstheorie* 20(2): 249–270.

Gavroglu, K., and A. Simões. 2012. *Neither physics nor chemistry: A history of quantum chemistry*. Cambridge, MA: The MIT Press.

Ginzburg, V.L., and L.D. Landau. 1950. On the theory of superconductivity. *Zhurnal Eksperimental'noi i Theoreticheskoi Fiziki* 20: 1064–1082. Reprinted in *L. D. Landau, Collected Papers*, ed. Ter-Haar, D, 1965. New York: Gordon and Breach.

Goodstein, D., and J. Goodstein. 2000. Richard Feynman and the history of superconductivity. *Physics in Perspective* 2: 30–46.

Gorkov, L.P. 2011. Developing BCS ideas in the former Soviet Union. In *BCS: 50 Years*, ed. L.N. Cooper and D. Feldman, 107–126. Singapore: World Scientific.

Gorter, C.J., and H. Casimir. 1934. On supraconductivity I. *Physica* 1: 306–320.

Haber, F. 1919a. Beitrag zur Kenntnis der Metalle. *Sitzungsberichte der Preussischen Akademie der Wissenschaften (Berlin)*, 1919: 506–518.

Haber, F. 1919b. Zweiter Beitrag zur Kenntnis der Metalle. *Sitzungsberichte der Preussischen Akademie der Wissenschaften (Berlin)* 1919: 990–1007.

Heitler, W., and F. London. 1927. Wechselwirkung neutraler Atome und homöopolare Bindung nach der Quantenmechanik. *Zeitschrift für Physik* 44(6): 455–472.

Hoddeson, L., and G. Baym. 1980. The development of the quantum-mechanical electron theory of metals 1900–1928. *Proceedings of the Royal Society of London. Series A: Mathematical and Physical Sciences* 371: 8–23.

Hoddeson, L., and V. Daitch. 2002. *True genius: The life and science of John Bardeen*. Washington, DC: Joseph Henry Press.

Hoddeson, L., G. Baym, and M. Eckert. 1987. The development of the quantum-mechanical electron theory of metals: 1928–1933. *Reviews of Modern Physics* 59: 287–327.

Hoddeson, L., E. Braun, J. Teichmann, and S. Weart (eds.). 1992. *Out of the crystal maze. Chapters from the history of solid-state physics*. Oxford: Oxford University Press.

Jaeger, G. 1998. The Ehrenfest classification of phase transitions: Introduction and evolution. *Archive for History of Exact Sciences* 53: 51–81.

Joas, C., and G. Waysand. 2011. Von Leitungsketten zur Paarhypothese. Die Entdeckung der Supraleitung und die wechselvolle Geschichte ihrer Erklärung. *Physik Journal* 10(6): 23–28.

Kaiser, W. 2001. Electron gas theory of metals: Free electrons in bulk matter. In *Histories of the electron. The birth of microphysics*, ed. J.Z. Buchwald and A. Warwick. Cambridge, MA: The MIT Press.

Kamerlingh Onnes, H. 1913. Nobel lecture. Available online: http://nobelprize.org/nobel_prizes/physics/laureates/1913/onnes-lecture.html. Last accessed 30 Jan 2013.

Kapitza, P. 1938. Viscosity of liquid Helium below the λ-point. *Nature* 141: 74.

Kuhn, T.S. 1978. *Black-body theory and the quantum discontinuity, 1894–1912*. Oxford: Oxford University Press.

Laesecke, A. 2002. Through measurement to knowledge: The inaugural lecture of Heike Kamerlingh Onnes (1882). *Journal of Research of the National Institute of Standards and Technology* 107: 261–277.

Landau, L. 1933. Zur Theorie der Supraleitfähigkeit. *Physikalische Zeitschrift der Sowjetunion* 4: 43–49.

Lindemann, F.A. 1915. XII. Note on the theory of the metallic state. *Philosophical Magazine, Series 6 (1901–1925)* 29(169): 127–140.

London, F. 1935. Macroscopical interpretation of supraconductivity. *Proceedings of the Royal Society of London. Series A: Mathematical and Physical* 152: 24–34.

London, F. 1938. The λ-phenomenon of liquid Helium and the Bose-Einstein degeneracy. *Nature* 141: 643–644.

London, F. 1939. *Une conception nouvelle de la supra-conductibilité*. Paris: Hermann. Commented edition: Matricon, J., and Waysand, J. 2005. Paris: Hermann.

London, F., and H. London. 1935. The electromagnetic equations of the supraconductor. *Proceedings of the Royal Society of London. Series A: Mathematical and Physical* 149: 71–88.

Matricon, J., and G. Waysand. 2003. *The cold wars. A history of superconductivity.* Trans. from French edition "La guerre du froid", éditions du Seuil, Paris 1994 by Charles Glasshauser. New Brunswick: Rutgers University Press.

Maxwell, E. 1950. Isotope effect in the superconductivity of mercury. *Physical Review* 78: 477.

Mehra, J., and H. Rechenberg. 2000. *The historical development of quantum theory. Vol. 6: The completion of quantum mechanics 1926–1941, Part 1.* New York: Springer.

Meissner, W., and R. Ochsenfeld. 1933. Ein neuer Effekt bei Eintritt der Supraleitfähigkeit. *Die Naturwissenschaften* 21: 787–788.

Reynolds, C.A., B. Serin, W.H. Wright, and L.B. Nesbitt. 1950. Superconductivity of isotopes of mercury. *Physical Review* 78: 487.

Sauer, T. 2007. Einstein and the early theory of superconductivity, 1919–1922. *Archive for History of Exact Sciences* 61: 159–211.

Schafroth, M.R., and J.M. Blatt. 1955. Phenomenological equations for superconductors. *Physical Review* 100: 1221–1222.

Schmalian, J. 2011. Failed theories of superconductivity. In *BCS: 50 Years,* ed. L.N. Cooper and D. Feldman, 41–55. Singapore: World Scientific. Preprint: arxiv.org/pdf/1008.0447.

Simon, F. 1927. Zum Prinzip von der Unerreichbarkeit des absoluten Nullpunktes. *Zeitschrift für Physik* 41: 806–809.

van Delft, D. 2007. *Freezing physics: Heike Kamerlingh Onnes and the quest for cold.* Amsterdam: Edita.

Chapter 6
Superfluidity: How Quantum Mechanics Became Visible

Sébastien Balibar

6.1 Introduction

Why is it that physicists keep trying to study matter at lower and lower temperature? One likely explanation is that, as temperature goes down, thermal fluctuations progressively vanish so that new phenomena appear as a landscape when clouds go up. When approaching the absolute zero (0 K = −273.15 °C) the behaviour of matter becomes sensitive to minute interactions that would be irrelevant at higher temperature when fluctuations are larger. This is how many new properties of matter have been discovered, not only superfluidity and superconductivity (Mendelssohn 1964; Dahl 1993).

In 1908, Kamerlingh Onnes succeeded in liquefying helium gas (Kamerlingh 1908; Delft 2005). This was actually natural helium, which Kamerlingh Onnes obtained by isolating the gas from monazite sand found in North Carolina, through the Office of Commercial Intelligence at Amsterdam where his brother was the director. Nowadays, it comes from oil wells. It contains mostly ^{4}He, which results from the radioactive alpha-decay of Uranium in the Earth. The light isotope ^{3}He was not available in large enough quantities until Tritium, which decays into ^{3}He, was used in the military nuclear industry after World War II. Kamerlingh Onnes wanted to see if it was possible to liquefy the last gas that had not yet been liquefied. The transition from gaseous ^{4}He to liquid ^{4}He occurred at 4.2 K under atmospheric pressure. In 1911 and at the same temperature (4.2 K) he made the major discovery for which he received his Nobel prize, that is the superconductivity of mercury (Kamerlingh Onnes 1911): at low temperature, mercury is a metallic solid and he discovered that its electrical resistance vanishes below 4.2 K.

S. Balibar (✉)
Laboratoire de Physique Statistique de l'Ecole Normale Supérieure,
associé au CNRS, à l'Université Pierre et Marie Curie et à l'Université
Denis Diderot, 75231 Paris Cedex 05, France
e-mail: balibar@lps.ens.fr

K. Gavroglu (ed.), *History of Artificial Cold, Scientific, Technological and Cultural Issues*, Boston Studies in the Philosophy and History of Science 299,
DOI 10.1007/978-94-007-7199-4_6,

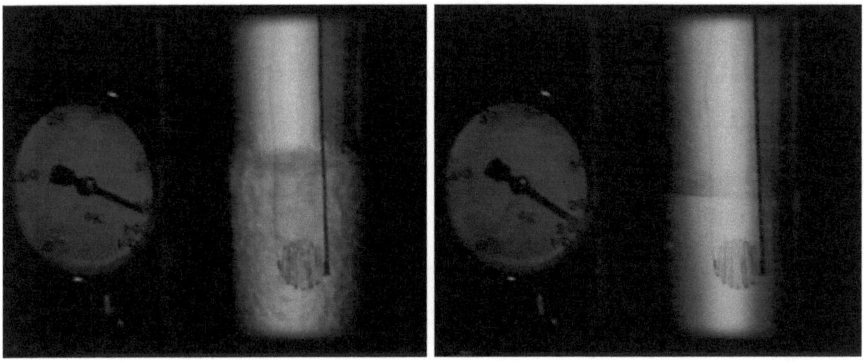

Fig. 6.1 As shown by these two images from a film by J.F. Allen and J.M.G. Armitage (Allen and Armitage 1982), superfluid helium stops boiling below $T\lambda$. This is due to its large thermal conductivity. The *left picture* is taken at 2.4 K as indicated by the needle of the thermometer on the *left*. The *right picture* is taken just below the lambda transition

In order to reach temperatures lower than 4.2 K, Kamerlingh Onnes simply pumped on liquid helium. This time, one could say that he really entered the era of artificial cold because the lowest temperature in the Universe is that of the cosmic background radiation, now known as 2.7 K (Fixsen 2009).

It is only a little further down in temperature – at –2.2 K and in December 1937 – that superfluidity was discovered simultaneously by Allen and Misener in Cambridge (Allen and Misener 1938) and by P. Kapitsa in Moscow (Kapitsa 1938).

Superconductivity and superfluidity are far from being the only two phenomena that have been discovered below 3 K, but they appear as the most important manifestations of quantum mechanics at the macroscopic scale. One can see with a naked eye that liquid helium is a normal liquid down to 2.17 K (–271 C) and that it changes into a different liquid below this temperature (Fig. 6.1). When pumping on any liquid, one reduces its vapour pressure so that its temperature goes down. Since the thermal conductivity of classical liquids is usually poor, the surface is usually colder than the inside, especially the walls of the container where the probability that bubbles nucleate becomes high. As result, a classical liquid is invaded by bubbles, which have grown on nucleation sites before being driven up by the buoyancy force. The result is the turbulent mixture of liquid and gas that everyone has in mind when talking about a boiling liquid. This is exactly what happens if one pumps on normal liquid helium above 2.17 K. But as soon as it is cooled below this temperature, boiling stops and the liquid only evaporates from the free surface, without any bubble formation (see Fig. 6.1). This property was discovered in 1932 by J.C. McLennan (McLennan et al. 1932) at the Toronto University. Some years later, one realized that this phenomenon is due to the unusually high thermal conductivity of superfluid helium below 2.17 K.

It is really London who understood for the first time that quantum mechanics shows up at the macroscopic scale, not only at the scale of atoms and molecules. When proposing that superfluids (London 1938a) and superconductors (London 1935)

could be represented by a single macroscopic wave function he made a giant step forward. More precisely he explained that this wave function had to be related to "Bose Einstein condensation", an astonishing phenomenon. This was in 1938. It took 20 Years to the famous trio BCS (John Bardeen, Leon Cooper and Robert Schrieffer) to understand in details that a similar condensation was taking place in superconductors so that these two phenomena were close to each other, as first imagined by London. Today, one usually considers that superconductivity is superfluidity in a charged fluid.

To judge the importance of superfluidity and superconductivity, one can look at the number of Nobel Laureates whose work was dealing with superfluidity (Landau in 1962, Kapitsa in 1978, Lee, Osheroff and Richardson in 1996, Cornell, Wieman, and Ketterlee in 2001, Leggett in 2003) and superconductivity (Kammerlingh Onnes in 1913, Barden, Cooper, Schrieffer in 1972, Giaever and Josephson in 1973, Bednorz and Mueller in 1987, Abrikosov and Ginzburg in 2003).

In this article, I wish to recall how superfluidity was discovered experimentally in liquid helium 4. Then how it was progressively understood. In a last part I describe shortly some important developments about the quantization of vortices (1949–1961) the discovery of superfluid helium 3 (1972), the superfluidity and the very clear evidence for Bose Einstein condensation in quantum gases (1995), and most recently supersolidity.

At the very end, I briefly describe some important applications of superfluidity and superconductivity, mainly very high field magnets leading to a revolution in medical imaging, and the impressive presence of superconductivity and superfluidity in the 27 km long LHC near Geneva.

6.2 Before the Discovery of Superfluidity

This discovery has already been described by various authors with various points of view (Donnelly 1995; Balibar 2007; Griffin 1995, 2005; Rubinin 1997; Matricon et al. 2003). Before describing it here, let us mention some of its prehistory and present the two main characters involved in the experimental discovery (Kapitsa and Allen).

In 1927 at Leiden, M. Wolfke and W.H. Keesom (Wolfke and Keesom 1927; Keesom and Wolfke 1927; Keesom and Keesom 1932) measured the specific heat of liquid helium as a function of temperature and they found that this quantity had a singularity in the shape of the Greek letter lambda (λ). This singularity had to separate two different liquid states and this observation must have been rather astonishing: how could it be that a liquid made of simple spherical atoms with no chemistry -only weak van der Waals long range attraction and a hard core repulsion at short distance – could liquefy in two different phases? Anyhow, W.H. Keesom called "helium I" the liquid above T_λ and "helium II" the mysterious liquid below T_λ. It took 10 years to realize –a little more to understand – that helium II was a fundamentally new state of matter.

Some years later, another striking property was discovered, this time by J.C. McLennan in Toronto (McLennan et al. 1932): when pumping on liquid helium I, one could see it boiling as any classical fluid would do. But below T_λ, liquid helium II was not boiling (Fig. 6.1). When any of us, researchers in low temperature physics, see this boiling stopping, it is some kind of evidence that quantum order sets in. We now know what it is: the quantum coherence of a macroscopic standing wave of matter.

In 1935 at the Toronto University Physics Department directed by E.F. Burton, an undergraduate student named Don Misener attempted to measure the viscosity of helium II with two technicians (J.O. Wilhelm who had already observed the absence of boiling with McLennan, and A.R. Clark). Burton actually published the results in Nature without the authors of the measurements (Burton 1935; Misener et al. 1935). Anyhow, Wilhelm, Misener and Clark measured the damping of the oscillations of a cylinder immersed in liquid helium and discovered that the viscosity of liquid helium dropped down as the system cooled below T_λ. Apparently, helium II had surprising mechanical and thermodynamical properties.

At the same time, a high thermal conductivity was found by B.V. Rollin in Oxford (Kuerti et al. 1936), followed by W.H. Keesom and his daughter Ania (Keesom and Keesom 1936) and by J.F. Allen, R. Peierls and Z. Uddin in Cambridge (Allen et al. 1937). As far as I know, Toronto, Cambridge, Oxford, Leiden and Kharkov were the only five places in the world where liquefiers produced liquid helium for such studies. It was not yet available in Moscow for reasons that will become clear after introducing Piotr Kapitsa and Jack Allen.

Piotr Leonidovitch Kapitsa had graduated as an electrical engineer in Saint Petersburg under the supervision of F. Ioffe in 1919. In 1921, Ioffe suggested that Kapitsa goes to Cambridge where he could work with Rutherford. There, he proved to be a brilliant experimental physicist. For example, he made the first detection of the bending of alpha-particle paths in a magnetic field thanks to a cloud chamber. Then, he built a pulsed magnetic field installation and a hydrogen liquefier with his student John Cockcroft. He was quickly elected Fellow of Trinity College (1925) and Fellow of the Royal Society (1929) "*a rare distinction for a foreigner, especially for one who became a Corresponding Member of the Soviet Academy of Sciences in the same year*", as explained by David Shoenberg (Shoenberg 1994). Then, Rutherford obtained from the Royal Society that part of the donation from Ludwig Mond was used to build the "Royal Society Mond Laboratory" where Kapitsa could develop his low temperature and high magnetic field studies. In this laboratory, he constructed a new type of helium liquefier, which produced its first drops of liquid helium on April 19, 1934, and made such experiments much easier (Rubinin 1997).

In the summer of 1934, Kapitsa went back to Leningrad where he could see his mother and participate in a symposium celebrating the centenary of Mendeleiev. However, on September 24, 1934, 5 months only after the first operation of his liquefier in Cambridge, he was not allowed to return to England from the Soviet Union (Rubinin 1997). The reasons for this are a little unclear but, according to D. Shoenberg (Shoenberg 1994), "*he had sometimes been rather boastful of his successes in England and gave the impression that his work could be of immense technological importance if only he were given the right support. The authorities, possibly Stalin himself, took him at his word and told him that he must in the future work for them, although in fact none of his work was secret and it was available to everyone"*. Later, Stalin would need Kapitsa for his nuclear program and conflicts with Beria triggered Kapitsa's disgrace. But in 1934, Kapitsa started a fight with Stalin and Molotov to obtain support for his research. Two years later, the "Institute for Physical Problems" was built for Kapitsa in Moscow. Thanks to the help of Rutherford, he could also arrange that part of his equipment be purchased from Cambridge and transferred to Moscow, so that he could start his research again.

At that stage, one major problem for Kapitsa was that Cambridge had kept his precious liquefier. But Kapitsa obtained the right to invite his student David Shoenberg and two technicians, E.Ya. Laurmann and H.E. Pearson, in order to build a new helium liquefier in Moscow. They made a better one, which produced liquid helium on February 22, 1937 (Rubinin 1997). Meanwhile, Cambridge had used Kapitsa's rather high salary (800 pounds a year) to hire two younger scientists, Rudolf Peierls and J.F. Allen who seemed satisfied with 400 pounds a year each (Allen 1988).

John Franck ("Jack") Allen was born in Winnipeg (Canada) and he had obtained his PhD on superconductivity in Toronto (1933). Then, he tried to join Kapitsa in Cambridge but when he arrived in the fall of 1935, Kapitsa was already detained in USSR. In 1936, he attracted A.D. Misener to work towards a PhD degree in Cambridge with him. We thus realize that Kapitsa was competing with two Canadian physicists who were using his former liquefier in his former laboratory where he was still in close contact with other people. Of course, this situation was very painful to him: "*I often see my laboratory in my dreams, and painfully want to work*" as he wrote to his wife (Rubinin 1997) in March 1935.

6.3 Who Discovered Superfluidity?

On January 8th, 1938 *Nature* published two articles side by side (Allen and Misener 1938; Kapitsa 1938). The first one, on page 74, received December 3, 1937, was entitled: "*Viscosity of liquid helium below the lambda point*" and signed by P. Kapitsa (Institute for Physical Problems, Moscow). The second one, entitled "*Flow of liquid Helium-II*" (page 75), received December 22, 1937, and signed by J.F. Allen and A.D. Misener (Royal Society Mond Laboratory, Cambridge, UK).

In 1937, Kapitsa tried to understand why the thermal conductivity of helium II was anomalously large. An explanation could be that convection in this liquid was important if its viscosity was small, as already proposed by Wilhelm, Misener and Clark. He thus tried to measure this viscosity by flowing liquid helium from a little reservoir through a slit about 0.5 μm thick, between two polished cylinders pressed against each other.

In his article (Kapitsa 1938), Kapitsa writes:

> The flow of liquid above the lambda-point could be only just detected over several minutes, while below the lambda-point the liquid helium flowed quite easily, and the level in the tube settled down in a few seconds. From the measurements we can conclude that the viscosity of helium II is at least 1500 times smaller than that of helium I at normal pressure.
>
> The experiments also showed that in the case of helium II, the pressure drop across the gap was proportional to the square of the velocity of flow, which means that the flow must have been turbulent. If, however, we calculate the viscosity assuming the flow to have been laminar, we obtain a value of order 10^{-9} cgs, which is evidently still only an upper limit to the true value. Using this estimate, the Reynolds number, even with such a small gap, comes out higher than 50,000, a value for which turbulence might indeed be expected.

These two paragraphs are a little difficult to understand. Kapitsa does not give any value for the flow velocity in the slit, nor for the height difference, which drove the flow. Since the Reynolds number is $R = UL/\nu$ where U is the velocity, L a typical length scale and ν the kinematic viscosity, I understand that he must have measured velocities U up to about 7 cm/s. As we shall see, this is comparable to what had been measured by Allen and Misener and confirmed by later work, although it depends on the size of the flow as reviewed in the book by Wilks, page 391 (Wilks 1967). As far as I know, Kapitsa's square law for the pressure drop has not been confirmed by any later work, but since he does not give much quantitative information on his measurements, it is difficult to appreciate the accuracy at which this square law could fit his data. Given what is known today, I suppose that he approximated the pressure dependence of the velocity -which shows a threshold – with a square law. Anyhow, Kapitsa finally proposes that:

> by analogy with superconductors, ... the helium below the lambda-point enters a special state which might be called superfluid.

This is a famous sentence where Kapitsa introduces the word "superfluid" for the first time. His intuition was quite remarkable because superfluids and superconductors are indeed analogous states of matter, but Kapitsa wrote this sentence long before the BCS theory of superconductivity was established, *a fortiori* before any demonstration of this analogy.

As for the article by Allen and Misener, it starts with the sentence:

> A survey of the various properties of liquid helium II has prompted us to investigate its viscosity more carefully. One of us [1] had previously deduced an upper limit of 10^{-5} cgs units for the viscosity of helium II by measuring the damping of an oscillating cylinder. We had reached the same conclusion as Kapitsa in the letter above; namely that, due to the high Reynolds number involved, the measurements probably represent non-laminar flow.

The note [1] refers to the article by E.F. Burton (Burton 1935) on the work by Wilhelm, Misener and Clark, which I mentioned above. More important for our present purpose is the reference to Kapitsa at the beginning of the article by Allen and Misener. We understand that they had read Kapitsa's article before writing their own, or at least that they had heard of its content. Together with the 19 days difference in the submission date, this has sometimes been taken as a proof that Kapitsa had some priority on Allen and Misener in the experimental discovery of superfluidity (Andronikashvili 1989). However, as we shall see, I do not agree with such a statement.

The Cambridge article contains a detailed study of the flow through two different capillaries with sections respectively equal to 6×10^{-4} and $0.8\ mm^2$. Measurements are given at two different temperatures (1.07 and 2.17 K) and at series of ten different pressures. Flow velocities range from 0.4 to 14 cm/s. Their main findings were that, contrary to Poiseuille's law, which describes laminar situations, the velocity was nearly independent of pressure. The measurements by Allen and Misener could obviously not be done in 19 days. I cannot imagine that they started their study after hearing of Kapitsa's article. If a proof is needed, it is in their notebook, which shows that Allen and Misener had obtained results already on November 24, 1937 (Griffin 2008).

When Kapitsa sent his letter to *Nature*, he wrote in the accompanying letter to the editor:

> I am sending herewith a short note: 'Viscosity of liquid helium below the lambda-point', which I hope you will kindly publish in your 'letters to the editor'. I think this is an important note and I should be glad if you could arrange it to be published as soon as possible, and with the day of dispatch. Please do not bother to send the proofs to me here to Moscow, it takes too much time. If necessary please send them either to Prof. P.A.M. Dirac, Dr. J.D. Cockcroft, or to Dr. W.L. Webster … All my good friends [are] sufficiently competent to make the necessary corrections. I hope you will kindly help me in publishing this note very soon …

As explained by Allen himself (Allen 1988) and by Shoenberg (Shoenberg 1994), it was John Cockcroft who took care of the proof-reading. He was the new director of the Mond Laboratory since Kapitsa had left. In 1951, he received the Nobel Prize for having verified Einstein's famous relation $E=mc^2$. In December 1937, he showed Kapitsa's letter to Allen and Misener and asked them to write down their own results as quickly as possible. He finally asked *Nature* to publish the two papers side by side. It is clear to me that the Cambridge work was independent of Kapitsa's work in Moscow. Nineteen days delay in the article submission is not a serious reason to doubt of that.

Now, was Kapitsa's work independent of the Cambridge work? After all, Kapitsa's insistence to be published with a mention of the date of receipt indicates that he probably knew that his competitors were working on the same subject. Furthermore, his letter presents qualitative ideas, which could have been written down quickly. Was Kapitsa aware of Keesom's work (Keesom 1930) on the ability of helium II to flow through narrow slits (the existence of "superleaks")? Could Kapitsa have written his letter *after* hearing of the progress made by Allen and Misener in Cambridge?

I have carefully inquired about this possibility (Balibar 2007). It appears that there were frequent contacts between Kapitsa in Moscow and his friends in Cambridge. Letters from one place took about 1 week to arrive at the other place. Kapitsa probably knew that he was competing with Allen and Misener but it is unlikely that he knew Allen and Misener's results. Anyhow, in December 1937, when Rutherford died, Kapitsa sent a letter to Niels Bohr, which proves that Kapitsa was already working on this problem in June 1937. In my opinion, there is no real priority in either way, the two works are independent. But they are not equivalent as we shall see now.

Let us summarize the content of the four experimental contributions to the problem of superfluidity. In 1930 Keesom had observed that helium II was highly fluid and in 1935 Wilhelm, Misener and Clark had measured in Toronto a sharp drop of the viscosity below the lambda-point. Then, in December 1937, Kapitsa claimed that the flow of liquid helium II was turbulent and that its viscosity could not be larger than 10^{-9} cgs units. As for Allen and Misener, they presented the results of a series of measurements, from which they concluded:

> the observed type of flow… in which the velocity becomes almost independent of pressure, most certainly cannot be treated as laminar or even as ordinary turbulent flow. Consequently any known formula cannot, from our data, give a value of the viscosity which would have much meaning.

In my opinion, it is Allen and Misener who discovered that, below T^{λ}, the hydrodynamics of helium required a totally new interpretation. Here is the real experimental breakthrough. At that time, everyone else kept considering that liquid helium was a liquid with a small viscosity. In 1941, Kapitsa published another article (Kapitsa 1941) where he kept considering the viscosity of superfluid (he found it at least 100 times lower than in 1938).

It would be very interesting to understand how Kapitsa had the intuition that helium II had something in common with superconductors. The idea that superconductors were quantum systems had to be described by a macroscopic wave function, had been put forward by Fritz London and his brother Heinz (London and London 1935) but, as we shall see now, London had not yet considered that it could be the case for liquid helium also. Furthermore, and as we shall see when considering Landau's work, this was not at all the way how one liked to think about liquid helium in Moscow.

6.4 Who Initiated the Understanding of Superfluidity?

Fritz London was born in Breslau (now Wroclaw in Poland) in 1900 and he had started studies in philosophy before switching to physics (Gavroglu 1995). He was educated at the universities of Bonn, Frankfurt, Göttingen and Munich where he graduated in 1921. Together with Walter Heitler in Zurich, he had devised the first quantum mechanical treatment of the hydrogen molecule in 1927 (Heitler and London 1927). He then joined Schrödinger in Berlin but in 1933, when the Nazis took power, he escaped to Oxford where Lindemann found support for him until 1936. Then, he was quite happy to find a position at the Institut Henri Poincaré in Paris where he was attracted by a group of intellectuals linked to the "Front populaire" (the coalition of political parties from the French left), namely Paul Langevin, Jean Perrin, Frédéric Joliot and Edmond Bauer.

Laszlo Tisza had arrived in Paris in 1937 for similar reasons. He was born in 1907 and he had studied in Budapest before attending Max Born's course in Göttingen. Later, he worked in Leipzig under Heisenberg and wrote his first paper with Edward Teller, just before being arrested by the Hungarian Nazi government under the accusation of being a communist (Teller 1998). In 1935, he was liberated and Teller strongly recommended him to his friend Lev Landau in Kharkov. There, Tisza entered as number 5 the famous school of theoretical physics that Landau had founded. But in March 1937, both Landau and Tisza left Kharkov. At this time, Tisza must have tried to protect himself from anti-Semitism, just like London. In September 1937, Paul Langevin and Edmond Bauer offered him a position at the Collège de France in Paris. This is where he met Fritz London; the Collège de France is about 300 m from the Institut Henri Poincaré.

London's first ideas on superfluidity (London 1938a, b) were triggered by the next article published by Allen in the same volume 141 of *Nature* on February 5, 1938 (Allen and Jones 1938). Entitled *"New phenomena connected with heat flow in helium II"*, this new letter described the discovery of what is now known as the fountain effect: together with Misener again for the experiments, but published with H. Jones, the new young theorist who replaced Peierls in Cambridge, Allen discovered that, when heat was applied to liquid helium II on one side of a porous plug, the pressure increased proportionally to the heat current so that the level of the free surface went up (it was later realized that the fountain pressure was in fact proportional to the temperature difference between the two sides). A liquid jet could even occur if the pressure was high enough. If the liquid had been classical, the vapour pressure would have been higher on the warm side so that, in order to maintain hydrostatic equilibrium in the liquid, its level would have had to go down. Allen and Jones explained that the opposite was observed. For London, it was no longer possible to doubt that this liquid had totally anomalous properties for which a radically new interpretation was needed. In previous work (London 1936), Fritz London had proposed that helium II was more ordered than helium I. This was probably because its specific heat decreased sharply below T_λ. Perhaps it had some kind of crystalline structure with a diamond lattice? However, on March 5, 1938, London sent a letter to *Nature,* which was published on April 9. There, he explained that liquid helium II was not crystalline before proposing that it was undergoing some kind of Bose-Einstein condensation at T_λ (London 1938a, b).

In the introduction of his first book (London 1950), London, writes:

> In 1924, Einstein developed a very strange concept of a gas of identical molecules, which were assumed to be indistinguishable... Einstein remarked that this removal of the last vestige of individuality from the molecules of a species would imply a statistical preference of the molecules for having the same velocity, even if any interaction between them were disregarded, and this preference would lead, at a well-defined temperature to a kind of change of state of aggregation; the molecules would 'condense' into the lowest quantum state, the state of momentum zero. Einstein did not give a very detailed proof, and his remark received little attention at the time. Most people considered it a kind of oddity which had, at best, an academic interest, for at the extremely low temperatures or high pressures in question there are no gases, all matter being frozen or at least condensed by virtue of the intermolecular interaction forces. In addition, doubt was cast on the mathematical correctness of Einstein's remark, and hence the matter was disposed of as if there were no 'Bose-Einstein condensation'.

On November 29, 1924, Einstein himself had sent a letter to his friend Paul Ehrenfest in Leiden, where he wrote: *"From a certain temperature on, the molecules 'condense' without attractive forces, that is, they accumulate at zero velocity.*

The theory is pretty, but is there also some truth in it?" (Pais 1982). By generalizing a calculation by the young Bengali physicist Satyendra Nath Bose (Bose 1924) to massive particles, Einstein had found (Einstein 1924, 1925) that, for an ideal gas of Bose particles, a macroscopic fraction of these particles accumulates in the ground state below a certain critical temperature. At that time, the theory of phase transitions was still in its infancy, and, in his PhD work, Uhlenbeck had argued against the BEC being a true phase transition by saying that it would not occur in a finite size system (Uhlenbeck 1927). Uhlenbeck was a graduate student of Paul Ehrenfest and, apparently, his criticism was generally accepted, even by Einstein himself (Griffin 1999). But in November 1937, a conference took place in Amsterdam in honour of van der Waals (Johannes Diderik van de Waals was born 100 years before, on November 23, 1837 in Leiden). Fritz London was there (Gavroglu 1995) and he must have heard discussions including Ehrenfest and Kramers about the thermodynamic limit in connection with phase transitions, also that Uhlenbeck had withdrawn his argument against BEC. This must be what triggered London's interest in Einstein's forgotten paper on BEC (Griffin 1999).

In a message that he sent me on the September 4th, 2001, Tisza made the following comment on the discovery of superfluidity:

> The novelty of the effect became strikingly apparent in the Allen and Jones fountain effect that started London and myself on our speculative spree…

In his letter to Nature (London 1938a, b), Fritz London first recalled that ^{4}He atoms were Bose particles, then that liquid ^{4}He was a quantum liquid because the quantum kinetic energy of the atoms was large, something he had explained in his previous article (London 1936). This large "zero point energy" was responsible for the absence of crystallization at low pressure, something which had been also noticed by Franz Simon (Simon 1934). Then London explained that, although BEC had "*rather got the reputation of having only a purely imaginary existence… it actually represents a discontinuity in the temperature derivative of the specific heat*", meaning that it was a phase transition of third order (according to the classification by Ehrenfest). Then he calculated the transition temperature at which an ideal Bose gas with the same density as liquid ^{4}He would condense in Einstein's sense and he found 3.1 K, a value close to T_λ. He further noticed that the singularity in the specific heat of helium resembled the cusp predicted for BEC. He then concluded that, "*Though the lambda-point resembles rather a phase transition of second order, it seems difficult not to imagine a connexion with the condensation of the Bose-Einstein statistics. The experimental values of the temperature of the lambda-point and of its entropy seem to be in favour of this conception*". Keeping this modest attitude, he also estimated that his model, "*which is so far from reality that it simplifies liquid helium to an ideal gas*", was a rough approximation, which could not give quantitative agreement with experimental measurements. To a modern eye, everything looks right in this letter to *Nature*. Shortly afterwards, he expanded his letter into a longer article published the same year (London 1938b).

London's new ideas created considerable interest (Gavroglu 1995; Griffin 1995), in particular from Laszlo Tisza. Laszlo Tisza told me in 2001 that they liked discussing physics together during long walks. On one such occasion, London explained his ideas about BEC to him and he had soon the intuition that, if BEC took place, there should be two independent velocity fields in liquid helium. One part would have zero viscosity and zero entropy; the other part would be viscous and would carry entropy; the proportion of each fluid would be related to temperature. He sent this as a short note to *Nature* on April 16, 1938 (Tisza 1938a), which introduced for the first time what is now known as the "two fluid model" (Tisza 1938a). He announced there more detailed publications which were presented in French by Paul Langevin at the Académie des Sciences in Paris on November 14th, 1938, and indeed published in its Comptes-Rendus (Tisza 1938b, c).

On the basis of his model, Tisza solved the apparent contradiction between different types of measurements of the viscosity of helium II: in the Toronto experiment (Misener et al. 1935), the damping of the oscillations of the cylinder was related to the viscosity of the whole fluid while in a flow through a thin capillary (Allen et al. 1937) or through a thin slit (Kapitsa 1938) only the non-viscous component of the fluid could flow. He further explained in this *Nature* note that the independent motion of the two fluids allowed one to understand the fountain effect. He eventually predicted an inverse phenomenon, namely that "*a temperature gradient should arise during the flow of helium II through a thin capillary*". The latter was to be named the "thermomechanical effect" by Fritz London (London 1938b) and his brother Heinz (London H. 1938c); evidence for its existence was soon found by Mendelssohn and Daunt in Oxford (Daunt and Mendelssohn 1938) and further studied by Kapitsa (Kapitsa 1941). In the following articles to the Comptes-Rendus (Tisza 1938b, c), Tisza predicted that, in helium II, heat should propagate as "*temperature waves*", another revolutionary idea. In July 1938, *Tisza "presented this prediction at a small low temperature meeting in London… and offered it to make or break [his] theory*" (Tisza 2000). His temperature waves were later renamed "*second sound*" by Landau, discovered by Peshkov in 1946 and were indeed taken as a crucial test of his theory (see below).

At least qualitatively, the 1938 papers by London and Tisza could explain all the experimental observations which had been already made at that time, namely the flow and heat conduction experiments, the heat capacity measurements, also the fast motion of films adsorbed on a wall by Rollin (Kuerti et al. 1936; Rollin and Simon 1939), confirmed by Daunt and Mendelssohn (Daunt and Mendelssohn 1938). But still, when London first heard about Tisza's two fluid model, he could not believe that, in a liquid, which was pure and simple, there could be two independent velocity fields (Tisza 1991). This was indeed quite a revolutionary idea. Later, Tisza wrote a more elaborate version of his theory, which he submitted as two articles (Tisza 1940a, b) to the Journal de Physique et du Radium on October 23, 1939, but he could not see them printed till the end of the war. In June 1940, part of Langevin's laboratory was evacuated to Toulouse, in the south part of France, which was not yet occupied by the Nazi army. In another e-mail (March 17, 2005), Laszlo Tisza told me that:

> Jacqueline Hadamard, the daughter of the mathematician Jacques Hadamard, was a member of the lab and she offered to me and my wife a ride to Toulouse. M. and Mme Hadamard travelled with their other daughter, but I had the privilege to travel as a virtual member of

> the Hadamard family. Just before leaving we had signed up for an American visa at the Budapest consulate without any definite plans for using it. By a fortunate coincidence the Clipper connection between Lisbon and New York started in the summer 1940 and suddenly we received airmail letters from friends and relatives in Cambridge in two days! We must have notified the Marseille consulate of our address and sometimes in October we got a telegram that our visa was authorized. After finishing all paper work we left Marseille early February 1941 for Madrid and Lisbon. Mid-March we sailed on a Portuguese boat to New York and joined friends and relatives in Cambridge. In a few months in September I was appointed instructor at MIT, to become eventually professor….

As for Fritz London, Frederic Joliot offered him a position of "Directeur de recherches" at the Institut Henri Poincaré in November 1938. He was strongly attached to France where his wife Edith had entered a group of painters led by André Lhote (Gavroglu 1995). But he found it wiser to accept an offer from Paul Gross, the head of the Chemistry Department at Duke University. He could escape just in time from France, in September 1939, on the boat *Ile de France* to New York, fortunately not on the *New Amsterdam*, which was destroyed by a submarine on September 3, three days after the beginning of the war (Meyer 2005). In October 1939, Fritz London was teaching at Duke as a professor of theoretical chemistry.

Nearly at the same time, Landau was coming out of Stalin's jails, and it is time now to introduce the fifth important character in our history.

Lev Davidovitch Landau was born in Baku on January 22, 1908. He graduated from the Physics Department of Leningrad in 1927, at the age of 19. He then travelled thanks to a Rockefeller fellowship to Germany, Switzerland, England and Copenhagen where he worked with Niels Bohr. From 1932 to 1937, he was the head of a theory group in Kharkov. There, Alexander S. Kompaneets, Evgueny M. Lifshitz, Alexander I. Akhiezer, Isaak Ya. Pomeranchuk, and Laszlo Tisza formed the first core of Landau's famous school. At the same time Landau was also teaching in Moscow and Kapitsa invited him to come in his new Institute in 1937. However, in March 1938, Landau was arrested by the NKVD, later called KGB (Pitaevskii 1992; Gorelik 1997). He had been accused of being one of the authors of a leaflet criticizing the Soviet regime (Gorelik 1997).

Kapitsa had already written some letters to Stalin in order to obtain the scientific equipment he needed for his research in Moscow. After Landau's arrest, Kapitsa started another fight to liberate him and eventually sent a letter to Molotov on April 6, 1939, where, as published in English by P.E. Rubinin (Rubinin 1997), he wrote:

> In my recent studies on liquid helium close to the absolute zero, I have succeeded in discovering a number of new phenomena… I am planning to publish part of this work… but to do this I need theoretical help. In the Soviet Union it is Landau who has the most perfect command of the theoretical field I need, but unfortunately he has been in custody for a whole year. All this time I have been hoping that he would be released because, frankly speaking, I am unable to believe that he is a state criminal… It is true that he has a very sharp tongue, the misuse of which together with his intelligence has won him many enemies… but I have never noticed any sign of dishonest behaviour… the Soviet Union and worldwide has been deprived of Landau's brain for a whole year. Landau is in poor health and it will be a great shame for the Soviet people if he is allowed to perish for nothing…

Three weeks later, Kapitsa was summoned to the NKVD headquarters where he tried to defend Landau as much as he could in a very hard discussion. Around 4 o'clock in the morning, it was said to him: "All right, Kapitsa, if you pledge your word for Landau, then give us a written guarantee. If anything happens, you will be held responsible" (Rubinin 1997). Kapitsa wrote a letter to Beria on April 26, and Landau returned to the Institute on April 28, 1939.

This allowed Landau to survive and to come back to work. On June 23, 1941, Kapitsa (Kapitsa 1941) and Landau (Landau 1941a) sent two letters together for publication in the Physical Review. They were published next to each other and Landau's letter announced a more elaborate paper to be published in the Journal of Physics of the USSR (Landau 1941b).

The two 1941 articles by Landau start with nearly the same sentence: "It is well known that liquid helium at temperatures below the lambda-point possesses a number of peculiar properties, the most important of which is superfluidity discovered by P.L. Kapitsa".

For Landau, superfluidity had thus been discovered by the man who had saved his life – P.L. Kapitsa – and only by him. Landau continues with:

> L. Tisza suggested that helium II should be considered as a degenerate ideal Bose gas… This point of view, however, cannot be considered as satisfactory… nothing would prevent atoms in a normal state from colliding with excited atoms, i.e. when moving through the liquid they would experience a friction and there would be no superfluidity at all. In this way the explanation advanced by Tisza not only has no foundation in his suggestions but is in direct contradiction to them (Landau 1941b).

Landau *never* cited Fritz London. Here as everywhere he attributes to Tisza instead of F. London the proposal that superfluidity is a consequence of Bose-Einstein condensation. Why is it that Landau never believed in the relevance of BEC? This is a major and quite interesting question. Moreover, why Landau needed to be so abrupt in his criticism of his former postdoc Tisza? This is a related question, which is no less interesting in my opinion.

After the above introduction, Landau's article starts with a first section entitled "*The quantization of the motion of liquid*". Everybody considers what follows as a brilliant breakthrough in the theory of quantum liquids. He quantizes the hydrodynamics of quantum liquids and arrives to the statement "Every weakly excited state can be considered as an aggregate of single elementary excitations", which he divides in two different categories: sound quanta which he calls "*phonons*" and elementary vortices which his friend I.E. Tamm suggested be called "*rotons.*"

Six years later (Landau 1947), Landau modified the roton spectrum and included them as part of the phonon spectrum. But already in 1941, Landau could calculate the specific heat of liquid helium and obtained a good fit of experimental measurements by W.H. and A.P. Keesom (Keesom and Keesom 1936). In his 1941 article, Landau then proposes that, for a superfluid flowing at a velocity V at zero temperature, dissipation can only result from the emission of either phonons or rotons, so that, from the conservation of energy and momentum in this process, dissipation is only possible if the fluid velocity V is larger than a critical velocity v_c, which is the phase velocity of rotons and today known as "Landau's critical velocity".

Landau has thus introduced a possible explanation why helium II flows at a velocity that is independent of pressure or capillary section: his critical velocity is a property of the helium itself. However, he also notices that the value he predicts for v_c is much larger than observed in experiments and "*left aside the question as to whether superfluidity disappears at smaller velocity for another reason*".

In the next section he calculates the properties of superfluid helium at finite temperature. For this he introduces a two fluid model: he distinguishes a "*normal component*", which is made of phonons and rotons, from a "*superfluid component*". The superfluid component carries no entropy and moves without dissipation, while the normal one is viscous and carries a non-zero entropy. The ratio of the respective densities of the two components depends on temperature since, at $T=0$, all the density of the fluid is superfluid while, at the lambda point the superfluid component disappears and all the fluid is normal. Given the values for the phonon and roton parameters, which he had adjusted to fit specific heat data, Landau calculates an approximate value for the lambda point temperature (2.3 K) also in agreement with experiment. He finally explains the thermomechanical effects – the fountain effect and the reverse phenomenon – and he predicts that heat should propagate as "*second sound*" instead of diffusing as in classical fluids.

Landau's theory is a remarkable success, and it is still in use nowadays. Its main features are common to Tisza's previous version but there is one major difference. The common features are: the existence of two independent velocity fields; the temperature variation of the two fluid densities; the non-dissipative flow of the superfluid component (but Tisza could not predict the existence of a critical velocity for it); the fact that all the entropy is carried by the normal component and the propagation of heat as a wave. When deriving the equations which describe thermomechanical effects, Landau writes: "*The formulae 6.1 and 6.4 were deduced already by H. London (Proceedings Royal Society 1939) starting from Tisza's ideas*". Let me remark that Landau cites Heinz London (London H. 1939), Fritz London's young brother, and it is very hard to believe that Landau had not noticed the work of Fritz London, whom he had met in 1932. The absence of reference to Fritz London must be intentional. He had perhaps personal reasons for this, but I have tried to understand why he never believed in the relevance of Bose Einstein condensation in the theory of superfluidity. The above sentence also means that Landau knew the existence of the two notes published in the Comptes-Rendus by Tisza (Tisza 1938b, c) in 1938, which are cited by Heinz London (London H. 1939).

The major difference between Landau's theory and Tisza's is in the nature of the normal component: according to Landau it is made of "quasiparticles", a new concept he introduces to quantize the elementary excitations of quantum fluids. In contrast, Tisza thinks in terms of ideal gases and proposes that the normal component is made of the non-condensed atoms. Shortly after the war, Peshkov did experiments to discriminate between the predictions by Landau and by Tisza (Peshkov 1946). Indeed, according to Landau, the second sound velocity should increase as temperature tends to zero, while Tisza predicted that it vanishes. At the low temperature meeting, which Allen organized in Cambridge in 1946, Fritz London was asked to give the opening talk (London 1947). He explained that Peshkov's preliminary results (Peshkov 1946) where not yet done at low enough temperature to discriminate between Landau and Tisza, but Peshkov's experiments soon showed that Landau was right (Peshkov 1948). In fact, Fritz London was very critical about Landau's theory: *"an interesting attempt to quantize hydrodynamics... based on the shaky grounds of imaginary rotons"*. London must have been rather upset by Landau's attitude, in particular by his rough rejection of Tisza's model. Some authors consider that the two fluid model has been independently discovered by Tisza and by Landau, but this is not true. I have demonstrated that Landau knew Tisza's work, whatever he declared later (Balibar 2007).

Landau's absence of reference to Fritz London is a different issue of greater scientific interest. At this stage, we have to realize that Landau's 1941 work never mentions Bose nor Fermi statistics. In fact he derives his quantization of hydrodynamics without making any difference between Bose and Fermi fluids. Today, of course, we know that degenerate Fermi liquids such as liquid ^{3}He are highly viscous while degenerate Bose fluids are superfluid. It means that there is a mistake or some misunderstanding somewhere in Landau's article. Where?

A weakness in Landau's view occurs when he claims that there is a gap between irrotational states and states where the circulation of velocity is non-zero. Landau does not justify this statement. It is the later work of Bogoliubov (Bogoliubov 1947), which showed for the first time that in a degenerate Bose gas with weak interactions, there is BEC and there are no individual excitations at low energy, only collective modes, that is phonons with a non-zero velocity. Bogoliubov showed that if dissipation results from the emission of elementary excitations, it can only occur beyond a certain critical velocity, (the sound velocity in this case), and that the motion of the condensate fraction is non-dissipative and irrotational below this critical velocity. In 1951, BEC was generalized by Penrose (Penrose 1951) as "off-diagonal long range order" (ODLRO) in the formalism of the density matrix. This approach was further developed by Penrose and Onsager in 1956 (Penrose and Onsager 1956). It allows the condensate fraction to be much smaller than one (the total mass) and irrotational dissipationless motion to occur below a certain critical velocity. One has also realized that in most macroscopic systems, the emission of quantized vortices is another mechanism, which is responsible for a critical velocity smaller than Landau's. In other words, the existence of superfluidity is really linked to BEC, at least to the Bose statistics and the quantization of vortices.

One could argue, of course that superfluidity exists in 2D-Bose fluids, where, strictly speaking, there is no BEC. But there are long range quantum correlations so that vortices are quantized, and dissipation cannot occur in practice below a certain velocity. In summary, the superfluidity is certainly linked to the Bose statistics, contrary to Landau's statement.

As for Fermi liquids, it is in fact the hydrodynamics itself, which breaks down. As Landau was to realize later (Landau 1957), the excitations of a degenerate Fermi liquid are Fermi quasiparticles which travel ballistically over a certain distance and which are responsible for the divergence of the viscosity in the low temperature limit. The existence of an energy gap between rotational and irrotational states in quantum fluids is simply not true in Fermi liquids. This takes us back to the already mentioned question: how can it be that Landau never referred to BEC nor mentioned Bose statistics in his theory of superfluidity?

Perhaps Landau could simply not believe that a theory of quantum ideal gases (BEC) could apply to liquids with strong interactions between atoms? This is the spirit of his criticism of Tisza's approach (there should be collisions between excited atoms and condensed atoms). Furthermore, as would show up later from Bogoliubov's work (Bogoliubov 1947), it is true that an ideal Bose gas with no interactions at all would have a sound velocity equal to zero, consequently a zero critical velocity: it would not be superfluid! Eventually, we now know that there is no continuous path from a low density helium gas to a higher density helium liquid: it has been predicted by H.J. Maris (Maris 1994, 1995) and experimentally verified in our research group (Caupin and Balibar 2001) that there is a range of densities for which helium is unstable, between two spinodal lines which respectively limit the range of possible metastability of either liquid or gaseous helium. For all these reasons, the most likely interpretation of Landau's absence of reference to BEC is just that he could not consider that a theory of quantum gases could apply to a liquid.

However the absence of reference to the Bose statistics needs a further explanation. In his 1992 article (Pitaevskii 1992) on Landau's theory of superfluidity, Lev Pitaevskii writes: *"Landau was only one step from the very interesting subject of macroscopic quantum phenomena. But he never made this step. And there is no sense now to guess why..."* Coming back to this issue with Landau, Lev Pitaevskii proposed to me another idea, which is the following. Kapitsa and most probably Landau as well considered superfluidity as a phenomenon analogous to superconductivity. This was long before the BCS theory and of course superconductivity occurs in a Fermi system of electrons. Since the same phenomenon occurred in both quantum fluids (Bose and Fermi), Landau could perhaps not admit that superfluidity was related to the quantum statistics. Whatever Landau really thought, a possible comparison of ^{3}He and ^{4}He progressively appeared as a crucial test. In his book, London insisted on the importance of such a test (London 1950). As soon as ^{3}He was available in large enough quantities, a test was made of the possible superfluidity of ^{3}He, which was found to be non-superfluid down to 1 K, in strong support to London's and Tisza's approach. This experiment was done by D.W. Osborne, B. Weinstock and B.M. Abraham in 1949 (Osborne et al. 1949).

As an aside, let me mention that B. Abraham had joined the Manhattan project during the war and owned a patent for the Lithium-Tritium compound to be used in H-bombs. Let me mention further that Landau also participated to the building of the H-bomb, but the Soviet one of course, and despite the severe conflict which opposed Kapitsa and Beria in this enterprise. Beria forced Kapitsa to leave his scientific position and activity at the Institute for Physical Problems because of their conflict. Landau kept working for the bomb, apparently because this was a way for him to be protected against any further problems with the Soviet regime (Gorelik 1997). Later, Beria was assassinated and Kapitsa recovered his position at the Institute for Physical Problems. When Stalin died, Landau left the Soviet H-bomb program (Gorelik 1997).

Coming back to superconductivity and the superfluidity of ^{3}He, we know that the BCS theory considers the condensation of Cooper pairs, which obey the Bose statistics, and that superfluidity was also discovered in liquid ^{3}He at a temperature low enough (about 2.5 mK) that ^{3}He atoms could form pairs (Osheroff et al. 1972; Osheroff 1997; Leggett 1972, 2004).

As for rotons, their existence was proven by inelastic neutron scattering experiments (Henshaw and Woods 1961). It also happens that, for my PhD work, I studied quantum evaporation and obtained the first experimental evidence that, at low enough temperature, a heat pulse decomposes into ballistic phonons and rotons, and that individual rotons can evaporate atoms with a minimum kinetic energy of 1.5 K (Balibar et al. 1978). This phenomenon had been predicted by P.W. Anderson as an analogue of the photoelectric effect (Anderson 1969). A.F.G. Wyatt and his group have performed a long quantitative study of this quantum evaporation (Hope et al. 1984; Brown and Wyatt 1990; Tucker and Wyatt 1999). Today, there is no doubt that rotons exist, only controversies remain on their physical picture. Landau had first proposed that they were vortices of atomic size and later considered them as part of the phonon spectrum. Surprisingly, Feynman insisted on Landau's first picture by considering that a roton could be an elementary vortex loop (Feynman 1955). In my opinion, rotons are phonons with a wavelength equal to the interatomic distance. Their low energy is a signature of the local order, which had already been mentioned by London. As expressed by Nozières, rotons are "ghosts of a Bragg peak" (Nozières 2004). This is because Feynman showed that, under certain approximations, the dispersion relation for elementary excitations is related to the static structure factor of liquid helium by the simple relation. As a consequence, if there is some short range order in this liquid, that is a large probability to find an atom at a distance, which is the average distance from another atom, in other words a large peak in the structure factor (Horner 1972), then there has to be a roton minimum in the dispersion relation of elementary excitations. One should not associate superfluidity with the existence of a roton minimum; Landau introduced rotons to calculate the specific heat of liquid helium and then explained that their existence limits the maximum value of the critical velocity. In reality rotons are precursors of solidification, and their existence works against superfluid order. In the superfluid gases which have been discovered in 1995 (Cornell and Wieman 2002; Ketterle 2002) there is superfluidity and no rotons because the system has weak interactions and consequently no short range order. Landau was right in a sense (rotons exist) but wrong concerning his first interpretation or physical picture (they are not elementary vortices, nor essential for superfluidity).

London and Tisza as well as Landau formulated theories, which were, in a way, complementary. Unfortunately, neither London nor Landau lived long enough to realize that a full theory should combine their respective approaches. Fritz London died of a heart attack in 1954. Landau was severely injured in a car accident shortly before receiving his 1962 Nobel Prize. The car accident occurred on January 7, 1960, he was in coma for a long time and suffered so much afterwards that he could never work anymore till he died in 1968. Of course he could not go to Stockholm and receive his Nobel Prize in person. Einstein had proposed London for the Nobel Prize. A few years before arriving to the famous BCS theory with Leon Cooper and Robert Schrieffer, John Bardeen also recognized the great importance of London's work on superconductivity (the introduction of a macroscopic wave function) as the basis of his work on the same subject; in a letter sent to London on December 9, 1950, he had written:

> Dear Prof. London,
> You may be interested in the enclosed manuscripts on superconductivity; they are both based on your approach (Meyer 2005)

Bardeen's admiration for London's work must be the reason why, when he received his second Nobel Prize in 1972 – he shared this one with Cooper and Schrieffer for the "BCS" theory of superconductivity but he had already shared one with Schockley and Brattain in 1956 for the discovery of the transistor -he decided to donate his part of the Nobel Prize to Duke University. The purpose was to create an endowment to enable funding a yearly lecture at Duke University in the honour of Fritz London and also to finance the Fritz London Prize for distinguished work in Low Temperature Physics. This Prize, which has become very prestigious, was given for the first time to N. Kuerti in 1958 for his work on nuclear magnetism. I was surprised to see that the second London Prize was given to Landau in 1960 (the third one was given to John Bardeen in 1962). Of course, Landau's exceptional achievements in physics deserved more than the London prize, but it means that the London Prize jury totally ignored the controversies and personal conflicts, which opposed London and Landau. In an e-mail, which he sent me on January 21, 2001, Tisza wrote:

> I know that Landau had no high regard for London. I think he was wrong and hurt his own science for yielding to his spite. London disliked Landau, and I did what I could to temper his feelings when writing his "Superfluids". I suspect that they had an unpleasant interaction in 1932 when Landau travelled in the West, but this will remain an unsolved mystery.

I am pleased to see that science is sometimes more important than personal conflicts. On June 17, 2005, I received another message from Laszlo Tisza where he commented on the London prize:

> Dear Sebastien,
> … Yesterday I was leafing through old correspondence and I found a letter in which I nominated Landau for the Prize. I am sure I was not alone. I was actually at LT-7 in Toronto when the Prize was announced. It is actually unconscionable of Landau not to have taken note of the remarkable Simon –London work on helium in Oxford 1934-35! I never heard a word of it while at UFTI. All he said was that London was not a good physicist. I am looking forward to your book to straighten out matters.
> With warmest regards,
> Laszlo

It is remarkable that Laszlo Tisza himself supported the nomination of Landau for the London prize. He had recognized the superiority of Landau's two fluid model on his early theory and he was never upset by any personal criticism, which he considered as secondary.

Kapitsa was awarded the Nobel Prize in 1978. This was 16 years after Landau and 41 years after he had sent his historical letter to *Nature*. In his speech, he noticed this surprising delay and talked about a different subject (nuclear fusion). I do not know if the Nobel Prize jury ever considered the possibility of dividing a Nobel Prize on superfluidity between Kapitsa and Allen. Some authors (Griffin 2008) believe that Kapitsa refused sharing it with Allen, which explains the 41 years delay, but I doubt that the Nobel committee asks potential laureates about their opinion. If some physicists considered that Kapitsa had some priority on Allen so that it was difficult to find agreement, I believe that this was unfair.

At the end of his talk for the hundredth anniversary of the Hungarian physical society in 1991 (Tisza 1991), Tisza wrote:

> "If history has a lesson, it is that the 'winner takes all' attitude deprives one of the pleasure of being the heir to the best of different traditions, even while avoiding their intolerance against each other."

Tisza was squeezed between London and Landau whose approaches of the theory of superfluidity were rather different (Tisza 2009a, b). In fact London considered the ground state of liquid helium and Landau its excited states. It took quite a long time to unify their respective points of view, even after Bogoliubov's work (Bogoliubov 1947). As for Tisza, some of his theory was wrong but he had introduced many of the fundamental ideas, which were later developed by Landau. Furthermore, Landau's theory was not entirely correct either. In conclusion, one should certainly attribute the discovery of the theory of superfluidity not only to London and Landau, but also to Tisza.

6.5 Conclusion: More Recent Developments and Discoveries, Applications

The discovery of quantized vortices is certainly one of the most important developments in the study of superfluidity. Onsager had predicted it first (Onsager 1949), followed by Feynman (Feynman 1955). It is a direct consequence of the existence of a macroscopic wave function: the superfluid velocity is the gradient of its phase so that, after integration, one finds that the circulation is an integer number of quanta h/m where h is Planck's constant and m the helium mass. It is the strict equivalent of the quantization of vortices in superconductors, which was predicted by F. London (London 1950). It explains why superfluids resist to rotation as superconductors resist to the penetration by a magnetic field. The vortex quantization was measured first by W.F. Vinen (Vinen 1961). Quantized vortices were later imaged by Yarmchuk et al. (Yarmchuk et al. 1979). As shown by Fig. 6.2, the same array of quantized

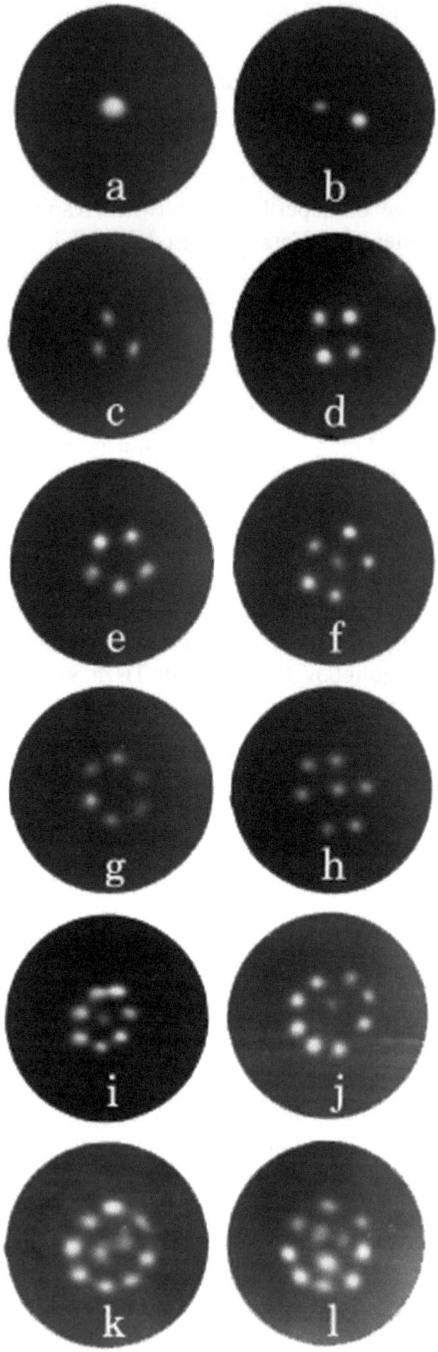

Fig. 6.2 Images of quantized vortices in liquid helium 4 on the *left* (Yarmchuk et al. 1979) and in a Bose-Einstein condensate of Rb atoms on the *right* (Madison et al. 2000). As the rotation speed increases, the system is invaded by more quantized vortices forming similar arrays in both cases

vortices was observed in the case Bose-Einstein condensates of Rubidium atoms by Madison et al. (Madison et al. 2000).

The Bose-Einstein condensation (BEC) in quantum gases of cold atoms was discovered in 1995 shortly before clear evidence for their superfluidity was demonstrated (Cornell and Wieman 2002; Ketterle 2002). The superfluidity of pairs of ^{3}He atoms was discovered in 1972 by Osheroff et al. at Cornell University (Osheroff et al. 1972) and shown to be a close analogue of the superconductivity of electron pairs in superconductors, whose famous "BCS" theory was established in 1957 by J. Bardeen, L.N. Cooper, and J.R. Schrieffer (Bardeen et al. 1957a, b). The continuity from BCS to BEC was also demonstrated with quantum gases of Fermi atoms as a function of the intensity of interactions between these atoms (Zwierlein et al. 2004). In summary, the connexion of superfluidity with Bose-Einstein condensation is well established, as well as its relation to superconductivity.

Superfluidity has been discovered in other examples of condensed matter, especially semiconductors where it concerns electron–hole pairs interacting with photons in cavities, that is so-called "polaritons" (Amo et al. 2011). Eventually, it is usually assumed to take place in the inside of neutron stars where it is invoked to explain glitches and cooling rates (see Chamel 2011 for a recent review). In the latter case it concerns pairs of neutrons, which may be superfluid even at temperatures of order a million degrees due to their very high density, as was predicted first by A.B. Migdal (Migdal 1959).

A future development might be the superfluidity of solids, called "supersolidity" (Balibar 2010). It is indeed possible that in a quantum crystal, where atoms are not totally localized on their crystal lattice sites so that they can exchange site with their neighbours, part of the total mass is superfluid while the rest ensures elasticity, an essential property of solids. The existence of supersolidity is presently debated in the case of solid helium 4 around 0.1 K since the experiments by Kim and Chan in 2004 (Kim and Chan 2004a, b). However its existence has been found unlikely to occur in solid helium-4 (Beamish 2012, Haziot 2013). Supersolidity might also be obtained in Bose-Einstein condensates of cold molecules where one could tune the interactions to the right value.

As concerns applications, the major application of superconductivity is in building high field electromagnets. The possibility to run very high amplitude currents without Joule heating has revolutionized medical imaging and MRI machines are now ubiquitous in modern hospitals. But if we restrict ourselves to superfluidity of neutral systems, its main application is as a cooling fluid. MRI machines contain superconducting magnets that are usually cooled down to 4 K with normal liquid helium 4. Only very high field experimental machines need superfluid helium as a coolant. The best-known electromagnet using superfluid helium is the 27 km long ring of magnets used in the Large Hadron Collider at CERN (Geneva) to accelerate protons and anti-protons and search for new elementary particles. From the 1930s when superfluid helium was available in quantities not much more than a few cubic centimetres to the giant LHC, it is obvious that superfluidity is no longer a curiosity but a phenomenon of possible use at the industrial scale. Still, its main applications are for low temperature research in laboratories where studying matter from 2 K down to mK temperatures would be practically impossible without superfluid helium. Not only superfluid helium has allowed to make fundamental progress in the understanding of electrical transport in matter and in magnetism, it has also allowed to cool down the very low noise detectors used in astrophysics to determine the Cosmic Background radiation in relation with the origin of the Universe.

Acknowledgements I acknowledge support from the ERC grant AdG247258-SUPERSOLID.

References

Allen, J.F. 1988. The beginning of superfluidity. *Physics World* 1: 29–31.
Allen, J.F., and J.M.G. Armitage. 1982. *VHS movie*, 5th ed. St Andrews: St Andrews University.
Allen, J.F., and A.D. Misener. 1938. Flow of liquid helium II. *Nature* 141: 75–78.
Allen, J.F., and H. Jones. 1938. New phenomena connected with heat flow in helium II. *Nature* 141: 243–244.
Allen, J.F., R. Peierls, and Z. Uddin. 1937. Heat conduction in liquid helium. *Nature* 140: 62–63.
Amo, A., S. Pigeon, D. Sanvitto, V.G. Sala, R. Hivet, I. Carusotto, F. Pisanello, G. Leménager, R. Houdré, E. Giacobino, C. Ciuti, and A. Bramati. 2011. Polariton superfluids reveal quantum hydrodynamic solitons. *Science* 332: 1167–1170.
Anderson, P.W. 1969. Interpretation of the Johnston-King experiment on evaporation of liquid He. *Physics Letters A* 29: 563–564.
Andronikashvili, F.L. 1989. *Reflections on liquid helium*. New York: American Institute of Physics.
Balibar, S., J. Buechner, B. Castaing, C. Laroche, and A. Libchaber. 1978. Experiments on superfluid 4He evaporation. *Physical Review B* 18: 3096–3104.
Balibar, S. 2007. The discovery of superfluidity. *Journal of Low Temperature Physics* 146: 441–470.
Balibar, S. 2010. The enigma of supersolidity. *Nature* 464: 176–182.
Bardeen, J., L.N. Cooper, and J.R. Schrieffer. 1957a. Microscopic theory of superconductivity. *Physical Review* 106: 162–164.
Bardeen, J., L.N. Cooper, and J.R. Schrieffer. 1957b. Theory of superconductivity. *Physical Review* 108: 1175–1204.
Beamish, J.R., A. Fefferman, A. Haziot, X. Rojas, and S. Balibar. 2012. Elastic effects in torsional oscillators containing solid helium. *Physical Review B* 85: 120501.
Bogoliubov, N.N. 1947. On the theory of superfluidity. *Journal of Physics USSR* 11: 23.
Bose, S.N. 1924. Planck's Gesetz und Lichtquantenhypothese. *Zeitschrift für Physik* 26: 178–181.
Burton, E.F. 1935. Viscosity of helium I and helium II. *Nature* 135: 265.
Brown, M., and A.F.G. Wyatt. 1990. The surface boundary conditions for quantum evaporation in 4He. *Journal of Physics: Condensed Matter* 2: 5025–5046.
Caupin, F., and S. Balibar. 2001. Cavitation pressure in liquid helium. *Physical Review B* 64: 064507. 1–10.
Chamel, N. 2011. A stellar superfluid. *Physics* 4: 14.
Cornell, E.A., and C.E. Wieman. 2002. Nobel lecture: Bose-Einstein condensation in a dilute gas, the first 70 years and some recent experiments. *Reviews of Modern Physics* 74: 875. For a review, see the Nobel lecture.
Dahl, P.F. 1993. *Superconductivity, its historical roots and development from mercury to the ceramic oxides*. New York: American Institute of Physics.
Daunt, J.G., and K. Mendelssohn. 1938. Transfer of helium II on glass. *Nature* 141: 911–912.
Donnelly, R. 1995. The discovery of superfluidity. *Physics Today* 48: 30.

Einstein, A. 1924. Quantentheorie des einatomigen idealen Gases I. *Sitzungsberichte Berlin Akad* 22: 261–267.
Einstein, A. 1925. Quantentheorie des einatomigen idealen Gases II. *Sitzungsberichte Berlin Akad* 1: 3–14.
Feynman, R.P. 1955. Application of quantum mechanics to liquid helium. In *Progress in low temperature physics*, vol. 1, ed. C.G. Gorter, 17–53. Amsterdam: North Holland.
Fixsen, D.J. 2009. The temperature of the cosmic microwave background. *The Astrophysical Journal* 707: 916–920.
Gavroglu, K. 1995. *Fritz London: A scientific biography*. Cambridge: Cambridge University Press.
Gorelik, G.E. August 1997. The top secret life of Lev Landau. *Scientific American* 277(2): 72–77.
Griffin, A. 1995. A brief history of our understanding of BEC: From Bose to Beliaev. In *Proceedings of the International School of Physics*, ed. E. Fermi, M. Inguscio, S. Stringari, and C.E. Wieman, 1–12. Amsterdam: IOS press.
Griffin, A. 2005. John C. McLennan and his pioneering research on superfluid helium. *Physics in Canada* 61: 33–40.
Griffin, A. 2008. Superfluidity: Three people, two papers, on prize. *Physics World* 21(8): 27–30.
Haziot, A., X. Rojas, A.D. Fefferman, J.R. Beamish, and S. Balibar. 2013. The giant plasticity of a quantum crystal. *Physical Review Letters* 110: 035301.
Heitler, W., and F. London. 1927. Wechselwirkung neutraler Atome und homöopolare Bindung nach der Quantenmechanik. *Zeitschrift für Physik* 44: 455–472.
Henshaw, D.G., and A.D.B. Woods. 1961. Modes of atomic motions in liquid helium by inelastic scattering of neutrons. *Physics Review* 121: 1266–1274.
Hope, F.R., M.J. Baird, and A.F.G. Wyatt. 1984. Quantum evaporation from liquid ^{4}He by rotons. *Physical Review Letters* 52: 1528–1531.
Horner, H. 1972. Scattering function S (Q, ω) for solid helium. *Physical Review Letters* 29: 556–558.
Kamerlingh Onnes, H. 1908. The liquefaction of helium. *Communications from the Physical Laboratory at the University of Leiden* 108: 3.
Kamerlingh Onnes, H. 1911. The disappearance of the resistance of mercury. *Communications from the Physical Laboratory at the University of Leiden*. 122b and 124c.
Kapitsa, P. 1938. Viscosity of liquid helium below the λ-point. *Nature* 141: 74.
Kapitsa, P.L. 1941. Heat transfer and superfluidity in helium II. *Physics Review* 60: 354.
Keesom, W.H., and M. Wolfke. 1927. Two different liquid states of helium. *Communications from the Physical Laboratory at the University of Leiden* 190b: 17–22.
Keesom, W.H., and J.N. van der Ende. 1930. *Proceedings of the Royal Academy of Amsterdam* 33: 24.
Keesom, W.H., and A.P. Keesom. 1932. On the anomaly of the specific heat of liquid helium. *Communications from the Physical Laboratory at the University of Leiden* 221d: 19–26.
Keesom, W.H., and A.P. Keesom. 1936. On the heat conductivity of liquid helium. *Physica* 3: 359–360.
Ketterle, W. 2002. Nobel lecture: When atoms behave as waves: Bose-Einstein condensation and the atom laser. *Reviews of Modern Physics* 74: 1131–1151.
Kim, E., and M.H.W. Chan. 2004a. Probable observation of a supersolid helium phase. *Nature* 427: 225–227.
Kim, E., and M.H.W. Chan. 2004b. Observation of superflow in solid helium. *Science* 305: 1941–1944.
Kuerti, N., B.V. Rollin, and F. Simon. 1936. Preliminary experiments on temperature equilibria at very low temperatures. *Physica* 3: 266–274.
Landau, L.D. 1941a. Theory of superfluidity of helium II. *Physics Review* 60: 356–358.
Landau, L.D. 1941b. The theory of superfluidity of Helium II. *Journal of Physics (USSR)* 5: 71–90.
Landau, L.D. 1947. On the theory of superfluidity of helium II. *Journal of Physics (USSR)* 11: 91.
Landau, L.D. 1957. The theory of a Fermi liquid. *Soviet Physics JETP* 3: 920–925.
Leggett, A.J. 1972. Interpretation of recent results on He^3 below 3 mK: A new liquid phase? *Physical Review Letters* 29: 1227–1230.

Leggett, A.J. 2004. Nobel lecture: Superfluid 3He: The early days as seen by a theorist. *Reviews of Modern Physics* 76: 999–1011.
London, F. 1935. Macroscopical interpretation of supraconductivity. *Proceedings of the Royal Society A* 152: 24–34.
London, F. 1936. On condensed helium at absolute zero. *Proceedings of the Royal Society A* 153: 576–583.
London, F. 1938a. The λ-phenomenon of liquid helium and the Bose-Einstein degeneracy. *Nature* 141: 643–644.
London, F. 1938b. On the Bose-Einstein condensation. *Physics Review* 54: 947.
London, H. 1938c. A ponderomotive effect associated with the flow of heat through liquid helium II. *Nature* 142: 612.
London, H. 1939. Thermodynamics of the thermomechanical effect of liquid HeII. *Proceedings of the Royal Society A* 171: 484–496.
London, F. 1947. The present state of the theory of liquid helium. In *International conference on fundamental particles and low temperatures*, 1. Cambridge: Taylor and Francis. Reprinted by R. Donnelly, Department of Physics, University of Oregon (1993).
London, F. 1950. *Superfluids I.* New York: Wiley.
London, F., and H. London. 1935. Supraleitung und Diamagnetismus. *Physica* 2: 341–354.
Madison, K.W., F. Chevy, W. Wohlleben, and J. Dalibard. 2000. Vortex formation in a stirred Bose-Einstein condensate. *Physical Review Letters* 84: 806–809.
Maris, H.J. 1994. Nucleation of bubbles on quantized vortices in helium-4. *Journal of Low Temperature Physics* 94: 125–144.
Maris, H.J. 1995. Theory of quantum nucleation of bubbles in liquid helium. *Journal of Low Temperature Physics* 98: 403–424.
Matricon, J., and G. Waysand. 2003. *The cold wars: A history of superconductivity.* New Brunswick: Rutgers University Press (Originally published as *La guerre du froid: Une histoire de la superconductivite*, 1994, Paris: Seuil).
McLennan, J.C., H.D. Smith, and J.O. Wilhelm. 1932. The scattering of light by liquid helium. *Philosophical Magazine* 14: 161–167.
Mendelssohn, K. 1964. Prewar work on superconductivity as seen from Oxford. *Reviews of Modern Physics* 36: 7–11.
Meyer, H. 2005. Fritz London aux Etats-Unis 1938–1954. Communication at the conference "Quantique… mais macroscopique, Hommage à Fritz London, physicien en exil", Institut Henri Poincaré, Paris, 11 mai 2005.
Migdal, A.B. 1959. Superfluidity and the moments of inertia of nuclei. *Nuclear Physics* 13: 655–674.
Misener, A.D., J.O. Wilhelm, and A.R. Clark. 1935. The viscosity of liquid helium. *Proceedings of the Royal Society A* 151: 342–347.
Nozières, P. 2004. Is the roton in superfluid 4He the ghost of a Bragg spot? *Journal of Low Temperature Physics* 137: 45–67.
Onsager, L. 1949. Nuovo Cimento 6, Suppl. 2, 249 (discussion on the article "The two fluid model of Helium II" by C.J. Gorter).
Osborne, D.W., B. Weinstock, and B.M. Abraham. 1949. Comparison of the flow of isotopically pure liquid He^3 and He^4. *Physics Review* 75: 988.
Osheroff, D.D., R.C. Richardson, and D.M. Lee. 1972. Evidence for a new phase of solid He. *Physical Review Letters* 28: 885–888.
Osheroff, D.D. 1997. Superfluidity in 3He: Discovery and understanding. *Reviews of Modern Physics* 69: 667–681. Nobel lecture.
Païs, A. 1982. *Subtle is the Lord*, 432. Oxford: Clarendon Press.
Penrose, O. 1951. On the quantum mechanics of helium II. *Philosophical Magazine* 42: 1373–1377.
Penrose, O., and L. Onsager. 1956. Bose-Einstein condensation and liquid helium. *Physics Review* 104: 576–584.

Peshkov, V.P. 1946. Determination of the velocity of propagation of the second sound in helium II. *Report of an international conference on fundamental particles and low temperature*, Cavendish Laboratory, Cambridge 22–27 July 1946, p. 19 (Taylor & Francis, London 1947). Reprinted by R. Donnelly, Department of Physics, University of Oregon (1993).

Peshkov, V.P. 1948. Skorost 2nd Zvuka OT 1,3 do 1.03-Degrees-K. *Zh Eksp Teor Fiz* 18: 951.

Pitaevskii, L. 1992. 50 years of Landau's theory on superfluidity. *Journal of Low Temperature Physics* 87: 127–135.

Rollin, B.V., and F. Simon. 1939. On the "film" phenomenon of liquid helium II. *Physica* 6: 219–230.

Rubinin, P.E. 1997. The discovery of superfluidity. Letters and documents. *Physics-Uspekhi* 40: 1249–1260.

Shoenberg, D. 1994. Kapitsa in Cambridge and Moscow. *Physics Uspekhi* 37: 1213–1216 (Kapitsa centenary symposium at the Cavendish laboratory, Cambridge, 8 July 1994).

Simon, F. 1934. Behaviour of condensed helium near absolute zero. *Nature* 133: 529.

Teller, E. 1998. Science and morality. *Science* 280: 1200–1201.

Tisza, L. 1938a. Transport phenomena in helium II. *Nature* 141: 913.

Tisza, L. 1938b. Sur la supraconductibiité thermique de l'hélium II liquide et la statistique de Bose-Einstein. *Comptes Rendus de l'Académie des Sciences (Paris)* 207: 1035–1037.

Tisza, L. 1938c. La viscosité de l'hélium liquid et la statistique de Bose-Einstein. *Comptes Rendus de l'Académie des Sciences (Paris)* 207: 1186.

Tisza, L. 1940a. Sur la théorie des liquides quantiques. Application à l'hélium liquide I. *Journal of Physique et le Radium* 1(5): 164–172.

Tisza, L. 1940b. Sur la théorie des liquides quantiques. Application à l'hélium liquide II. *Journal of Physique et le Radium* 1(8): 350–358.

Tisza, L. 1991. *The history of the two-fluid concept.* Centenary meeting of the Eötvös Society, Budapest, Hungary, 19 Oct 1991.

Tisza, L. 2000. E-mail to S. Balibar, 12 Sept 2000.

Tisza, L. 2009a. Adventures of a theoretical physicist, part I: Europe. *Physics in Perspective* 11(1): 46–97.

Tisza, L. 2009b. Adventures of a theoretical physicist, part II: America. *Physics in Perspective* 11(2): 120–168.

Tucker, M.A.H., and A.F.G. Wyatt. 1999. Direct evidence for R – Rotons having antiparallel momentum and velocity. *Science* 283: 1150–1152.

Uhlenbeck, G.E. 1927. *Over Statistische Methoden in de Theorie der Quanta.* PhD dissertation, University of Leiden.

van Delft, D. 2005. *Heike Kamerlingh Onnes. Een biografie.* Amsterdam: Bakker (to be translated in English).

Vinen, W.F. 1961. Detection of single quanta of circulation in liquid helium II. *Proceedings of the Royal Society A* 260: 218–236.

Wilks, J. 1967. *The properties of liquid and solid helium.* Oxford: Clarendon Press.

Wolfke, M., and W.H. Keesom. 1927. Über zwei verschiedene Flüssigkeitszustände. *Proceedings of E Amsterdam* 31: 81.

Yarmchuk, E.J., M.J.V. Gordon, and R.E. Packard. 1979. Observation of stationary vortex arrays in rotating superfluid helium. *Physical Review Letters* 43: 214–217.

Zwierlein, M.W., C.A. Stan, C.H. Schunck, S.M.F. Raupach, A.J. Kerman, and W. Ketterle. 2004. Condensation of Pairs of fermionic atoms near a Feshbach resonance. *Physical Review Letters* 92: 120403.

Chapter 7
The Physics of Cold in the Cold War—"On-Line Computing" Between the ICBM Program and Superconductivity

Johannes Knolle and Christian Joas

Superconductivity—the loss of resistance in various materials close to absolute zero temperature—was a hot topic after World War II. Advances in nuclear reactor technology led to the discovery of the isotope effect in 1950 (Maxwell 1950; Reynolds et al. 1950), which brought about crucial insights about the role of electron-lattice interactions in superconductors that ultimately led to the formulation of a microscopic theory of this phenomenon. Generations of physicists had been struggling to find an explanation of superconductivity ever since its discovery in 1911 by Heike Kamerlingh Onnes.[1] While the sudden loss of resistance below a critical temperature close to absolute zero in some materials was long deemed the unique feature that a successful theory would need to explain, in the mid-1930s it also became clear that perfect diamagnetism below the critical temperature (i.e., the complete expulsion of external magnetic fields from the inside of the material, up to a certain field strength) was a constitutive feature of superconductivity (see, e.g., Gavroglu and Goudaroulis 1984). After the missing piece—the isotope effect—was added, John Bardeen, Leon Cooper, and Robert Schrieffer in 1957 managed to formulate a microscopic theory of superconductivity, which came to be known as the BCS theory after the initials of their last names (Bardeen et al. 1957; see also Matricon and Waysand 2003). Their work was based on recent advances in the quantum-field theoretic description of many-body systems. The theory was built on a simplified electron-lattice interaction, yet was able to capture many key features of superconductivity qualitatively. However, the elaboration of BCS theory towards a

[1] See contribution by Joas and Waysand, Chap. 5 in the present volume. See also (Schmalian 2011).

J. Knolle
Max Planck Institute for the Physics of Complex Systems, Dresden, Germany

C. Joas (✉)
Ludwig Maximilians University, Munich, Germany

Fritz Haber Institute of the Max Planck Society and Max Planck Institute for the History of Science, Berlin, Germany
e-mail: christian.joas@lmu.de

K. Gavroglu (ed.), *History of Artificial Cold, Scientific, Technological and Cultural Issues*, Boston Studies in the Philosophy and History of Science 299,
DOI 10.1007/978-94-007-7199-4_7,

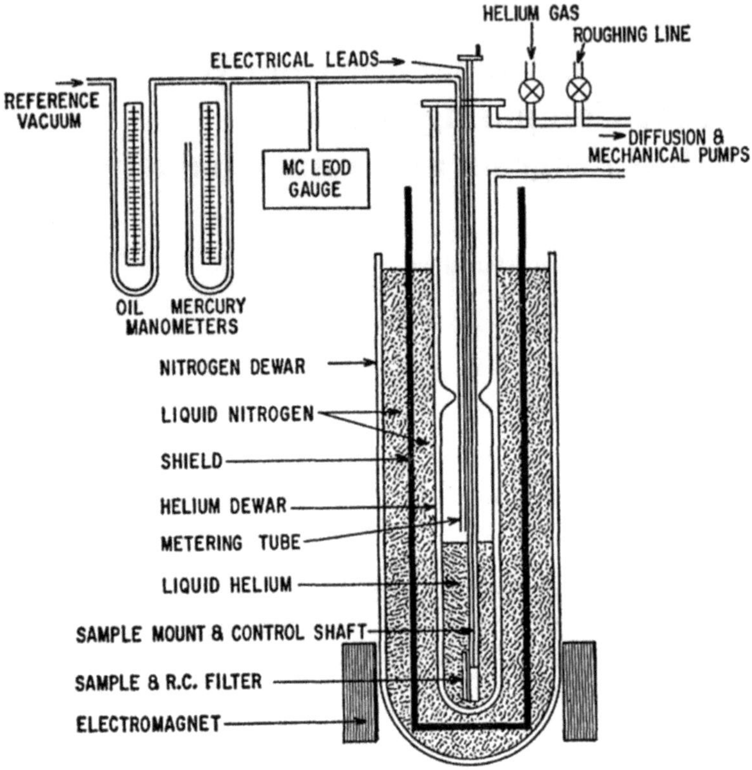

Fig. 7.1 A schematic drawing of Ivar Giaever's apparatus for his tunneling experiments which showed deviations from the expected BCS behavior. Giaever measured the tunneling current between two superconductors separated by a thin normal-conducting barrier and related it to the energy gap of the superconducting state. In their 1961 paper, Giaever and Karl Megerle described the basic principle of the apparatus as follows: "The apparatus [...] consists basically of a liquid helium Dewar with provisions for pumping on the helium, and an outer Dewar containing liquid nitrogen which acts as a radiation shield for the helium. The helium Dewar has a constriction in its diameter to minimize creep losses of the superfluid helium when the temperature is below the lambda point" (Giaever and Megerle 1961:1101–1102)

truly quantitative theory of superconductivity only came in the early 1960s: In order to explain tunneling experiments (see Fig. 7.1) by Giaever (1960a, b) that showed deviations from the predictions of BCS theory, physicists began to use computer technology that had been developed in the context of the US program to build intercontinental ballistic missiles (ICBM).

7.1 Cold War Physicists

World War II and the Cold War affected not only the relations between military, industry, and science, but also the characteristic practices of scientists in general and of physicists in particular. In the case of the development of ICBM in the 1950s, highly complex problems arose whose solution necessitated the expertise of scientists. Military contractors created their own research and development (R&D) units and hired engineers and scientists both as consultants and as staff. In 1954, Ramo-Wooldridge (RW)—according to one of its founders, "the company that developed the U.S. missile" (Ramo 1988)—came to be the most important contractor for the US Air Force's program to build ICBMs (Gruntman 2004:233). Intriguingly, a 1963 technical report on the use of computers in weapons control, issued by TRW, the successor company of RW, acknowledges the help of eminent physicists, among them both J. Robert Schrieffer, co-developer of BCS theory, and Richard P. Feynman:

> We are greatly indebted to the Data Processing Laboratory at Rome Air Development Center for support of this work and for the AN/FSQ-27 portion of the equipment; [...] and to Professor R. P. Feynman for suggestions concerning the possible extension to an algebra machine. We acknowledge with special thanks the efforts of Professor J. R. Schrieffer, Professor Karl Menger, Professor H. W. Wyld, Jr., Professor K. A. Johnson, Fred Dion and Martin Schultz who spent much of the summer of 1962 as *cooperative guinea pigs*, using the on-line system for research problems in their own fields, notwithstanding its then somewhat raw and rough-edged character, thus contributing greatly to its present state of development and to our understanding of the user's needs and desires in an on-line system [our emphasis].[2]

As the quote indicates, the collaboration between the theoretical physicist Schrieffer and the military contractor RW revolved around an innovative computer technology—the so-called "on-line computing." Initially, in the late 1950s, on-line computing was developed within the ICBM program to be used for command-and-control applications such as weapons control, rocket steering, and radar-based air control. In the early 1960s, RW developed the method further, thus adding a novel tool to the arsenal of physicists. Schrieffer played a major role in this process. He not only served as a "cooperative guinea pig," he also applied the new technique to his own research interests: It was by using on-line computing that Schrieffer eventually succeeded in turning the microscopic BCS theory of superconductivity into a quantitative theory. An efficient and precise numerical solution of the convoluted integral equations which appeared in BCS theory became possible only through the use of on-line computing.

Whereas the role of computers in the history of high-energy physics has been studied rather extensively (Galison 1996, 1997), the same does not hold true for other fields of physics. The present contribution focuses on the history of the development of a quantitative theory of superconductivity, in which on-line computing played a pivotal role. During the course of the 1960s, this computer technology enabled Schrieffer and others to reach agreement between their theoretical computations and ever more precise empirical results. At a deeper level, the story told here

[2] Culler, G. J., and Fried, B.: An on-line computing center for scientific problems. M19-3U3. Revised June 1963. TRW Computer Division. Canoga Park, California. Online: http://www.bitsavers.org/pdf/trw/trw-85/Culler_Fried_An_On-Line_Computing_Center_for_Scientific_Problems_Jun63.pdf (accessed on Jul 12, 2011).

illustrates some of the momentous changes in scientific practice that were brought about by the use of computers at the intersection of military, industry, and science during the Cold War.

7.2 Research and Development at Ramo-Wooldridge

After World War II, the military-industrial sector became an increasingly important actor in scientific research (Forman 1987). Scientists played important roles in the planning and the realization of large-scale military projects, acting as external consultants for industry, military, or government, or serving as managerial or scientific staff in companies that conducted research on behalf of the military or the government (Kevles 1988).

Ramo-Wooldridge was founded by Simon Ramo and Dean Wooldridge, who had earned their PhDs in physics at the California Institute of Technology in Pasadena. From 1946 on, they had helped to transform the R&D department of Hughes Aircraft from a small unit into a highly-profitable company of its own (Dyer 1998:171; Sheehan 2009:207–215). In the fall of 1953, they left Hughes Aircraft in order to found a new type of company, one headed by scientists. Their goal was first and foremost to serve as military contractors, but they also aimed, in the medium term, to market the technology developed in these projects for civilian use (Dyer 1998:170).

In the spring of 1954, a group of consultants to President Eisenhower headed by John von Neumann, to which Ramo and Wooldridge also belonged, suggested a "radical reorganization" of the United States ICBM program. This so-called Tea-Pot Committee believed that the aeronautical industry, which had grown to great strength during the War, was unable to cope with the complex technological challenges involved in the building of ICBMs (Sheehan 2009:215). In the fall of 1954, RW was chosen to serve as "Systems Engineering and Technical Direction Contractor" for the ICBM program of the US Air Force (Neufeld 1993:63–64; Marcus 2006:26). Within a very short period, just a few years in fact, RW thereby grew into a company with thousands of employees (Dyer 1998:186). A report issued by the "Technological Capabilities Panel" in early 1955 propelled the development of ICBMs to the top of the national research agenda, owing to the perceived race between the US and the USSR (Wang 2008:49–50). The 1957 "Sputnik shock" then catapulted the ICBM program, headed by RW, into the center of public attention.

7.3 "On-Line" Computing

During the 1950s, the growing complexity of military research projects necessitated the development and application of new computer technologies. Gradually, transistor-based computers replaced the older vacuum-tube computers (Laplante et al. 1995). Like IBM, RW stood on the cutting edge of this development. In July

Fig. 7.2 The civil variant of the *AN/FSQ-27* computer was marketed as a "polymorphic" computer. It consisted of many different modules (such as processing units, data storage units, or input and output devices) that could be combined at will (Photograph, Ramo Wooldridge Division, Thompson Ramo Wooldridge, Inc., online: http://ed-thelen.org/comp-hist/BRL61-0842.jpg)

1957, RW completed the construction of their first transistor-based computer (Dyer 1998:205). One of its successors, which was constructed in the following years for applications in rocket- and weapon-steering, bore the military name "AN/FSQ-27." For civilian use, it was marketed as the "RW-400," a "polymorphic" computer which allowed one to variably and rapidly connect different modules like processing and data storage units, or input and output devices, to conform to the specific demands of the tasks at hand (Bell and Newell 1971:470–488) (Figs. 7.2 and 7.3).

The most remarkable innovation of the "RW-400" was a novel man–machine interface which allowed the user to affect program execution "on-line," i.e., during the execution of code. RW had developed a new peripheral device, the so-called "Display and Analysis Console (DAC)." It consisted of a keyboard and a light gun equipped with a photo cell for input, and two tube monitors for the graphical display of data as well as intermediate results (Helfinstein 1961:77). Instead of feeding the system with code that is executed automatically in full, the user was able to issue commands step-by-step, to check intermediate results, and to determine further appropriate actions during program execution (including the possibility to "undo" previously-issued commands). The key accomplishment of the "on-line" system thus was not a faster processor, but the direct man–machine interaction during the runtime of a program—very much similar to later programmable pocket calculators or to modern "interactive" computer programs.

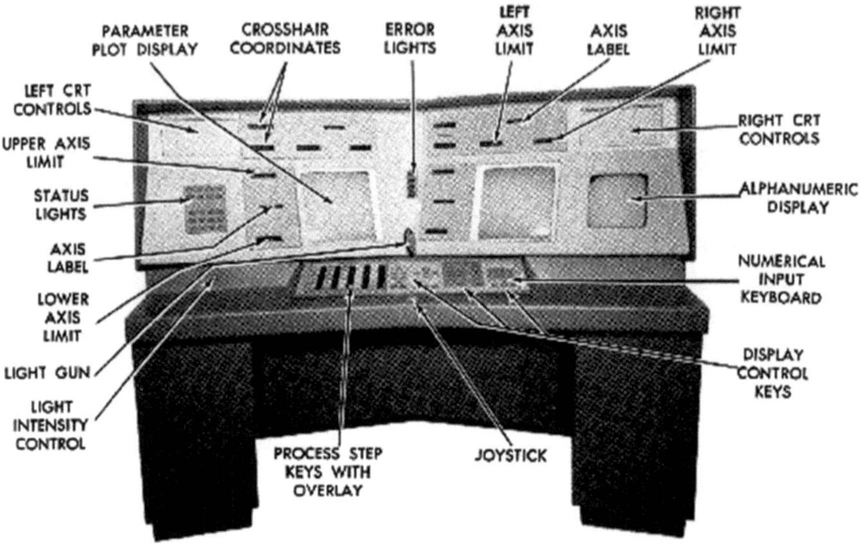

Fig. 7.3 The *Display and Analysis Console* (*DAC*) of the *RW-400* on-line system allowed users to display intermediate results graphically during the program's runtime, to issue or take back commands at the press of so-called *Process Step Keys* input data via the *Numerical Input Keyboard* or the so-called *Light Gun*, a precursor of the modern computer mouse (Helfinstein 1961)

The Sputnik shock of 1957 led to an extensive restructuring of RW (Dyer 1998:198). Among other changes, the computing and electronics unit was moved to Canoga Park, California. In 1959, RW also created the "Intellectronics Research Laboratory" to explore the potentialities of man–machine interactions (Dyer 1998:242). Its first director was the theoretical physicist Burton Fried, who had joined RW in 1954 after completing his PhD with Gregor Wentzel in Chicago and working briefly at the Lawrence Radiation Lab (Morales and Coroniti 2003). Fried convinced the experienced programmer and professor of mathematics Culler to join him at the new lab.[3] They had previously collaborated at RW and, in 1957, had published a joint article on the numerical computation of rocket trajectories (Culler and Fried 1957).

During the restructuring of RW, the "Guided Missile Research Division" in late 1957 was renamed (Dyer 1998:198) the "Space Technology Laboratories" (STL). At the STL, a new plasma physics research group was established.[4] Its main objective was to study the ionization of air taking place during the re-entry of ICBMs into the atmosphere (Dyer 1998:199). In 1960, Fried published a joint article with the

[3] Interview with Culler by George Michael, 12/7/1997. Online: http://www.computer-history.info/Page1.dir/pages/Culler.html (accessed on 12 Jul 2011).

[4] Interview with Roy W. Gould by Shirley K. Cohen (Pasadena, California, March–April 1996). Oral History Project, California Institute of Technology Archives. Online: http://resolver.caltech.edu/CaltechOH:OH_Gould_R (accessed on 21 Jan 2011).

renowned physicists Murray Gell-Mann, John David Jackson, and Henry William Wyld. They reported on their study of electromagnetic phenomena in plasmas and discussed analytical approximation methods for the solution of the nonlinear integral equations governing these phenomena (Fried et al. 1960). Soon thereafter, Fried and Culler also began to work on numerical solutions of such integral equations and quickly realized the potential of "on-line" computing when it came to solving mathematical problems of this sort.

Nonlinear integral and differential equations arise in many areas of science. An exact solution is often impossible. An iterative numerical solution is hampered in most cases by singularities and slow convergence. This constituted a very practical problem for the then-prevailing punch-card controlled computers, as they allowed for neither the inspection of intermediate results nor the user's direct intervention during program execution. If a punch-card controlled computer was fed initial conditions that led to singularities in the numerical solution procedure, the user would only find out after the full program had finished running—a process that frequently took hours, if not days. Valuable computing time was thus frequently wasted on meaningless calculations. But "on-line" computing was different. It allowed the user to assess convergence during program execution, to abort the program if necessary, and to modify either the code or the initial conditions before restarting the computation.

During the years 1961 and 1962, Culler and Fried developed a method for solving nonlinear integral equations on the RW-400. They decided to demonstrate the advantages of on-line computing to a wider audience by applying it to scientific problems of high relevance:[5]

> An on-line digital system allowing an unusually direct coupling between the user (physicist, mathematician, engineer) and the computer is described. This system, which has been successfully operated during the past six months, was designed principally to provide assistance for problems whose structure is partially unknown (and frequently surprising). These typically require the development of new methods of attack, and hence an amount of program experimentation not feasible with classical computer center organizations. With the system described here, the interaction between user and computer is close enough to permit effective use of a scientist's intuition and of his detailed understanding of techniques appropriate to his special field. He is able to construct, with ease, and with no necessity for a knowledge of conventional programming techniques and procedures, machine representations of those tools he considers essential to his area, and then use these, on-line, to study or solve problems of interest.[6]

[5] TRW encouraged their technical staff to entertain close relations with researchers at universities and to publish in scientific journals (Dyer 1998, p. 175). This policy at least partly appears to have been driven by TRW's interest in diversifying their customer base by marketing products outside of the military sector, since the cyclicality of the latter was perceived as a potential threat to the company's future (see Dyer 1998, p. 244).

[6] Culler and Fried, Burton: An on-line computing center for scientific problems. M19-3U3. Revised June 1963. TRW Computer Division. Canoga Park, California. Online: http://www.bitsavers.org/pdf/trw/trw-85/Culler_Fried_An_On-Line_Computing_Center_for_Scientific_Problems_Jun63.pdf (accessed on Jul 12, 2011).

Besides the potential savings in computing time, the on-line system had the additional advantage that scientists themselves, with little training in programming, were able to operate the system directly, and therefore could use the system to build up an intuitive understanding of the structure of the problem at hand. This intuitive understanding could then immediately be translated into changes in the solution procedure, e.g., when it came to the choice of appropriate initial conditions or of a specific mathematical technique. In order to generate attention for their new approach to numerical computation, Culler and Fried actively sought out unsolved scientific problems whose structure suited the capabilities of their on-line system:

> To explore the potentialities and characteristics of on-line scientific computing, it is clear that one must attempt to solve a variety of suitably chosen, difficult problems. [...] We imposed the following criteria:
>
> a. the problem must be one which presents real difficulties for conventional computer techniques and mathematical analysis;
> b. the problem must be one as yet unsolved [...]; and
> c. the problem should be one of significance for some area of current scientific research (Culler and Huff 1962:129).

Culler and Fried invited prominent physicists to Canoga Park as consultants, hoping for their help in identifying suitable problems that fit the above criteria. The first scientific application of on-line computing, suggested by and realized in collaboration with Schrieffer, was the solution of the so-called gap equation in the theory of superconductivity in 1961 (Culler and Huff 1962:130). While this problem without question satisfied criteria "a" and "b," its accordance with criterion "c" must have had an enormous appeal to Culler and Fried. The solution of the gap equation stood at the center of heated debates about the validity of the recent microscopic theory of superconductivity, a phenomenon that had lacked a consistent microphysical explanation for almost half a century.

7.4 Computers Enter Superconductivity Research

The gap equation is the key element of the BCS theory of superconductors. It is a nonlinear integral equation of exactly the form that Culler and Fried had already studied in the context of plasma physics. Intriguingly, the progress in the quantum field theory of many-body systems, which had been crucial for the BCS work, had also been fueled by developments in plasma physics. While most experimentalists rather quickly accepted the validity of BCS theory, several prominent theorists remained skeptical, if not hostile (Hoddeson et al. 1992:558ff). Besides issues concerning the gauge invariance of the theory, which could be resolved quite rapidly, it was mostly the impossibility of quantitative predictions that led to the chagrin of theorists. In the early 1960s, ingenious tunneling experiments by Ivar Giaever confirmed the qualitative predictions of the BCS theory to a remarkable degree (Giaever 1960a, b). Not long thereafter, however, Giaever discovered notable deviations from the BCS behavior in a small number of superconductors (Giaever et al. 1962). To explain them, he pointed to the numerical solution by J. C. Swihart of a refined version of the gap equation (Swihart 1962). Swihart had used a more realistic model for the electron-lattice interaction than BCS, which had made the analytical solution of the gap equation impossible and thus had necessitated a numerical solution, which he undertook on a (conventionally-programmed) IBM 7090 computer.

The original BCS theory had assumed an idealized effective interaction between the electrons in a superconductor. The microscopic treatment of the combined Coulomb and electron-lattice interactions had not yet been achieved, despite Swihart's refinement. Little noticed in the West, Lev Gorkov (1958) in 1957 had reformulated BCS theory using propagator techniques (i.e., so-called Green functions, a well-known tool in the theory of integral equations). This had laid the foundation to Gerasim Eliashberg's 1959 generalization of Arkady Migdal's equations for the electron-lattice interaction in the normal-conducting state (Migdal 1958) to the superconducting state (Eliashberg 1960). These methods spread but slowly in the West[7]:

> The eventual quantitative theory of energy gaps had, as an important beginning, an informal collaboration between Schrieffer and [Philip W.] Anderson. During a Utrecht many-body theory meeting in June 1960, [...] Schrieffer told of his study of the Green's function formalism, developed by Russians, especially Eliashberg, which was the correct way to do this calculation and to express the tunneling current (Hoddeson et al. 1992:561).

What Eliashberg had achieved was the correct treatment of the electron-lattice interaction. As expected, the resulting gap equation was highly complicated, yet its structure resembled that of the BCS gap equation. First attempts to solve Eliashberg's gap equation were undertaken in the fall of 1961, almost at the same time as Swihart published his numerical work. While Morel and Anderson (1962) used an interaction that was insufficiently physically motivated but analytically solvable, Schrieffer, together with Culler, Fried, and R. W. Huff, took on the full complexity of the Eliashberg equation numerically using on-line computing on the RW-400 during Schrieffer's stays at Canoga Park (Culler et al. 1962). Schrieffer must have been thrilled at the possibility of demonstrating the compatibility between the predictions of BCS theory and the numerical treatment of Eliashberg's theory, as it opened the door to a truly quantitative theory of superconductivity that would take the wind out of the sails of BCS theory's critics.

7.5 On-Line Computing and Eliashberg's Gap Equation

In the 1962 Proceedings quoted above, Culler and Huff described the solution of nonlinear integral equations on the on-line system using as their example the solution of the gap equation of superconductivity (Culler and Huff 1962). They

[7] This also had to do with the lack of communication between East and West in the beginning Cold War (see, e.g., Matricon and Waysand 2003; Kaiser 2006).

included photographs of the output screen of the DAC which documented input and output after each successive iteration step. This allowed them to demonstrate in practice how the intuition of the scientist helped to speed up the solution procedure and to reach convergence after just a few iteration steps. Without extensive experience in programming, scientists working with the on-line system could develop a feeling for the structure of mathematical problems and, consequently, solve them in shorter time and with dramatically reduced computational effort. Everyone involved was enthusiastic. Fried later described the atmosphere as follows:

> The Computer Division engineers needed the RW-400 for testing during the day, leaving the machine for us to use at night. I have vivid memories of working with Glen at the console until the wee hours of the morning, and coming out of the building into the clear night air feeling elated and exuberant. We were solving a difficult problem using an approach which was, as far as we knew, unique in the world. [...] Our enthusiasm was tempered somewhat by some sense of frustration, because it was virtually impossible to tell anyone else what we were doing: it was so at variance with the standard approaches and capabilities with which people were familiar that it was exceedingly difficult to explain to others.[8]

The Proceedings article by Culler and Huff is structured more like a user's manual than like a scientific paper. The scientific results were published in a separate paper in *Physical Review Letters*, co-authored by Schrieffer (Culler et al. 1962).

In 1962, the young experimentalist John Rowell of Bell Labs published the first of his many articles on tunneling experiments (Rowell et al. 1962). For the first time, he was able to measure the tunneling current as a function of voltage directly using lock-in amplifiers. Rowell discovered structures in the tunneling current that were finer than those discovered by Giaever, and that could neither be explained by the work of Morel and Anderson, nor by the on-line solution of the gap equation. The conviction that Eliashberg theory really was the key to a quantitative understanding of superconductivity, and the discrepancies between experiment and theory observed by Rowell, in the following years led to ever more ambitious attempts to solve the Eliashberg equations numerically. Often, theorists and experimentalists worked hand-in-hand and published articles jointly, or published articles that directly succeeded one another in the same journal. Computer numerics was key for both sides, as a means for the confirmation of theories through the computation of involved integral equations on the one hand, or for the understanding of empirical data on the other. Computer numerics—and the people trained in using computers—thus opened up a new channel for the exchange between theory and experiment, very much reminiscent of their role in high-energy physics (see Galison 1996, 1997).

[8] B. Fried: Online Adventures with Glen Culler. Online: http://www.kathy.kramer.net/kk.engr.ucsb.edu/culler/stories/fried.html (accessed on 14 Jul 2011).

Volume 10, number 8 of the Physical Review Letters in 1963 contains two back-to-back articles whose connection is outlined at the beginning of the second article by Schrieffer and his co-authors:

> In the preceding letter (Rowell et al. 1963), Rowell, Anderson, and Thomas present the results of improved experiments [...] Below we summarize the results of a theoretical determination of the tunneling characteristic which is in good agreement with these experiments (Schrieffer et al. 1963).

The theorist Anderson, just like Rowell, worked at Bell Labs at the time. During a visit of Schrieffer's to Bell Labs, Anderson helped Rowell to extract a model of the phonon spectrum (i.e., the lattice vibration spectrum) from his tunneling data, on which Schrieffer could base his theoretical work (Rowell 1987). While earlier only oversimplified models of the phonon spectrum had been used, this new, more realistic model led to a "remarkably good agreement [...] considering the simplicity of the model we used" between the numerical calculations on the on-line system and experiment (Schrieffer et al. 1963:338).

This result dispelled all doubt that the extension of BCS theory by Eliashberg was the foundation of a quantitative theory of superconductivity.[9] Ever since, computer numerics has been an integral part of the arsenal of physicists dealing with superconductivity. Additional important developments followed. For example, using an improved version of the numerics, William McMillan and Rowell in the late 1960s managed to employ tunneling experiments for phonon spectroscopy (McMillan and Rowell 1969).

7.6 Conclusion

The surprising connections between military research in the context of the development of ICBM and fundamental research in the field of superconductivity are an example for the changing scientific practices and the changing cycles for the usage of knowledge in the Cold War. On-line computing, a computational technique that was developed for applications in weapons control—most notably the steering of intercontinental ballistic missiles—contributed the key innovation leading to the numerical solution of complicated quantum-field theoretic equations and laid the foundation for the quantitative theory of superconductivity.

As this case study demonstrates, the amalgamation of military and academic research in the days of the Cold War reached much farther than the mere exchange and application of technologies and techniques. Scientists themselves became military contractors, like Ramo and Wooldridge, they conducted research and developed technologies within military projects, like Culler and Fried, or they

[9] Anderson, Interview of P.W. Anderson by Alexei Kojevnikov on May 30, 1999, Niels Bohr Library & Archives, American Institute of Physics, College Park, MD USA, www.aip.org/history/ohilist/23362_2.html

served as consultants to the military or its contractors, like Feynman and Schrieffer. Each and every one maneuvered in the potential landscape between military, industry, and science. In solid-state physics, much like in high-energy physics, the application of computer technology promoted exchange between theory and experiment. Physicists mastering the new numerical techniques served as catalysts for exchanges between theory and experiment. They were part of a new generation of physicists that Parks describes as follows in his 1969 textbook that has attained the rank of a standard reference:

> The explosion in superconductivity was ignited by the BCS theory, but a good share of the credit must go to the new breed of experimenters who were incubating in the fifties. They converged on the scene with their lock-in amplifiers, stacks of computer cards, and repertory of Feynman diagrams. They were schooled in the new tradition and could speak the theorist's language. Consequently, there occurred a cross-fertilization between the two groups, and an unforetold escalation of scientific discovery (Parks 1969: v).

Acknowledgments The authors wish to thank R. Joseph Anderson, Alexander S. Blum, Jeremiah James, Christoph Lehner, Jean Matricon, Jürgen Renn, Skúli Sigurðsson, and Georges Waysand for helpful comments and suggestions, as well as the Project on the History and Foundations of Quantum Physics (a collaborative project of the Max Planck Institute for the History of Science and the Fritz Haber Institute of the Max Planck Society, Berlin) for its generous financial support. Substantial parts of the present contribution are based on Knolle, J., Joas, C. 2013. Supraleitung und Interkontinentalraketen. "On-Line Computing" zwischen Militär, Industrie und Wissenschaft. In *Physik im Kalten Krieg. Beiträge zur Physikgeschichte während des Ost-West-Konflikts*, ed. D. Hoffmann and C. Forstner. Wiesbaden: Springer.

Bibliography

Bardeen, J., L.N. Cooper, and J.R. Schrieffer. 1957. Theory of superconductivity. *Physical Review* 108: 1175–1204.

Bell, C.G., and A. Newell. 1971. *Computer structures: Readings and examples*. New York: McGraw-Hill.

Culler, G.J., and B.D. Fried. 1957. Universal gravity turn trajectories. *Journal of Applied Physics* 28: 672–676.

Culler, G.J., and R.W. Huff. 1962. Solution of nonlinear integral equations using on-line computer control. *Proceedings of the Western Joint Computer Conference*, 129–138. May 1962.

Culler, G.J., B.D. Fried, R.W. Huff, and J.R. Schrieffer. 1962. Solution of the gap equation for a superconductor. *Physical Review Letters* 8: 399–402.

Dyer, D. 1998. *TRW. Pioneering technology and innovation since 1900*. Boston: Harvard Business School Press.

Eliashberg, G.M. 1960. Interactions between electrons and lattice vibrations in a superconductor. *Soviet Physics JETP* 11: 696–702.

Forman, P. 1987. Behind quantum electronics: National security as basis for physical research in the United States, 1940–1960. *Historical Studies in the Physical and Biological Sciences* 18: 149–229.

Fried, B.D., M. Gell-Mann, J.D. Jackson, and H.W. Wyld. 1960. Longitudinal plasma oscillations in an electric field. *Journal of Nuclear Energy. Part C: Plasma Physics* 1: 190–198.

Galison, P. 1996. Computer simulations and the trading zone. In *The disunity of science*, ed. P. Galison and D.J. Stump. Stanford: Stanford University Press.

Galison, P. 1997. *Image and logic: The material culture of microphysics*. Chicago: University of Chicago Press.

Gavroglu, K., and Y. Goudaroulis. 1984. Some methodological and historical considerations in low temperature physics: The case of superconductivity 1911–1957. *Annals of Science* 41: 135–149.
Giaever, I. 1960a. Energy gap in superconductors measured by electron tunneling. *Physical Review Letters* 5: 147–148.
Giaever, I. 1960b. Electron tunneling between two superconductors. *Physical Review Letters* 5: 464–466.
Giaever, I., and K. Megerle. 1961. Study of superconductors by electron tunneling. *Physical Review* 122: 1101–1111.
Giaever, I., H.R. Hart, and K. Megerle. 1962. Tunneling into superconductors at temperatures below 1 K. *Physical Review* 126: 941–948.
Gorkov, L.P. 1958. On the energy spectrum of superconductors. *Soviet Physics JETP* 7: 505–508.
Gruntman, M. 2004. *Blazing the trail. The early history of spacecraft and rocketry*. Reston: American Institute of Aeronautics and Astronautics.
Helfinstein, B. 1961. *Programming manual AN/FSQ-27 (RW-400)*. 2nd edn., February 1, 1961. Canoga Park: Data Systems Project Office, Ramo Wooldridge. Online: http://www.bitsavers.org/pdf/trw/rw-400/AN-FSQ-27_RW-400_Programming_Man_Feb61.pdf. Accessed 17 Sep 2011.
Hoddeson, L., H. Schubert, S.J. Heims, and G. Baym. 1992. Collective phenomena. In *Out of the crystal maze. Chapters from the history of solid-state physics*, ed. L. Hoddeson, E. Braun, J. Teichmann, and S. Weart. Oxford: Oxford University Press.
Kaiser, D. 2006. The physics of spin: Sputnik politics and American physicists in the 1950s. *Social Research* 73(4): 1225–1252.
Kevles, D.J. 1988. R&D and the arms race: An analytical look. In *Science, technology, and the military*, vol. XII, ed. E. Mendelsohn, M.R. Smith, and P. Weingart. Dordrecht: Kluwer.
Laplante, P.A., E.P. Rose, and M. Gracia-Watson. 1995. An historical survey of early real-time computing developments in the U.S. *Real-Time Systems* 8: 199–213.
Marcus, G. 2006. The pioneer rocket. *QUEST The History of Spaceflight Quarterly* 13(4): 26.
Matricon, J., and G. Waysand. 2003. *The cold wars. A history of superconductivity*. New Brunswick: Rutgers University Press.
Maxwell, E. 1950. Isotope effect in the superconductivity of mercury. *Physical Review* 78: 477.
McMillan, W.L., and J.M. Rowell. 1969. Tunneling and strong-coupling superconductivity. In *Superconductivity*, vol. I, ed. R.D. Parks. New York: Marcel Dekker Inc.
Migdal, A.B. 1958. Interaction between electrons and lattice vibrations in a normal metal. *Soviet Physics JETP* 34: 996–1001.
Morales, G.J., and F.V. Coroniti. 2003. Burton David Fried (Obituary). *Physics Today* 56(7): 68–70.
Morel, P., and P.W. Anderson. 1962. Calculation of the superconducting state parameters with retarded electron–phonon interactions. *Physical Review* 125: 1263–1271.
Neufeld, J. 1993. *Reflections on research and development in the United States air force: an interview with General Bernard A. Schriever and Generals Samuel C. Phillips, Roger T. Marsh, and James H. Doolittle, and Dr. Ivan A. Getting. Conducted by Richard H. Kohn*. Washington, DC: Center for Air Force History.
Parks, R.D. (ed.). 1969. *Superconductivity*, vol. 1. New York: Marcel Dekker Inc.
Ramo, S. 1988. Memoirs of an ICBM pioneer. *Fortune* 117(9): 309.
Reynolds, C.A., B. Serin, W.H. Wright, and L.B. Nesbitt. 1950. Superconductivity of isotopes of mercury. *Physical Review* 78: 487.
Rowell, J.M. 1987. Superconducting tunneling spectroscopy and the observation of the Josephson effect. *IEEE Transactions on Magnetics* 23(2): 384.
Rowell, J.M., A.G. Chynoweth, and J.C. Phillips. 1962. Multiphonon effects in tunneling between metals and superconductors. *Physical Review Letters* 9: 58–61.
Rowell, J.M., P.W. Anderson, and D.E. Thomas. 1963. Image of the phonon spectrum in the tunneling characteristics between superconductors. *Physical Review Letters* 10: 334–336.
Schmalian, J. 2011. Failed theories of superconductivity. In *BCS: 50 years*, ed. L.N. Cooper and D. Feldman. Singapore: World Scientific.

Schrieffer, J.R., D.J. Scalapino, and J.W. Wilkins. 1963. Effective tunneling density of states in superconductors. *Physical Review Letters* 10: 336–339.

Sheehan, N. 2009. *A fiery peace in a cold war: Bernard Schriever and the ultimate weapon.* New York: Random House.

Swihart, J.C. 1962. Solutions of the BCS integral equation and deviations from the law of corresponding states. *IBM Journal of Research and Development* 6(1): 14–23.

Wang, Z. 2008. *In Sputnik's shadow: The President's Science Advisory Committee and Cold War America.* New Brunswick: Rutgers University Press.

Part II
Industries of Cold

Chapter 8
Domestic Ice-Making Machines 1830–1930

Simón Reif-Acherman

To my Mother, who was present at the beginning of this work, and who would have been very happy and proud to see it published

8.1 Introduction

The permanent evolution of everyday needs, as well as the changing customs that have characterized the civilizations' behavior along its history, have determined the dependence towards natural resources in several, specific and different temporal ways, throughout many ages. Refrigeration is a very good example, and ice supply becomes essential to the global way of life since the mid-nineteenth century. The circulation of cold air to cool foods by melting blocks of ice placed in the compartments of the so-called ice boxes used in the early part of the century, very quickly became a great improvement over previous methods of storing perishable foods. The ice obtained during the winter months was stored with this purpose in specially constructed buildings located near to the supplying river or lake, and later delivered to customers in horse-drawn wagons and railroads and, a great quantity of it, exported in the holds of sailing ships. Slowly but surely, natural ice supply became an industry into itself, to such an extent that, in the United States for example, it became the second most important export, after cotton, by dollar volume in the second half of the century. The dramatic growth of cities, the increasing distances between the consumers and the several sources of food, and the improvement in economic status of the general population fueled, not only greater demands for fresh food and the initial worry about its preservation, but promoted, in some way, other uses of ice as a natural resource too, such as for fermentation at breweries, in

S. Reif-Acherman (✉)
School of Chemical Engineering, Universidad del Valle, Cali, Colombia
e-mail: sireache@univalle.edu.co

K. Gavroglu (ed.), *History of Artificial Cold, Scientific, Technological and Cultural Issues*, Boston Studies in the Philosophy and History of Science 299,
DOI 10.1007/978-94-007-7199-4_8, © Springer Science+Business Media Dordrecht 2014

comfort ice-type cooling systems, crystallization of salts in chemical manufactures, several medical uses, and in intercontinental transport of frozen meat, to mention just a few. As an example, in New York City, over a 36-year period after 1843, the cyphers of consumption of natural ice increased from 12.000 tons to 1,000.000 tons.

Two general facts would negatively affect, however, the commercial success of the new business. On one hand, the very uncertain supply of natural ice, its difficult transportation, its limited availability to certain locations, and the subsequent product's increasing prices due to factors of diverse order, such as the repeated failure of the ice-crops during some years because of unseasonably warm weather, or the interruption of shipments because non-natural circumstances as it was the case, for example, with the northern states' supplies to the south during the Civil War in the United States around the 1860s. On the other hand, the constantly increasing contamination of the water sources in the vicinity of large cities from which ice-crops are harvested became a threat to public health and induced a growing demand among consumers for a purer material than the one then supplied by nature. Artificial methods that allowed machines to make ice on demand and freed the market from dependence on the weather and other external factors, became an indispensable and very useful way to give a right solution to the above mentioned difficulties.

The higher importance ice acquired day by day was not limited, however, to industrial purposes. Machines capable of producing ice in small quantities, quickly and cheaply, were also of considerable utility for domestic purposes, such as preserving food and cooling beverages during the warm seasons at homes and small business of small towns in hot countries where natural ice could not be easily procured, and modern ice manufacturing and distributing methods still did not exist. The purpose of this article is to show the most relevant developments in low-capacity ice-making machines between the nineteenth century and the first decades of the twentieth century, with methods to effect the temperature reduction, most of which were very similar to those simultaneously or later used on an industrial scale.

8.2 Refrigerating Mixtures and the "glacières des familles"

The discovery that water is cooled when salts are dissolved in it became a key element for the definitive establishment of technical methods for obtaining artificial reductions of temperature, and thereby producing ice. The method, later called *the refrigerant mixtures*, is based on the energetic balance established in the melting process of mixtures of some salts and water or acids, or some salts with ice, with freezing points below their original temperatures. The temperature decrease exhibited by such mixtures, because of the significant difference between the heat required for their melting and that externally available, was an empirically well-exploited scientific fact to chill drinks and food from early times in Mesopotamian, Chinese and Persian cultures. The first written record of the making of artificial ice or snow by the addition of saltpetre to cold water seems to be found in a passage of a work by the thirteenth century Arab physician and historian Ibn Abi Usaibia

(1203–1270) (Needham 1976; Thevenot 1979). The difficulty in identifying and purifying the respective salts could very probably be the reason for this technique not to be extensively used for cooling purposes. The diffusion of this knowledge into the West goes back to the beginning of the sixteenth century, and it seems that it was in Italy where people learned that immersing a container of wine in a vessel filled with cold water mixed with potassium nitrate, or saltpetre, would freeze the water (Lippmann 1898). Although the Italian Marco Antonius Zimara was probably who, in 1530, first mentioned the endothermic effect produced by dissolving certain salts and crystalline hydrates in water (Beckmann 1846), it was the Spanish physician Blausius Villafranca who first described the full method in a tractate published in Rome (Villafranca 1553). Some decades later, the also Italian physician and professor Latinus Tancredus stated that the addition of snow or powdered ice to saltpetre strengthened the cooling effect, and the mixtures reached even lower temperatures (Tancredi 1607). Probably, the best known, although very short, reference to this technique is found in a book by the Italian natural philosopher and scholar Giambattista Della Porta (1535–1615), devoted to explore "the secrets of nature", by proposing to teach people how to cool and freeze wine (de La Porta 1589). Decades later, the German physicist Daniel Gabriel Fahrenheit (1686–1736) additionally using vacuum (Fahrenheit 1724), and the British scientist Henry Cavendish (1731–1810) (Cavendish 1783), also discussed the obtaining of low temperatures and the freezing of water and mercury respectively, by using different refrigerating mixtures. The property by which salt and saltpeter added to ice and snow produced a depression of the freezing point, called the interest of scientists of different nationalities and various academic inclinations in the following century. As an example, remarked men of science as the British natural philosophers Francis Bacon (1561–1626) (Bacon 1765) and Robert Boyle (1627–1691) (Boyle 1665–1666, 1725), and even the physician and mathematician Gabriel Philippe de la Hire (1640–1718) (de La Hire 1694), among others, also reported related experiments by which, each one successfully produced ice by using the cited method, although using different salts and in diverse proportions.

The first significant work on the subject was carried out by the British apothecary Richard Walker over more than four decades. The first part of this research was devoted to the examination of several refrigerating mixtures containing different salts and in various proportions, in order to identify the most appropriate one and to, not only freeze water for summer in hot climates, but to also provide ways for cooling wines and other alcoholic beverages and for preparing frozen creams (Beddoes 1787; Walker 1788, 1789). The mixtures he studied could be classified in three general types: those composed by water and one salt, where the only thing that happened was the fusion of the compound; others that involved the dilution of one salt in an acid medium, with the occurrence of some amounts of heat produced due to the chemical affinity between both substances, and the subsequent implications in the degree of temperature reduction reached; and one third class, composed by ice, or snow, and some specific chemical compound, in which the effect on temperature reduction was doubled by the simultaneous fusion of both substances. Chemical compounds such as salt ammoniac

Fig. 8.1 Walker's apparatus (Walker 1828)

(ammonium chloride), Glauber salt (sodium sulphate decahydrate), nitrated volatile alkali (ammonium nitrate), vitriolated magnesia (magnesium sulphate), alum (hydrated potassium aluminum sulfate) (potassium alum), Rochelle salt (potassium sodium tartrate), and muriate of lime (calcium chloride) were some of the salts studied by Walker. Compounds used for the corresponding solution included spirit of nitre (nitric acid), nitrous acid, diluted nitrous acid (sulphuric acid) and marine acid (hydrochloric acid), besides obviously water and ice itself. In the second part of his work, Walker was devoted to the design and construction of an apparatus for freezing water, which, after some modifications, could be also used, either for cooling wine, or for preparing frozen creams (Walker 1795, 1801). The structure of the basic apparatus (Fig. 8.1), with which he was able to reach cooling up to 47 °C, was relatively simple and consisted of a tube for receiving the water (or other substance) to be congealed (or simply cooled), placed in a tin larger vessel which contained the freezing mixture. Then, the refrigerant mixture filled the void space, and some class of stirring arrangement had to be provided in order to get an indirect thermal contact as perfect as possible. Once the water was frozen, the cover was taken off, the tube wiped up, and the ice that had been detached by the heat after inverting the apparatus.

Numerous machines based on this principle were designed and built since that time, some of them even before preliminary theoretical studies that could describe the behavior of the refrigerating mixtures were carried out and published (Reif-Acherman 2009). The general functioning was similar for all, differing in the nature of the refrigerating mixture used, materials and capacities. The machines were also provided with mechanisms, which varied according to the specific model and its corresponding functioning system of mechanisms to remove the ice and to upgrade the refrigerant potential of the mixtures.

The use of these new developed apparatuses very quickly extended to applications different to the housework-related gastronomic tasks. In France for example, the pharmacists Michel de Courdemanche (de Courdemanche 1825), Pierre Hippolyte Boutigny (Boutigny 1834), and Pierre-Prosper Malapert (Malapert 1836) of Caen, Evreux and Poitiers, respectively, developed in the 1820s and

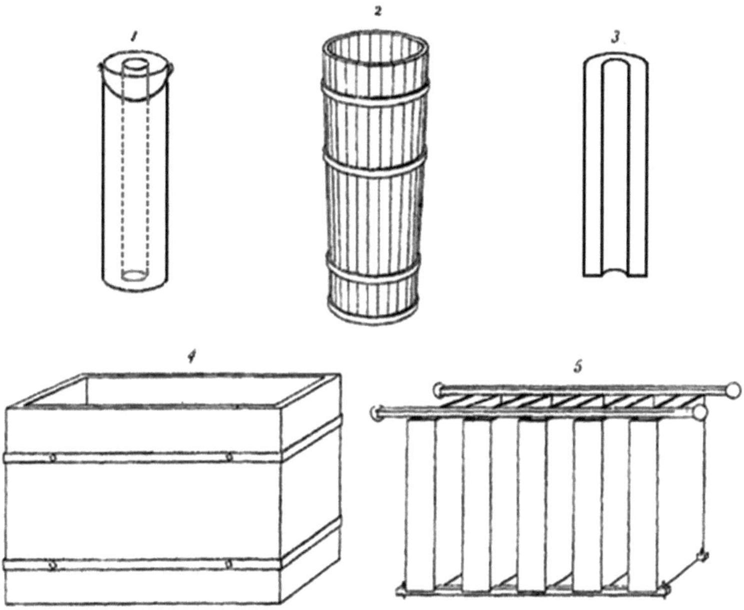

Fig. 8.2 de Courdemanche apparatuses to make ice for therapeutic purposes (de Courdemanche 1825)

1830s, some designs in order to obtain ice and cold water for hygienic and therapeutic uses. The apparatuses did not exhibit significant differences among them, except for modifications in their shape, the vessels used for the indirect thermal contact arrangement and the molds where ice was formed. The usual refrigerating mixtures used were those studied by Walker, particularly those composed by sulphuric acid (in concentrations between 41 ° and 56 °Bé) and sodium sulphate in proportion 4:5. Figure 8.2 shows details of two apparatuses designed by de Courdemanche; the three elements above, correspond to a low-capacity model, 3.0 kg of ice per hour, using something similar to two concentric tube system, while the elements appearing below refer to a model of higher capacity, 18 kg of ice per hour, where the product was formed inside tin cans suspended and maintained by two rods from below. The approximate consumption of refrigerating mixture in this class machines was generally reported to be in a 4:1 ratio, distributed in three charges along the specific batch. The refrigerant mixtures in both apparatuses must be treated by evaporation, in order to correct the permanent crystallization of the salts occurred along the process, but it must be periodically, and in any event, renewed after a definite number of batches.

The first reported trial on the introduction, from a commercial viewpoint, of the recently discovered alternative for freezing in the market, at least in Great Britain, goes back to 1844 with the works by Thomas Masters, which lasted for almost one decade, to include cooling of alcoholic and non-alcoholic beverages and, even, the human body (Masters 1844a, 1846, 1853). The basic apparatus, also used to freeze,

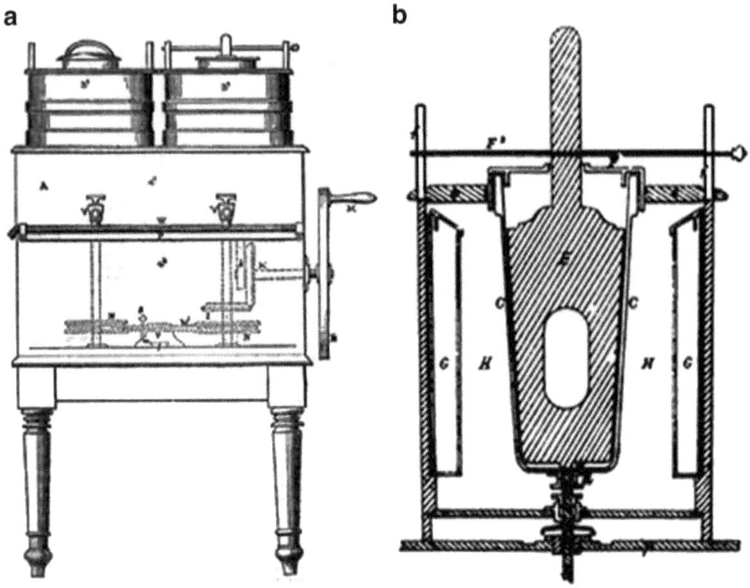

Fig. 8.3 Master's apparatus with two pails (**a**) Basic arrangement (**b**) enlarged sectional front view (Anonymous 1844a, b; Masters 1844b, 1850)

consisted of a rotary churn supported on four legs that had a case divided in two compartments, the upper one containing generally two circular, although different pails (Fig. 8.3a). In order to duplicate the freezing effects, the original design of each pail had in turn two vessels, the inner one made of pewter and of conical shape for holding water, cream or other material wanted to be frozen (or churned), and an outer one, made of zinc or other appropriate metal, and of a specific shape (generally circular) for holding the water or any other material that was to be frozen. The space that surrounded the vessels was filled up with refrigerating mixtures, and the inner vessel was rotated at high-speed, while certain number of spatulas which remained steady in the middle of it, carried out the churning part of the process. Once the freezing process concluded, the block of ice was removed by dipping the exterior of the vessel in cold or lukewarm water. There was no preference for any specific refrigerating mixtures, being several (snow with common salt and ammonium chloride, potassium nitrate, or calcium chloride, among others) that were claimed to be appropriate for the process, although, of course, in different weight proportions. Figure 8.3b, shows the section of one half of the apparatus on an enlarged scale.

The number of pails in the original design could be conveniently increased to three, four, six, or any number accordingly to the quantity of ice required, by merely enlarging the apparatus and connecting the spindles of the additional pails to the common mover, in the same way as the mechanism used for the basic two-pail model. A version of the apparatus having only one pail, in which case the parts

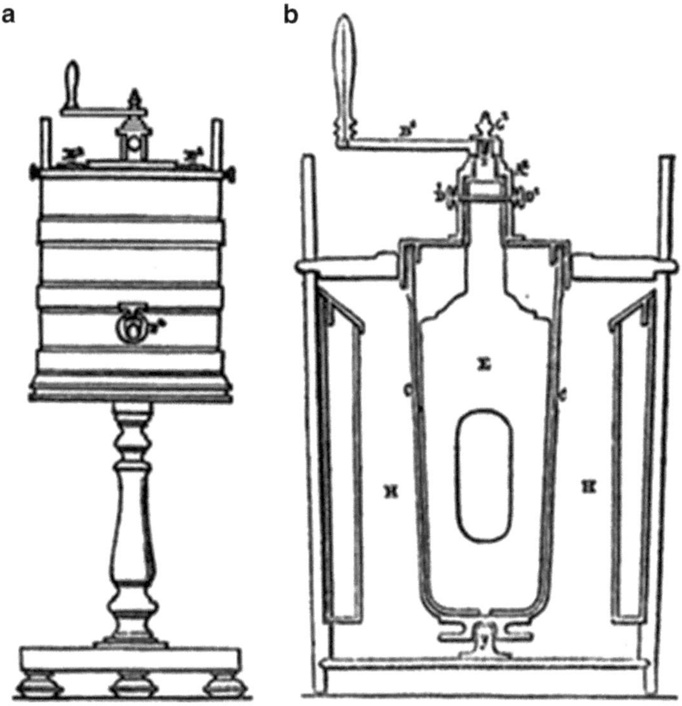

Fig. 8.4 Domestic Master's apparatus with one pail (**a**) basic scheme, (**b**) enlarged section (Anonymous 1844a, b; Masters 1844b, 1850)

worked from the top, and of its corresponding enlarged sectional front view, specially adapted for familiar domestic uses, are represented in the Fig. 8.4a, b, respectively. Besides being of a cheap and simple construction, easy, rapid and efficient to work with, the functioning of the machine was versatile. It could alternatively be used in some cases, to prepare ice creams and as a butter churn, or any of the previous and making ice if a division was provided. It is interesting to note that, although the developments received wide diffusion at the time by remarked technical journals (Anonymous 1844a, b) and by the inventor himself (Masters 1844b, 1850), there are few references in the different sources to production capacities, and any one explicit to the corresponding required consumption of chemicals for the refrigerating mixtures. The only available information for this apparatus mentions that a typical-size apparatus as described was able to produce between 5 and 6 kg of ice, or 2–2.5 kg of ice cream per hour of work (Masters 1844b, 1850).

In 1844 the French physician and professor of the Ecole Préparatoire de Médecine at Marseille, Étienne Villeneuve (1800–1882) presented an apparatus to the Académie des Sciences, which this institution considered for domestic economy rather than a scientific machine, and that first received the name of *congélateur* later known as *glacières des familles* (Dusserre et al. 1844). It consisted of four concentric metallic vessels alternatively containing water and a refrigerating mixture.

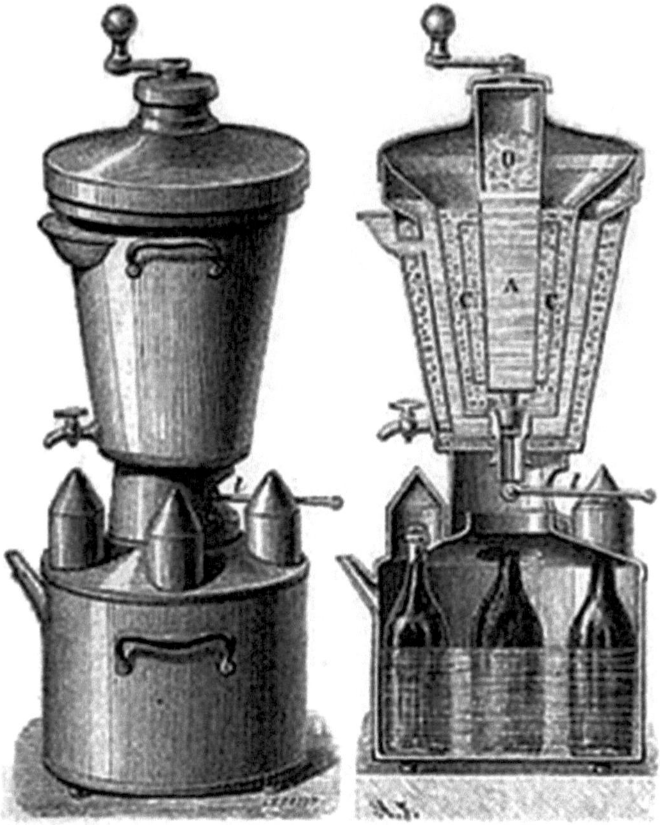

Fig. 8.5 Villeneuve's *glacières des familles* (Anonymous 1867)

The arrangement (see Fig. 8.5) included a compartment A, called *sorbetière*, where water (or other substance to be frozen) was placed, which resulted completely surrounded by the refrigerant mixtures filling the compartments O and C. The inner vessel, composed by the *sorbetière* and the compartment O, was provided on the top with a crank to make rotate the assembly, and on its sides with oblique fins for stirring the mixture. A correct functioning allowed the fulfilling of two objectives. On one side, it was possible to get a hollow cylinder of glass that easily detached with a little knock; on the other, and once the refrigerant mixture had lost his cooling efficiency, the action of a little lever could move a valve allowing the very cold, almost frozen, water filling the compartment located between the outer vessel and that one containing the refrigerating mixture, to discharge into a lower compartment where bottles filled with wine or champagne were cooled.

Two refrigerants were used with this apparatus: a first one composed by chlorhydric acid and sodium sulphate in proportion 2:3, and a second one with water and ammonium nitrate in equal quantities. Although this last mixture led to a slower cooling, and eventually freezing, it was preferred because it could be used almost indefinitely, given that, unlike the other, its salt could be recovered by evaporation, reducing the associated costs. Working on a room temperature of 15–20 °C, the typical apparatus was able to produce between 3 and 4 kg of ice per hour (Pouillet et al. 1845).

A significant contribution to the subject appears in 1845 as a response to the campaign led in France by the Société d'Encouragement pour l'Industrie Nationale in order to call the attention of the public on the necessity for the construction, not only of little ice-houses for preserving enough amounts of ice in warm seasons, but above all, of economic apparatuses for production of artificial ice that could satisfy the needs of people in remote areas or little communities. The prize of 200 francs offered by this institution for the construction of an apparatus that met basic requirements of a minimum production of 5 kg each 30 min, a cost price less than 30 centimes per kilogram, and an initial temperature of the water to be frozen higher than 12 °C, aroused the interest of several empirical men of science, but, above all, of empirical inventors (Silvestre 1845). A very simple, both in its construction and mode of working, apparatus proposed by Goubaud was maybe one of the most interesting offered for consideration of the Jury. It consisted of a great vessel containing an assembly of 12 slightly conic tubes closed at the bottom and joined at the top to a cylindrical chamber (allowing the simultaneous and very quick filling of all tubes) provided with an orifice to pour the water and a closing system (Anonymous 1846; de Silvestre 1846). A sheet of helical shape surrounding the whole assembly facilitated the required agitation and prevented salt sedimentation. The reported refrigerant mixture in use was an aqueous solution of equal amounts of ammonium chloride and ammonium sulphate. Figure 8.6a–d show several views for this apparatus. Figure 8.6e–g show similar views for a specific version of the apparatus, especially adapted for the preparation of sorbets and ice creams.

The main characteristics that highlight Goubaud's design regarding the previous one, were, the effort for multiplying the contact surfaces between water and the refrigerant mixtures, and the avoidance for using dangerous acids and salts in the mixture. The use of the mentioned mixture prevented, according to the inventor, the decomposition of the salts through the necessary successive evaporations made in order to get their maximum recovery, their possibility for becoming acid, and the loss of some of their properties. The 15–18 min manual rotary movement of the handle provided the appropriate thermal contact for obtaining 0.5 kg of ice per each batch of 2.5 kg of a mixture of salts added to 2 l of water. The loss of salts after each evaporation reported by the inventor, reached about 1.5 % and the cost price per kilogram of ice produced was estimated only slightly less than that reported by using the Villeneuve's apparatus.

Three years later, the also Frenchman Claude-Félix Fumet presented to the same Société d'Encouragement, a less expensive apparatus in its functioning following the same previously-described basic elements but with a simpler design (Fumet 1849; Anonymous 1849; de Silvestre 1849). The apparatus consisted of two concentric cylindrical vessels: a pail to receive the freezing mixture (conveniently isolated from the external view by woolen rags), and the inner vessel

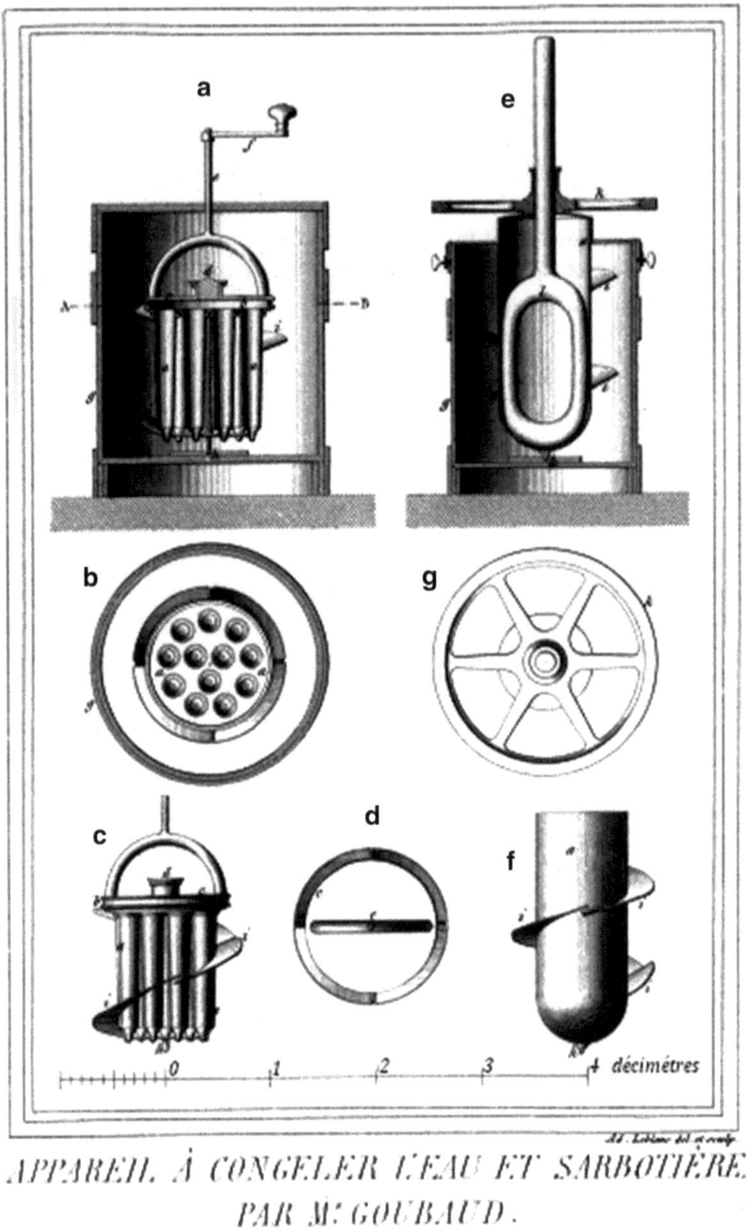

Fig. 8.6 Different views for the domestic Goubaud's apparatus for making ice and its adaptation to prepare sorbets and ice creams (de Silvestre 1846)

inclosing a conical tube open at the bottom, in order to improve the respective contact surface. There is no provision, unlike previous descriptions, of mechanisms to improve the thermal contact, as they are the rotatory movement and the extended fins mentioned for the Goubaud's apparatus. Each batch of production

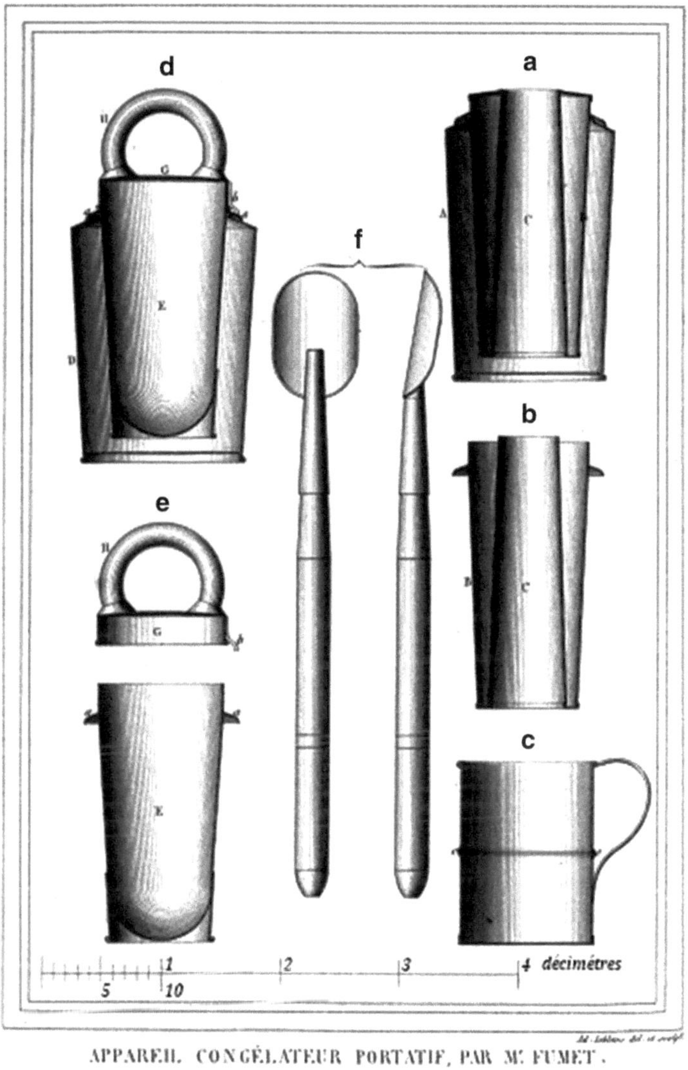

Fig. 8.7 Different views for the domestic portative Fumet's apparatus for making ice and its adaptation to prepare sorbets and ice creams (de Silvestre 1849)

took about 15 min to get ice if the mixture was fresh, and around 30–45 min if it required recovery by evaporation. Two kilograms of mixture were usually required to produce 1 kg of ice. Figure 8.7a–c show a vertical view of the apparatus. The vessel for the water, separately, and a sample of the measuring vessel used for preparing the mixture, respectively. On the other hand, Fig. 8.7d–f represent views of the adaptation of the Fumet's apparatus for preparing iced foods, meaning the basic functioning, the apparatus itself, and two views of the spatula usually used for revolving the preparation, respectively. Although the unitary cost price of the ice produced was very similar to those corresponding to the

Fig. 8.8 Toselli's first design for an ice-making apparatus (Toselli 1866)

previously described machines, this specific style received wide acceptance in the common citizens and for therapeutic purposes mainly because its size and portability (Anonymous 1850).

Maybe the maybe best-known models of domestic apparatuses in the second half the nineteenth century, were the so-called "Italienne Glacières", proposed by the French civil engineer of Italian origin, Jean-Baptiste Toselli (1807–1883), based on the same refrigerating mixtures principle. His first design worked by the combined action of two different cycles; one with vessels for the refrigerant mixture and the corresponding recovery of the vaporizable liquid connected by a tube, that simultaneously served as axis of vertical rotation in order to improve the heat exchange, and the other for the cooling system of those vapors (Fig. 8.8) (Toselli 1866). Although the temperature decrease reached with this assembly was close to 30 °C and allowed the owner to simultaneously obtain ice and sorbets, the high costs associated with the refrigerant mixture used with his machine, an aqueous solution of ammonium nitrate and sodium carbonate, as well as the opaque appearance of the manufactured ice, quickly forced Toselli to improve the design in order to, not only correct these details but to build a more efficient and practical apparatus.

Toselli's most popular design, using only ammonium nitrate as dissolved salt, consisted of two concentric compartments of cylindrical shape placed and properly isolated from the external medium, provided with two opposite openings which closed hermetically (Fig. 8.9a–c) (Toselli 1868a, 1873). Water, or any other liquid to be frozen, was added to the apparatus through one of the openings, whereas a

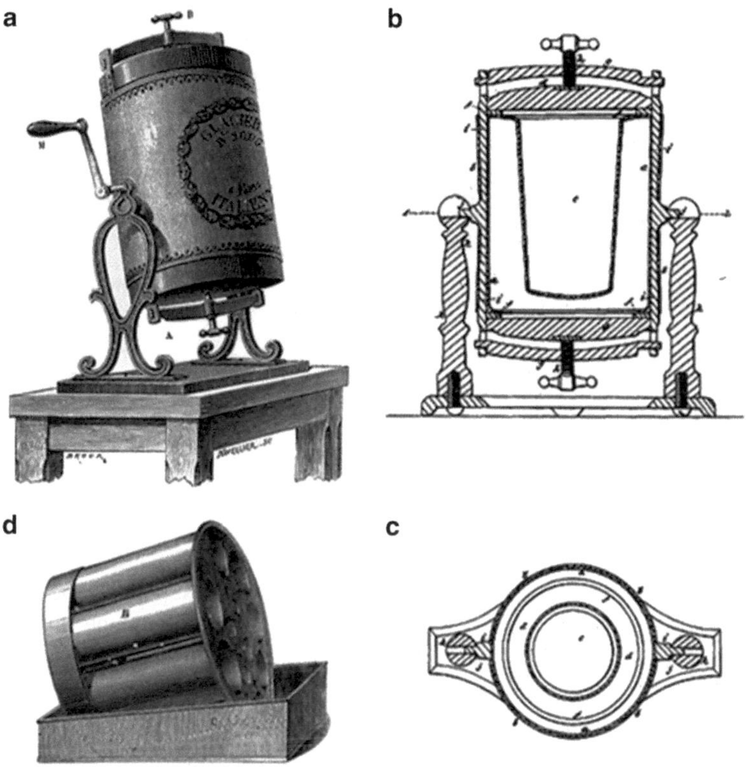

Fig. 8.9 Toselli's apparatus: (**a**) general view, (**b**) central vertical section, (**c**) transversal section, (**d**) tin tube with tapered moulds of varying sizes (Toselli 1868a; Figuier 1873–1877)

solution of ammonium nitrate entered through the other. The functioning implied now a rotary horizontal movement around its own axis for short time, after which compact and transparent ice blocks were formed. A specific model included a suspended ice tin-tube containing several slightly tapered moulds of circular cross section and of varying sizes (Fig. 8.9d). Only 5 min were enough for approximately 0.5 kg of ice to be formed around the edges to the thickness of nearly 3 mm. The diameters of these molds were so carefully calculated that when the corresponding rings or tubes of ice were removed and placed one inside the other, they formed a small compact stick of ice that remained solid for a very long time. He reported, for example, that 10 kg out of a 23 kg block of ice, prepared in only 18 min under the above described procedure, remained in a firm solid condition after its transport from France to Algeria (Anonymous 1869). The more than 50 different designs Toselli presented during 15 years, for which he was granted with several foreign patents, allowed him to progressively reduce the required time for each batch, from 18 to 20 min with his 1866 designs to about 5 min with the 1877 version without significantly increasing their performance (Toselli 1868b, 1876). The high requirements of water and coal for the corresponding regeneration of the mixture (about 4.0 and 0.5 kg for each kg of ice, respectively) were the main inconveniences

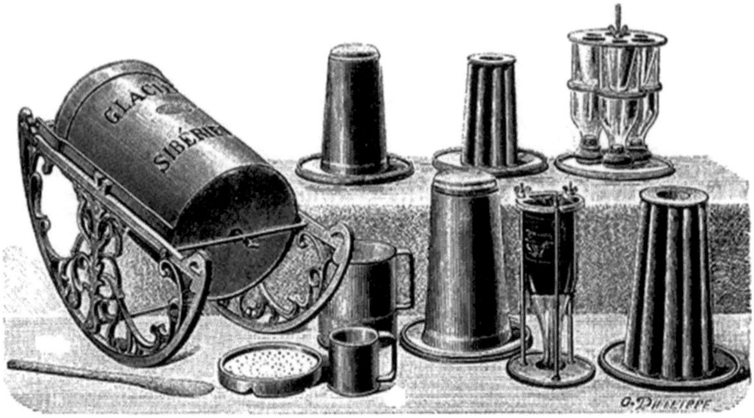

Fig. 8.10 "Glacière Siberienne" with accessories (Brewer Frères 1910)

of this apparatus. Besides the previously mentioned applications, Toselli's machines were widely used by armies in the field in therapeutic and supplies issues.

A model derived from that by Toselli, called "Glacière Siberienne" or "Glacière a bascule", and proposed by Penant, showed a different disposition in order to increase the contact surface between the refrigerant mixture, composed by sodium sulphate and chlorhydric acid, and the water to be frozen. It was essentially constituted by a hermetically closed, metallic cylinder containing the mixture, and a ribbed or plane mold consisting of two concentric surfaces with water in its inside space (Fig. 8.10). The whole apparatus was then placed in a rocker carriage, whose movement activated the cooling process (Héraud 1890). It took about 20 min and two batches of mixture to produce one block of ice of 0.8 kg.

The step by which the ice formed in the internal surface of the *sorbetière* (usually included in several ice-making machines), generally a long and a tedious one, was replaced by a wooden spatula. An ingenious mechanism for ice to be easily and very quickly removed from ice making machines using refrigerating mixtures containing open vessels, was proposed by Loefz (Peclet 1878). Figure 8.11 shows the schematic view of a typical working system apparatus with two identical *sorbetières* actuated by only one crank. By using this machine, and with the *sorbetière*'s work of only one man equivalent to 300 revolutions per minute, it was possible to obtain between 4.0 and 5.0 kg of ice in a 5-min period.

The interest for the application of refrigerating mixtures remained latent along all the nineteenth century. A full inventory of possible mixtures and their diverse applicability was published in 1886 (Lightfoot 1886), at the same time as studies about their behavior with temperature continued appearing (Rudorff 1873; Pfaundler 1875). Continuous improvements in applications of refrigerant mixtures allowed the creation of little factories which specialized in the manufacturing of multifunctional apparatuses that could be used to make ice and ice cream, as well as to chill wines. More sophisticated mounts used included coils with one end connected to a beer or ale pump in the restaurant, and the other end with the barrel

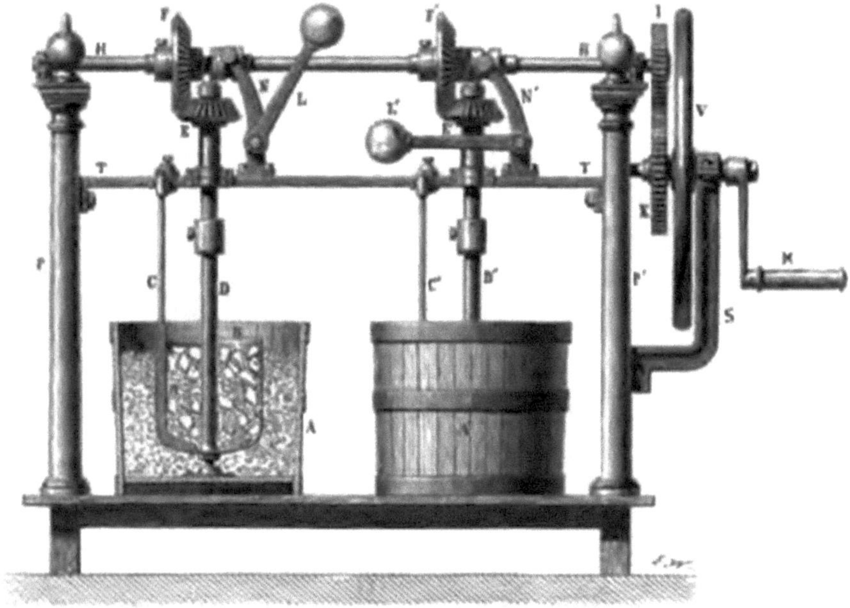

Fig. 8.11 Loefz's apparatus (Peclet 1878)

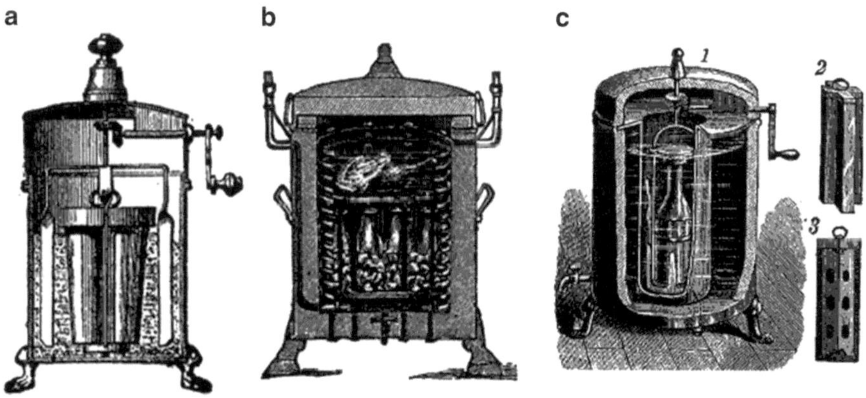

Fig. 8.12 Different models of family ice machines (Anonymous 1889, 1890)

in the vault. These compact and easy-to-work machines were serviceable, not only for family use as refrigerator for the cold storage of fish and vegetables, but also for restaurants, cafes, confectionaries, hotel keepers, etc. Figure 8.12a, b represent small and large sizes of some particular model, capable of turning out a block of ice from 0.4 to 9 kg at each batch, requiring for that purpose from 10 min to 1 h, or from 2 to 20 cups of ice cream, or from 1 to 4 carafes of ice for the table (Anonymous 1889). A larger capacity apparatus as that depicted in Fig. 8.12c, for example,

a b

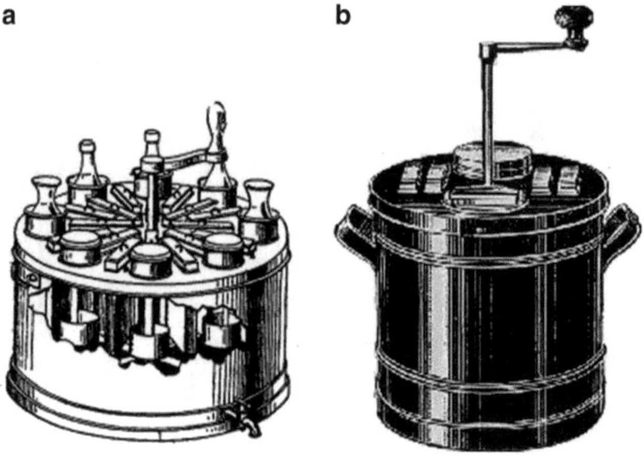

Fig. 8.13 Twentieth century domestic ice-making machines (**a**) "La Rayonnante" (Anonymous 1908) (**b**) "Frigoria" (Anonymous 1924)

exclusively devoted to domestic affairs and provided with the shown ice and ice cream molds, either one to be coupled as desired, was able to make all the ice required for an ordinary household daily use in only 15 min (Anonymous 1890).

Few new models of ice-making machines for household use based on this principle, driven either by hand or an electric motor, continued appearing in the market even in the first decades of the twentieth century. The usual refrigerant mixture was water and ammonium nitrate in equal proportions. They were generally cylindrical in shape and supplied with a number of molds which, when filled with water produced briquettes of ice from 10 to 25 min (Fig. 8.13). The machines were made in different sizes between 0.6 and 22 kg at once (Anonymous 1908, 1924).

Different design trials of industrial, large-scale machines for making ice in continuous processes using refrigerant mixtures were proposed. Designs as those carried out between 1855 and 1858 by the British civil engineer Charles Williams Siemens (1823–1883), for example, using an aqueous solution of calcium chloride (Siemens 1855, 1857, 1858), or others introducing mechanisms for automatic continuous feeding of the corresponding mixtures, proved to be unsuccessful because of the associated high functioning costs, compared to those of contemporary machines producing ice by purely mechanical methods.

8.3 Evaporation and Absorption Systems

Primitive techniques for producing cold by evaporation were known from the antiquity. Cooling of water contained in porous pottery, and eventually production of ice, by the simultaneous evaporation of small quantities filtering through the wall and the removal of heat required from the remaining contents of the vessel, were a usual practice all over the East (Forbes 1966). The extension to which this process was carried out determined whether the water was cooled to the freezing point or not. Some specific conditions, such as the relative humidity and the speed of the outside atmosphere could strongly accelerate the process by the displacement of the saturated layers of air immediately adjacent to the evaporative surface, and the bringing of larger and new amounts of air to be placed in contact with the pottery. Shallow and unglazed water vessels constructed of porous earthenware and having a small neck and a spout, or bottles wrapped with wet clothes or leaves, laid on straw and exposed to the night air were used in earlier times for the purpose of procuring ice. Moisture evaporation on the outside of the vessels produced intense cold within, and the water froze solid in the course of the night, being possible to observe in the morning, if the sky had been clear, the formation of thin plates of ice, which were carefully gathered and stored away.

The scientific basis for producing cold by evaporation are due, however, to the researches carried out in the 1750s by the Scottish professor of medicine and chemistry William Cullen (1710–1790) by the use of an arrangement consisting of a vessel containing nitrous ether immersed into a larger one containing water and placed under a receiver. After exhausting the air by means of a vacuum pump and allowing the whole apparatus to remain for a few minutes in vacuum, Cullen was able to observe the ether boiling and most part of the water frozen. As the vacuum forced the evaporation without giving any heat, this ascending vapor absorbed the latent heat required for its formation from the water, and at last lowered thus its temperature to such a degree as to cause its freezing (Gamgee 1877; Cullen 1756). It seems that Cullen's purpose was only a theoretical one, because, neither he made an attempt to recover the ether by compression and reuse it, nor he went beyond with practical applications for his researches. This ice-production principle by forced evaporation of water, but driven by the remote condensation of its vapor, was the one that jointly with the notions of specific and latent heat, became the basics of the ingeniously simple cryophorus mechanism (Fig. 8.14a), invented by the English chemist and physicist William Hyde Wollaston (1766–1828), in order to demonstrate the efficacy of the mechanism of the vapor pump acting over a considerable distance (Wollaston 1813).

The inefficiency showed, however, by Cullen's apparatus because of the considerable quantities of water vapor that must be removed at low suction-pressures, was later corrected by the British optician and scientific instrument maker Edward Nairne (1726–1806). His improved design accelerated the freezing process by including an additional little vessel with sulphuric acid, in order to, not only allow the absorption of the aqueous vapor formed thanks to the existing affinity between these substances, but also to prevent the saturation of the space (Nairne 1777). Nairne's proposal became the first example of a technique that involved the combination of evaporation and absorption principles in order to get a common objective, which this time was the mechanical production of ice. Absorption involved the use of two substances, which showed an affinity for one another, so that one tended strongly to be combined or dissolved into the other one when they were in cold state,

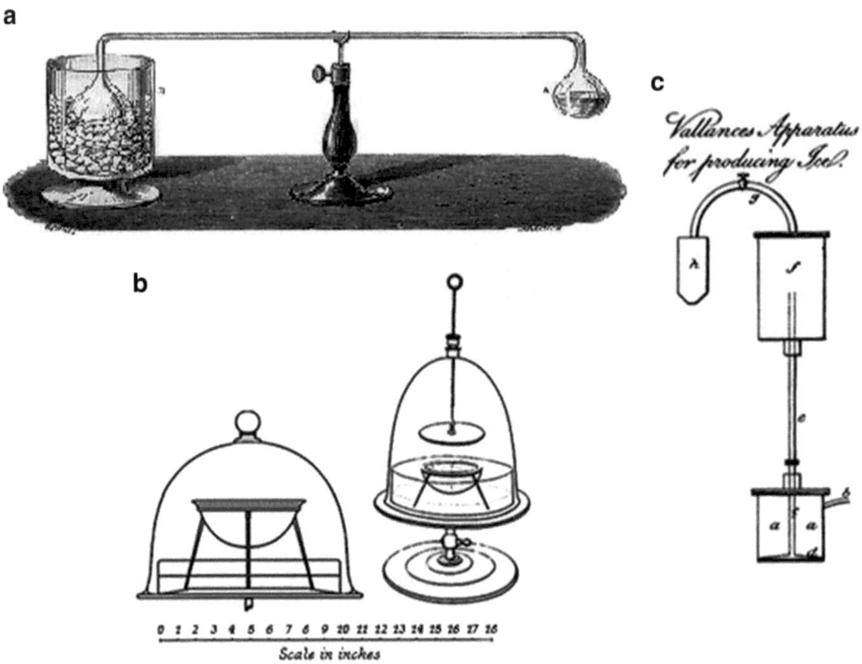

Fig. 8.14 Ice-making apparatuses for producing ice (**a**) Wollaston's cryophorus (Wollaston 1813) (**b**) Leslie's assembly (Leslie 1813) (**c**) Vallance's arrangement (Vallance 1824)

but that can be separated by applying heat. Then, once evaporation reached its own purpose, cold, and eventually freezing could be produced by absorption as a direct utilization of high temperature heat without requiring an intermediate conversion of heat into mechanical work.

The first domestic-level apparatus for producing ice based on these principles was however reported to be built by the Scottish mathematician and physicist John Leslie (later Sir) (1766–1832), who was able to get, in 1811, about 0.6 kg of ice after 45 min of partial vacuum (Fig. 8.14b) (Leslie 1813). A continuous version of this model (Fig. 8.14c), which allowed to the obtaining of larger quantities of ice, became the first patent of an ice-making machine granted in 1824 to the British brewer John Vallance (Vallance 1824).

The first successful machine based on this principle was made in 1866 by the French civil engineer Edmond Carré (1833–1894). The apparatus he designed was made up of a simple and inexpensive pneumatic machine, a flask of resistant glass for the water to be frozen and a reservoir for sulphuric acid, made with an alloy of lead and antimony to resist chemical attack at ordinary temperatures (Fig. 8.15a). This reservoir communicated on one side with a carafe of water, and on the other with the pneumatic machine. A rod connected to the pump handle sets in motion an internal stirrer, which, by a continuous renewing of the liquid surface, made easy the

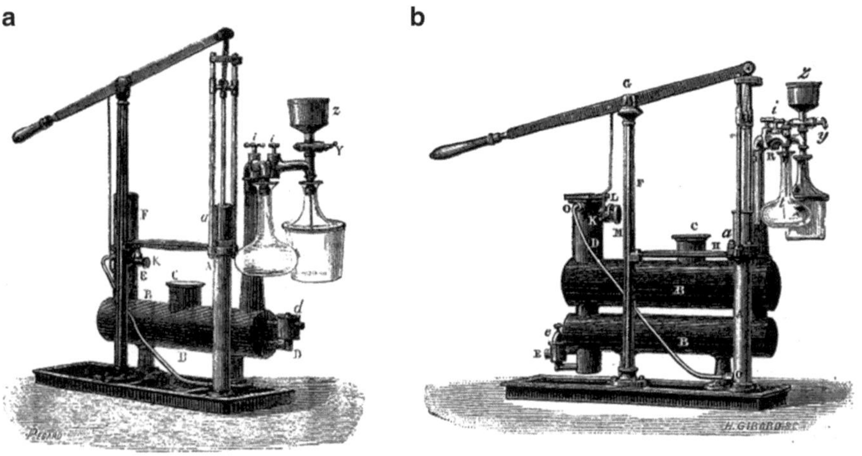

Fig. 8.15 Carré's machines for making ice by evaporation under reduced pressures (**a**) with only one tank for sulphuric acid (**b**) with two tanks for acid circulation (Fontaine 1891)

absorption of vapors. As the pressure was reduced, the water in the carafe boiled, thanks to its own heat, and the sulphuric acid absorbed the water vapor, lowering the pressure still further. The heat required for the vaporization was then taken from the water in the carafe until this removal of heat caused ultimately the remaining water to be frozen. A small time (about 4 min) before the vacuum started, a crust of ice formed in the flask, could be seen, which increased more and more until the whole, which fills about half the flask, is frozen. Taking into account the unavoidable losses of cold, or rather unavoidable absorptions of heat, it was possible to obtain 0.34 kg of ice per each 0.06 kg of water evaporated. A small size model of the apparatus, with a new charge of 2.5 kg of acid of 66 °Bé, had the capacity to freeze 15–20 demijohns of 0.34 kg each one. This performance decreases, however, as the acid grew hot and became diluted, making its replacement necessary when the concentration went down to 52 °Bé (Carré 1867; Lhéritier 1875). Carré modified the apparatus in 1883 by housing the acid in an assembly of two tanks, the one above provided with a series of diaphragms designed to slow the acid circulation, in order to get a stronger cooling (Fig. 8.15b).

Carre's machines were successfully manufactured until the conclusion of World War II, being highly appreciated in France and Great Britain because of their portability and little weight. The main inconveniences were associated with the vacuum pumps originally utilized, and specifically with the deformations in the cylinder and the piston, resulting from the natural wear of the materials, and the subsequent higher suction works required. The development by the British inventor Henry Albert Fleuss (1851–1932) of an improved, almost frictionless double-action pump machine, able to produce a practically perfect vacuum and to avoid air to leak back, helped to overcome these difficulties and provided the evaporation technique with a better accessory (Ewing 1908). The domestic hand ice-machines

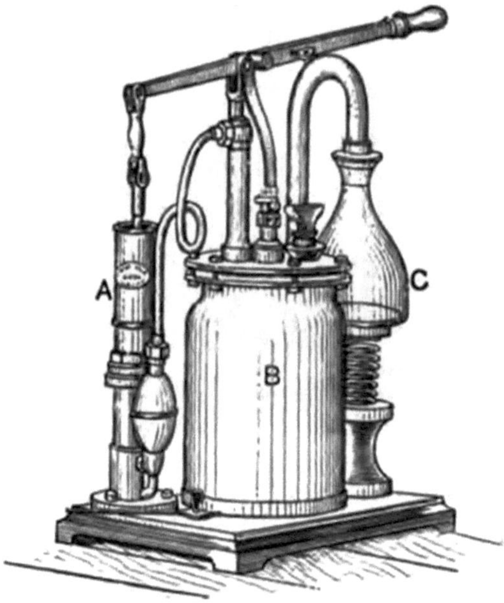

Fig. 8.16 Fleuss's modified evaporation ice-making domestic machine (Ewing 1908)

constructed with this pump derived in a simpler, cheaper and more compact apparatus than anything else of its kind in the market. The same way, the batches of sulphuric acid lasted considerable longer in the equipment on account of the high vacuum used (Fig. 8.16). These machines received wide acceptance for cold drinking water, table butter preparation, and making of ice creams and ice blocks in 3 min, depending on the given model.

A different evaporation – absorption machine had been previously proposed, in 1858, by the French civil engineer Ferdinand Philippe Édouard Carré (1824–1890), brother of Edmond, by using a solution of ammonia in water (Carré 1860). The main difference regarding the previously described apparatus lays on the non-existence of vacuum (because of the higher volatility of ammonia compared with that of water), but on the contrary, pressures higher than the atmospheric throughout the whole cycle. The apparatus consisted of two vessels capable of resisting considerable pressure, free from air and connected firmly by means of a tube (Fig. 8.17). Vessel *A* contained a strong solution of ammonia in water, and vessel *B*, initially empty, was surrounded by cold water *C*. A first phase where pure ammonia vapors were driven off from the liquid mixture by placing the vessel *A* over charcoal fire was followed by the passing over pressure into the other vessel where condensed, giving up its latent heat to the water outside. The process continued until most of the ammonia had passed over the apparatus, and the solution in the first tank was subsequently highly diluted. The apparatus was then taken up and both vessels' function changed. At the same time that the vessel which initially served as condenser of the anhydrous ammonia became the new refrigerator, the other vessel, which originally functioned as generator, changed its role to serve as absorber in the second stage of the process.

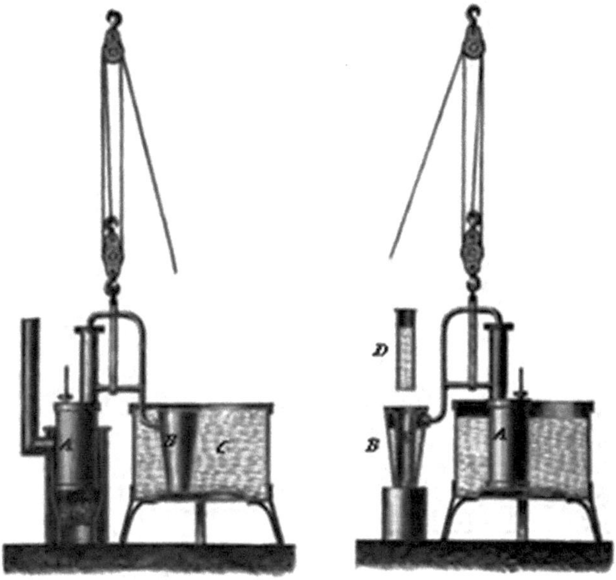

Fig. 8.17 F. Carré's intermittent ice-making machine (Lami 1885)

A tin cylinder *D*, containing water, was then placed in the concavity of the vessel containing the liquid ammonia. The refrigerating effect caused by the re-evaporation of the latter and the subsequent great reduction of temperature produced was then the responsible of the water freezing (Anonymous 1862).

Widely different ciphers had been reported in relation with the performance of the machine. An acceptable mean value informed that it was possible to obtain around 5.0 kg of ice per each kg of coal burned (Carré 1860). The main inconveniences with this apparatus were associated with the long time required for each batch of production (between one and one and half hour for the heating step depending on the production capacity of the apparatus, and almost 2 h for the freezing) and the high pressures utilized (up to 20 atmospheres).

Although the functioning was relatively simple, these machines, usually constructed for 0.5–2 kg of ice per batch capacities, produced ice more slowly (35–70 min for each step) than those working by evaporation, and were initially used only to chill bottles and carafes in cafes and restaurants. It was usual to observe many vehicles loaded with carafes of refrigerated water and ice prepared with this class of machine running on the streets of Paris during summer times to distribute them to private houses, cafes and restaurants. The vessels were however so heavy that their use was considered to be more appropriate for restaurants than for homes, and only the regularly appearing improvements gradually facilitated their later use in households. Improved machines working under the combined principles of evaporation and absorption continued appearing in the following decades. Figure 8.18 shows, for instance, details of one of those improved designs, then called "Minimax"

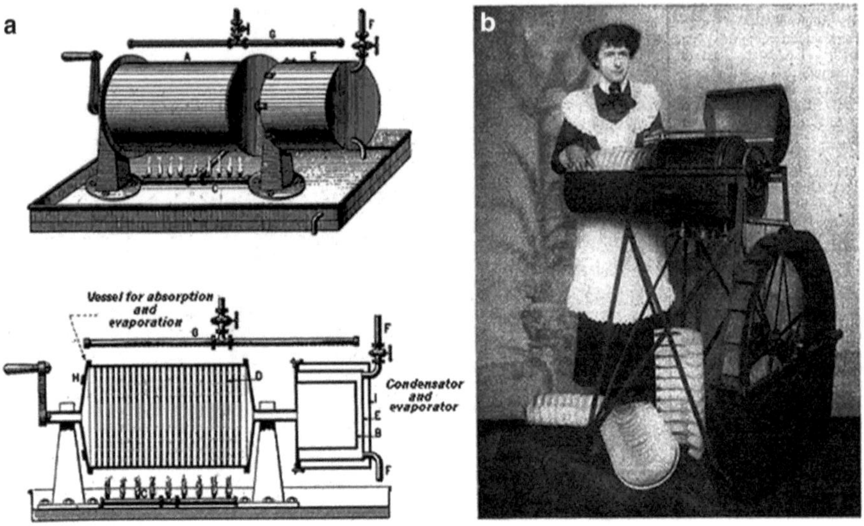

Fig. 8.18 'Minimax' Machine working on evaporation and absorption principles (Serbar 1914)

(Serbar 1914), which worked under vacuum conditions and with a solution of zinc chloride as an absorbent liquid replacing ammonia. It consisted of two cylindrical vessels connected by a hollow tube in a whole assemble which rotated on bearings around a longitudinal axis, in order to contribute to the absorption step (see Fig. 8.18a). Both vessels fulfilled two different functions, in the same way as it was mentioned above for the F. Carré's machine. When water was evaporated from vessel A to B, the first one functioned as a boiler and the other as a condenser; when the process was the opposite and the refrigerating effect was produced, A acted as an absorber and B as a boiler. Figure 8.18b shows a machine of this class with a production capacity of 3 kg of ice per batch.

8.4 Compression Systems with Liquefiable Compounds

Refrigerating machines working upon compression principles introduced a completely new element on the subject: a separate refrigerating agent of a more or less volatile nature, able not only of abstracting heat for its own evaporation, but of being subsequently restored to its original physical condition by mechanical compression and cooling. Only those liquids whose vapors were able of being liquefied under pressure at ordinary temperatures could be used as refrigerant agent. The process made use of the principle by which all gases during the process of passing from a liquid to a gaseous state are bound to absorb a certain amount of heat, and whilst returning from a gaseous to a liquid state to give up or throw off the same amount of heat. The four phases which conform the general closed cycle of these machines included vaporisation, compression, condensation and expansion, steps in which the refrigerant alternatively absorbs and releases energy experiencing changes in its pressure, temperature and physical state. The absorption of heat from the material to be cooled or converted to ice by the refrigerating agent during its transit through the evaporator is the responsible of the desired refrigeration effect.

Early trials carried out in the United States, Great Britain and Australia for producing ice by using compression systems, focused in commercial machines of medium to large capacities and involved sulphuric ether and ethyl ether as refrigerant agent. It was the French civil engineer and professor at the Central School of Arts and Manufactures, Camille Vincent (1839–1910), the responsible for constructing, in 1878, the first little-scale machine working on this principle, using methyl chloride as refrigerant (Crispin 1879). This chemical compound, previously produced by the French engineer and also enthusiast on refrigeration issues, Charles Tellier (1828–1912), had efficiently been obtained in a condition of high purity by Vincent himself, distilling the residues of the mother liquors of beet vinasses. The apparatus (Fig. 8.19a) consisted of a double-cased copper vessel surrounded by isolating materials, being the central space (M) filled with some non-freezing liquid at working conditions, such as alcohol. Taking advantage of an ingenious mechanism, methyl chloride in liquid condition was allowed to enter from a cylindrical reservoir by a screw tap (B) into the space between the two cases (A). A screw (S) was left open in order to allow the escape of gas, fact which favored the boiling of the liquid. A few minutes were required for the temperature to go below −23.8 °C, the boiling point of methyl chloride. The ebullition ceases and the screw (S) could be replaced. Liquid could be placed in connection with a good air-pump if lower temperatures were required; the lowering in pressure accelerated the evaporation, making temperatures of about −55 °C possible to be reached. This way the alcohol could provide the refrigerant effect by keeping itself for several hours at temperatures that could range between the above mentioned numbers (Anonymous 1879a).

Besides its low production costs and chemical stability, methyl chloride showed other several properties that placed the corresponding ice-making machines in advantageous position compared to similar contemporaneous machines. Their construction was simpler, their parts were not exposed to corrosive effects and did not require lubrication, and the apparatuses worked at less demanding pressure conditions (up to 3 or 4 atmospheres according the temperature of the ether used for refrigeration, and vacuums up to 0.5 atmospheres depending on the lowering of temperature required) than those required, for example, with the above mentioned F. Carré's absorption machine (Anonymous 1879b). It was these factors, among other technical considerations, which allowed this class of hand driven machines, usually constructed for productions that ranged between 1 and 3 kg per hour, to receive wide acceptance, mainly at houses in little colonies located far from big towns for domestic issues, and at hospitals, and laboratories for manufacturing and as storage auxiliary equipment. Figure 8.19b shows an improved model of similar capacities, but provided of a little compressor and accessories for recovering the refrigerant and feeding it back to the process. A bigger apparatus, capable of producing between 5 and 20 kg of ice per-hour, provided with a steam engine, is shown

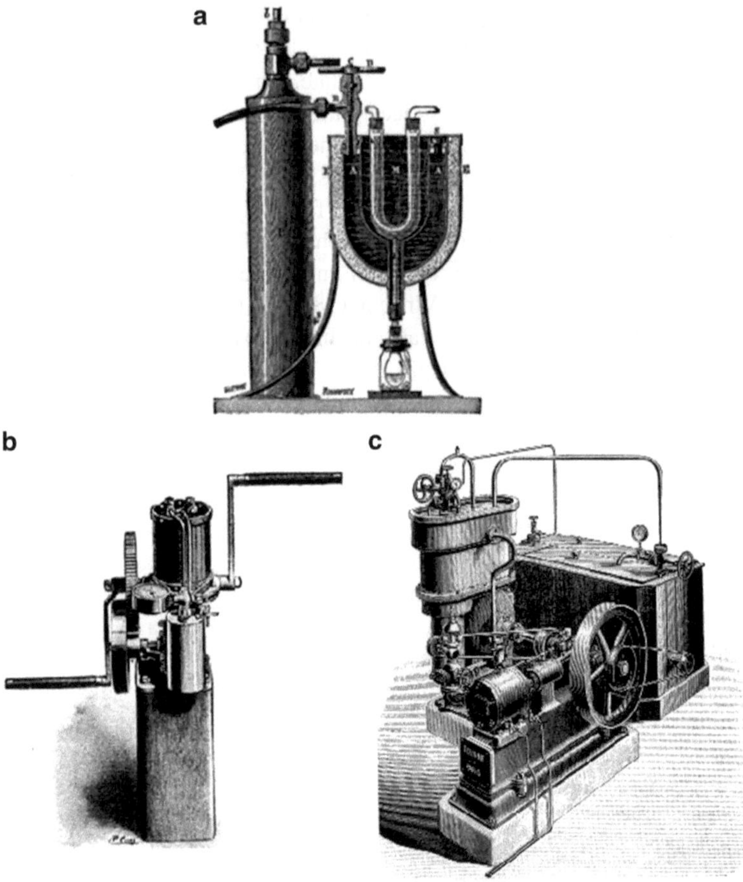

Fig. 8.19 Freezing machine by compression of methyl chloride (**a**) Vincent manual model (**b**) Manual model provided with a refrigerant recovery mechanism (**c**) Manual model adapted to a steam engine (Anonymous 1879a; Jacomet 1904)

in Fig. 8.19c (Jacomet 1904). These apparatuses were used until the first decades of the twentieth century, with the main modifications introduced being related to the compressor's design (Anonymous 1930).

Another representative machine constructed during this period, on the basis of the compression system, was that proposed in 1882, by the Swiss physicist Raoul-Pierre Pictet (1846–1929), involving the use of sulphurous dioxide as the volatile liquid (Anonymous 1882). The little apparatus (with a square shaped base of side length of 50 cm and height not exceeding 1.5 m) was very similar to those at large-scale constructed also by Pictet, in order to satisfy industrial needs (Fig. 8.20). It included a compression pump, a refrigerator where the molds filled with water to be frozen were placed and the sulphurous dioxide was evaporated in order to produce cold, and a condenser where the vapor was cooled and condensed again to then

Fig. 8.20 Pictet's ice-making machine by compression of sulphurous acid (Anonymous 1882)

be returned to the refrigerator, this way, closing the cycle. The Pictet's apparatus exhibited similar favorable advantages to those previously mentioned for the methyl chloride machine, but also exhibited simultaneous disadvantages such as high toxicity of the refrigerant, unpleasant smell, and the possibility to form highly corrosive sulphurous acid in the presence of water. Several years later, Pictet himself, overcame this latter drawback in the different scales by enclosing the motor and the compressor in an air-tight cover. Between 12 and 18 min were required to produce a block of 1 kg of ice and about 15 additional minutes to be ready for the next batch.

8.5 New Century, New Tendencies

First uses of refrigerating machinery in most countries, where it was available at the end of the nineteenth century, were to make ice, and one of the most important uses for that ice, at least in its beginnings, was surely the handling of foods. Blocks of ice were initially produced in a centralized way and distributed to different storage places. The fact that industrial users of the new techniques very quickly realized that the storage facility could be cooled in-situ, with the collaboration of a refrigerating machine, accelerated their progress. The situation for domestic machines was, however, different, and the changeover from ice blocks distributed by special enterprises to small refrigerating machines was definitely slower. The arrival of the new century was, to some extent, a turning point for these developments. Reasons for this shift were diverse, including the continuation of the theoretical supporting studies on the subject, introduced by the German engineer Carl von Linde (1842–1934) in 1876, the development of new motive powers for the machines, the striking changes in population growth and distribution in mainly developed industrial countries, the appearance of an always-increasing interest on the behavior of foods and the determination of the optimum conditions in which they must be held, among others. Some details associated with this new stage of the story need to be described in order to have a full picture of the subject in the period under review.

One of the most influential facts in the observed changes was strongly related to the worldwide technological leadership transitions, a subject to which refrigeration technology was not strange. Although the responsibility of the proposition corresponded to talented people of different nationalities, the first rank on the whole subject of refrigeration to be considered was initially occupied by two European countries. First, Germany, around 1875, with the scientific contributions of Linde and its compressor manufacturing industry, and second, 10 years later, Great Britain, thanks to its domination of the world's production of refrigerating machinery, mainly regarding the requirements of the cold chain of refrigerated meat. The preeminence very quickly changed, however, to other continent. The development in the United States of diverse refrigeration applications, which complemented the evident strong growth of related equipment construction, began in the 1880s. This very quickly led this country to assume the global leadership on the subject. Two clear reasons in order to explain the facts which motivated this transition and how they were associated with the development of domestic refrigerators, joined the usually supplied explanations referring to the American community, such as the diverse climatic conditions, great extension of territory, accelerated economic progress and growth of large city complexes, and remarked entrepreneurship and subsequent great industrial expansion. First of them was the increasing problems associated with the urban distribution of an also ever-growing artificial ice consumption, which in 1909 was estimated to be around 12 million tons around the country, and whose solution included from horses and automobile trucks to deliver ice, until the establishment of portable ice stations of small size that provided the always greater requirements (Thevenot 1979). The second reason was related to the diet variations of American population arising as a consequence of the introduction of the new technology and the subsequent requirements for applications, in order to preserve and transport perishable products over great distances.

Striking technological developments characterized this new stage. By 1885, the compression system of liquefiable gases seemed to have gained ground over the absorption machines. The abandonment of ether, and the introduction of some safer and with better refrigerant quality compounds, such as those mentioned in the previous section, allowed this system to be dominant in the refrigerating machinery market. The compressor then in use, of "open type", showed, however, several important difficulties. One of the most significant was associated with the fact that compressor and motor were mounted separately, with the great possibility of significant refrigerating leakage and the subsequent requirements of periodic servicing

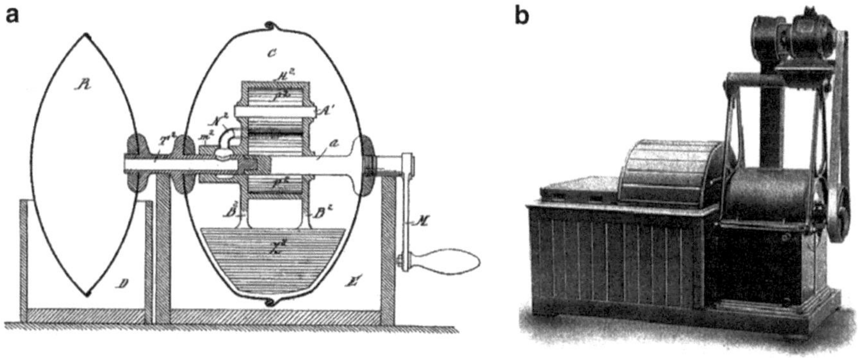

Fig. 8.21 Audifffren's machine (**a**) Sectional plan of a manual apparatus (**b**) View of machine of a later model adapted with an electric motor (Audiffren 1894; Audiffren and Singrun 1907)

and maintenance. First efforts were focused on solving one of the most typical and significant difficulties associated with the then used class of compressors. Solutions focused on the reduction and later elimination of the space located between the piston and the end of the cylinder, as well the introduction of a rotary stuffing-box in order to improve the prevention of leakage along the machine's moving parts allowing the increasing output of large compressors, were not, however, enough to solve the exaggerated amount of energy absorbed in the corresponding latter accessory for small scale machines. Since the providing of continuous maintenance on small systems was not a logical alternative, domestic machines required different solutions to this inconvenience. It was the French priest, physicist, and inventor Marcel Audiffren, who made the first proposal in this regard as a consequence of researches he carried out on the subject since 1893. Audiffren, who served as the abbot of a Cistercian monastery at Epinal (France), received in 1894 and 1904, patents for the development of a hand or machine-cranked unit to keep liquids cool and accessories for oil removal in this class of machines (Audiffren 1894, 1895, 1904). An improved model of the machine, resembling a large dumbbell in appearance, was constructed in 1907 in cooperation with Albert Singrun. Because of its hermetic condition, the apparatus became the first entirely self-contained and sealed machine constructed in the area of refrigeration (Fig. 8.21). It consisted of two containers connected by a hollow shaft, used to rotate a belt pulley at the shaft's end. One of the containers was the condenser, which also contained the reciprocating compressor hanging, cylinder down, on the crankshaft, which turned with the unit. This container turned in a tank of circulating cooling water. The other container acted as the expansion or cooling chamber, and was semi-immersed in a tank filled with the water or brine to be chilled (Audiffren and Singrun 1907; Lallié 1908; Anonymous 1909). The most remarkable fact exhibited by the new design laid on the immediate transformation of the power force into refrigerating effects without the apparent collaboration of mechanical accessories. Although several of the known refrigerants could be used, the usual procedure with this and other compression machines of the century was to work with sulphur dioxide.

The machine, with production capacities ranging from 5 to 25 kg of ice per hour, could be used for decades requiring only very few adjustments. Between 150 and 200 machine units based on this principle were manufactured every year in the United States by General Electric being the first unit sold in 1911, with costs exceeding US $1,000. Ten thousand of these machines, manufactured in France, Germany, Great Britain, and the United States were in use around 1920s and continued being very successful in the interwar period (Thevenot 1979). They were still of current use in some non-developed countries until the 1950s. The rudimentary conditions of the still incipient electric motor models of that time, did not allowed, however, for Audiffren's contribution to become a successful commercial idea, and it was not until after the War World I, with the incorporation of a completely enclosed electrically driven machine by the Australian, Douglas Henry Stokes, with the design in 1918 of an improved refrigerating apparatus, that the idea of a hermetic or sealed compressor took definitive form (Stokes 1920). The new development, with the motor, compressor and cooler contained in a hermetically closed case, very quickly received wide acceptance and became one of the strong roots that led to the definitive entry of refrigeration systems to households.

Small scale compression systems became so popular in the first decades of the twentieth century, that some literature intended for amateurs with medium abilities on the subject, offered detailed instructions for their construction (Anonymous 1920). By the end of War Word I, the appearance of the new models changed significantly, and it was a common fact to observe machines composed of two main parts, a food cabinet to store the material to be frozen or cooled, and the machinery, placed in the basement, being both parts linked by a series of pipes that circulated the refrigerant. The machines were only allowed to work a definite number of hours per day, hours that varied according the specific model. Figure 8.22a–c shows the different pieces of a Kelvinator brand model with a production capacity of 1.75 kg of ice and a daily functioning period of 4 h (Fougeret 1929). Figure 8.22d, e show similar apparatuses of other contemporaneous brands, with mean productions of 3 kg of ice for both models and daily functioning periods from 4 to 8 h, and from 3 to 4 h, respectively (Fougeret 1929; Redenbacher).

Compression machines exhibited other significant and, in some way, inter-related innovations. One of them had relation with new possible refrigerants and the others with details of the compressors in use. Indeed, the appearance of several types of rotary air compressors coincided with the introduction, in the 1930s, of several halogenated hydrocarbons, which received the trade mark of "freons" and that were mainly used, at least in their beginnings, in small domestic compressors. These new thermodynamic agents slowly but surely began to replace the largely-used in small domestic compressors, sulphur dioxide. This element is, however, outside of the subjects and period covered here and it will not be discussed in this essay.

Absorption machines were the other type of domestic refrigerating machine then constructed. A comparative study of machines corresponding to different systems and at different scales of production, carried out in the last quarter of the nineteenth century and included in the 1889 Official Report of the *Exposition Universelle* in Paris, remarked similar outputs for compression and absorption

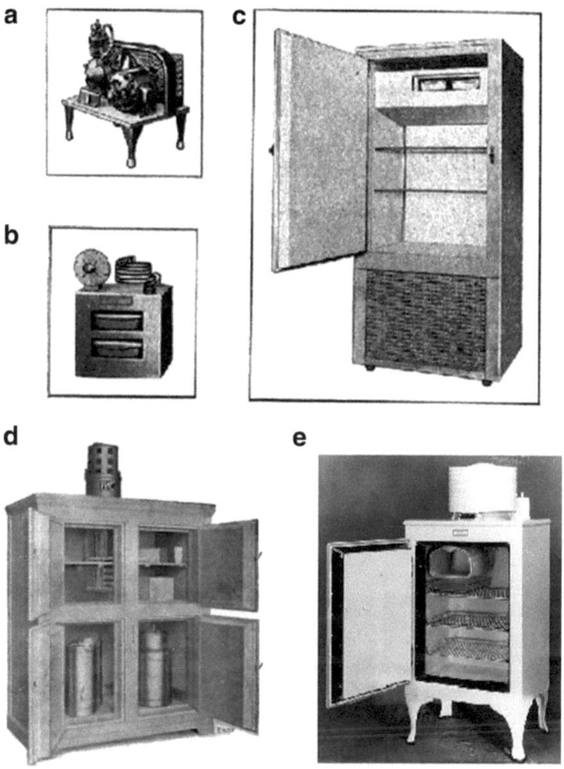

Fig. 8.22 Ice-making (or refrigeration) machines in the 1920s (**a**) *Kelvinator*'s refrigerating unit (**b**) *Kelvinator*'s assemble of compressor-motor-condenser, placed in the lower part (**c**) Whole *Kelvinator*'s apparatus (**d**) *Autofrigor* apparatus (**e**) *Frigéco-Thomson* apparatus (Fougeret 1929; Redenbacher n.d.).

machines in a scale of high production capacities, but slight superiority of the latter one for small machines (Barrier 1891). Production of cold by absorption-system machines had especially been developed for large capacities, but facts related with cost, size, complexity of the refrigeration system and toxicity of ammonia, prevented its use at homes.

Two main innovations favored the construction of small machines based on this principle. The first one was purely related with mechanical issues. On one side, the involved expansion valves were necessarily small and the orifice was continually clogging up with dirt; on the other side, the mechanical pumps were a source of constant maintenance and had to be driven with auxiliary independent power. Machines were then required to have neither expansion valves nor pump and that could work with the only heat supply used for the evaporation of the ammonia – water solution. A first idea in order to reach this aim was proposed by the German inventor H. Geppert by introducing an inert gas, air, in the evaporator, closing this

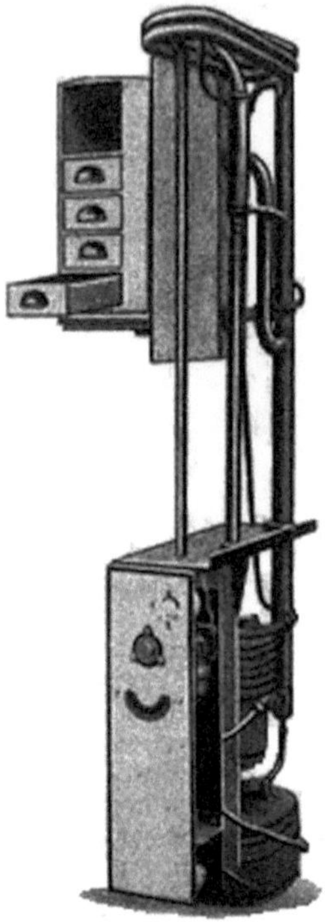

Fig. 8.23 Elements disposition in the Munters-von Platen absorption refrigerating machine (Fougeret 1929)

way the pressure gap between this equipment and the condenser and eliminating the pump. His research, carried out between 1899 and 1901, was based on the theoretically right idea that a pressure drop requiring throttle valves and pump, became unnecessary in a refrigerating machine if the liquid ammonia in the evaporator meets with a gas in whose presence the ammonia evaporated. This simple modification was not enough and Geppert's machine did not work. A new trial, creating a closed loop for the gas between the evaporator and absorber, led to a very low efficiency machine that yet did not solve the problem of circulating the gas. It was not until 1922 when two Swedish students of the Royal Swedish Institute of Technology, Baltzar Carl von Platen (1898–1984) and Carl Georg Munters (22 March 1897–1989) invented a three-fluid system that did not require a pump and designed a working model which dispensed with all moving and mechanical parts (Villers 1926). The third fluid was hydrogen and the first sample of the new workable and sealable refrigerating unit of this diffusion-absorption cycle was made 3 years later. The new concept avoided the use of the previously known toxic gases. The heat source that initiated the process could be fueled by electricity, gas or kerosene, making the system extremely flexible. The apparatus, with an air cooled condenser, was commercially introduced in 1929, and industrially manufactured from 1931 by the Electrolux Aktiebolaget in Europe and the Electrolux Servel Corporation in the United States, becoming in the latter case as the sole gas refrigerator in the market until the 1950s (Fig. 8.23) (Fougeret 1929). The second significant innovation in this class of machines was the replacement of the ammonia-water pair by other combination. Trials with lithium bromide as the absorbent and water as the refrigerant became successful and were industrially implemented after 1940.

8.6 Concluding Remarks

Writing a short essay on the history of refrigerating techniques, even when its emphasis is referred to a clearly-defined period of time, is a difficult task because of the complexity of the issue, the random succession of the events, the multiple countries and personalities involved, and the warp and the woof on which both, simple and elaborated designs, were continuously superimposed. The simple but also ingenious mechanisms that characterized almost all achievements in refrigeration along the nineteenth century, arose as a consequence of long, mainly empirical, developments of some explicit and implicit applications of physical principles, some of them in a still incipient evolution, with the only purpose to satisfy imminent needs of varied population in an always greater number of countries.

Manufacturing of ice, brewing, and transport of meat by sea, were undoubtedly the three motive forces behind the advances in refrigeration, but it was the first one the definitive only common element that paved the roads for developments on all scales of production. But, although it were the same physical principles which were usually applied for their developments, advances in industrial and domestic refrigerating machines included, as it has been explained above, distinct circumstances and showed subsequently different dynamics.

The idea of a domestic refrigerator, as it is known today, originated in the United States, and the interwar period was definitive for its consolidation. The increased reliability and the reduced costs of the great variety of small machines appearing in this period, transformed the domestic life of the American population after 1916, and of citizens of other countries after 1945 (Thevenot 1979). Statistical ciphers clearly showed that household refrigeration was no longer the privilege of rich families but became a daily necessity. The impressive growth of domestic mechanical refrigerators, from 5,000 units manufactured in 1921 to about 850,000 units 9 years later, and the simultaneous decrease in prices, from US 600 to US 250 dollars in the same period, are clear evidence of the social impact of the new technological development. The fact that it was in the same period that electric motor was confirmed as best choice for a prime mover was not a coincidence. The remarked change from a null percentage participation of electric motors in the different powers used in ice-making machines in Great Britain, for example, at the beginning of the century, to more than 50 % almost two decades later, reveals the existence of an obvious relation between these two subjects. Around 1930, 22 manufacturers competed for the American customer's preference with all class of commercial offerings (Fig. 8.24). This situation contrasted with what happened in European countries during the same dates. The above mentioned cipher of 850,000 refrigerators manufactured in the United States in 1930, differs markedly with the 30,000, 27,000 and 17,000

Fig. 8.24 Refrigerator advertisements during the 1920s and 1930s

units sold in Great Britain, Germany and Sweden, respectively in the same year. The comparing numbers of 49 % of American households enjoying domestic refrigerators in 1937, versus no more than 1 % in France 3 years later is conclusive (Thevenot 1979). Although industrial application developments in ice-making machines and refrigeration in general usually obscure those at domestic scale, there should not be any doubt that the historical evolution of this latter becomes one of the most striking events in the history of this subject.

References

Anonymous. 1844a. Masters' patent ice-making apparatus. *Mechanics Magazine* 40: 81–86.

Anonymous. 1844b. Master's patentirter eisbereitungs-apparat. *Polytechnisches Journal* 93: 206–208.

Anonymous. 1846. Goubaud's apparat zur bereitung von eis und gefrorenen sorbets. *Polytechnisches Journal* 102: 223–225.

Anonymous. 1849. Fumet's apparat zur eisbereitung für haushaltungen. *Polytechnisches Journal* 114: 302–303.

Anonymous. 1850. Application aux usages médicaux d'une nouvelle glacière portative. *Gazette Médicale* 5: 817.

Anonymous. 1862. Carré's apparate zur künstlichen eserzeugung. *Polytechnisches Journal* 163: 180–184.

Anonymous. 1867. La science apliquée a l'économie domestique. Le *Magasin Pittoresque* 35: 227–230, on 229.

Anonymous. 1869. Note. *Comptes Rendus Hebdomadaires des Séances de l'Académie des Sciences* 69: 421.

Anonymous. 1879a. Le chlorure de méthyle et la production du froid: Recherches de M. Vincent. *La Nature* 316: 43–46.

Anonymous. 1879b. Application industrielle du chlorure de méthyle a la production du froid. *La Nature* 330: 263–266.

Anonymous. 1882. Petite machine a glace: Système Raoul Pictet. *La Nature* 474: 80.

Anonymous. 1889. Family ice machines. *The Manufacturer and Builder* 21: 200.

Anonymous. 1890. The Dormigny domestic ice machine. *The Manufacturer and Builder* 22: 19.

Anonymous. 1908. La machine à glace "La Rayonnante". *La Nature* 1830(supplement): 19.

Anonymous. 1909. Machines frigorifiques nouvelles. *Revue de mécanique* 25: 179–195.

Anonymous. 1920. Construction d'une machine à glace domestique. *La Nature* 2406–2407 (supplement): 148–151.

Anonymous. 1924. Machine à glace "Frigoria". *La Nature* 2624(supplement): 19.

Anonymous. 1930. La fabrication de la glace par le chlorure de methyle. *Recherches et Inventions* 11: 223–225.

Audiffren, M. 1894. *Appareil frigorifique*. French patent 238,845, 31 May.

Audiffren, M. 1895. *Refrigerating apparatus*. United States patent 551,107, 10 December.

Audiffren, M. 1904. *Oil-separator for rotary refrigerating appliances*. United States patent 764,515, 5 July.

Audiffren, M., and A. Singrun. 1907. Le frigorigène Audiffren. *Comptes Rendus Hebdomadaires des Séances de l'Académie des Sciences* 145: 1268–1270.

Bacon, F. 1765. *The works of Francis Bacon*, vol. I. Sylva Silvarum: or a natural history in ten centuries; Century I, On induration of bodies, 158. London: Millar.

Barrier, M. 1891. Machines à produire le froid et la glace. In *Rapports du Jury International Exposition Universelle Internationale de 1889*; Groupe VI, Outillage et procédés des industries mécaniques (2e partie), classes 50 et 51, ed. Picard, A., 159–331, especially on 242–252. Paris: Imprimerie Nationale.

Beckmann, J. 1846. *A history of inventions, discoveries and origins*, 4th ed., 51–154. London: Henry G. Bohn. Adapted in: Anonymous. 1893. Refrigeration in olden times. *Ice and Refrigeration* 5: 100–102, 181–183.

Beddoes, T. 1787. An account of some new experiments on the production of artificial cold. *Philosophical Transactions of the Royal Society of London* 77: 282–287.

Boutigny, P.H. 1834. Note sur la préparation de la glace artificielle. *Journal de pharmacie et des sciences accessoires* 20: 698–699; *Bulletin général de thérapeutique* 7: 271–272.

Boyle, R. 1665–1666. An experimental history of cold. *Philosophical Transactions of the Royal Society of London* 1: 8–9.

Boyle, R. 1725. The mechanical origin of heat and cold. In *The philosophical works of the honorable Robert Boyle*, Esq., vol 1, ed. Shaw, P., 550–559. London: Innys, Osborn and Longman.

Brewer Frères. 1910. *Catalogue general: Machines à produire la glace*, 607. Paris: Brewer Frères.

Carré, F.P.E. 1860. Note sur un appareil propre à produire du froid. *Comptes Rendus Hebdomadaires des Séances de l'Académie des Sciences* 51: 1023–1027.

Carré, E. 1867. Sur de nouveaux générateurs de froid. *Comptes Rendus Hebdomadaires des Séances de l'Académie des Sciences* 64: 897–898; Ueber neue Kältegeneratoren. *Polytechnisches Journal* 185: 77–79.

Cavendish, H. 1783. Observations on Mr. Hutchins's experiments for determining de degree of cold at which quicksilver freezes. *Philosophical Transactions of the Royal Society of London* 73: 303–328.

Crispin, A. 1879. *An improved process and apparatus for manufacturing ice and for other refrigerating or cooling purposes* [being a communication from abroad by Camille Vincent], British patent 470, 6 February.

Cullen, W. 1756. Of the cold produced by evaporating fluids and of some other means of producing cold. In *Essays and observations, physical and literary*, 145–156. Read before a Society in Edinburgh and published by them, 2. Edinburgh: Hamilton and Balfour.

de Courdemanche, M. 1825. Mémoire sur la congélation artificielle de l'eau. *Journal de pharmacie et des sciences accessoires* 11: 584–592.

de la Hire, P. 1694. Explications des principaux effets de la *glace* et du froid. In *Memoires de mathematique et de physique*, 97–213 [113], on 97–203 [103]. Paris: Imprimerie Royale. Also published in *Mémoires de l'Académie Royale des Sciences*, depuis 1666 à 1699 (1730), pp. 475–500.

de La Porta, G. 1589. *Magiae Natyralis*, XIV book (Of cookery), chapter XI (Of diverse confections of wine). Available at http://homepages.tscnet.com/omard1/jportac14.html#bk14XI. Accessed Oct 2011 from an English translation published in London in 1658.

de Silvestre, A. 1846. Rapport sur un appareil à faire de la glace présenté par M. Goubaud. *Bulletin de la Société d'Encouragement pour l'Industrie Nationale* 45: 259, 401–403, 423–424, 459–460.

de Silvestre, A. 1849. Rapport sur un appareil congélateur à l'usage des familles. *Bulletin de la Société d'Encouragement pour l'Industrie Nationale* 48: 361–363.

Dusserre, Pagnerre, and Villeneuve. (1844). Appareil proper à faire de la glace, et à glacer toutes sortes de liquides et substances, dit congélateur – glacière des familles, French patent 16517, 5 June.

Ewing, J.A. 1908. *The mechanical production of cold*, 50. Cambridge: The University Press.

Fahrenheit, D.G. 1724. Experimenta & observationes de congelatione aquae in vacuo. *Philosophical Transactions of the Royal Society of London* 33: 78–84.

Figuier, L. 1873–1877. *Les merveilles de l'industrie ou description des principales industries modernes*: 3. Industries chimiques, 601–604. Paris: Furne, Jouvet et Cie.

Fontaine, G. 1891. *Catalogue des appareils et utensiles de laboratoire*, 3me. fascicule, 367–369. Paris: Maison G. Fontaine.

Forbes, R.J. 1966. *Studies in ancient technology*, Heat and heating, refrigeration, light, vol. VI, 109–111. Leiden: E.J. Brill.

Fougeret, H. 1929. Les machines frigorifiques du type domestique. *La Nature* 2809: 440–453.

Fumet, C.F. 1849. Pour une glacière artiicielle. French patent 4257, April 6; (1852). Appareil congélateur portatif. *Le Génie Industriel* 2: 41–43.

Gamgee, J. 1877. On artificial refrigeration. In *United States Fish Commission*. Rep. 5, 903–904.

Héraud, A. 1890. *Les secrets de l'alimentation à la ville et à la champagne*, 208. Paris: J.-B. Baillière et fils.

Jacomet, Ch. (ed.). 1904. Machines a chlorure de methyle. In *Revue Technique de l'Exposition Universelle de 1900*, sixth part, vol. 2, 104–119. Paris: E. Bernard

Lallié, N. 1908. Une nouvelle machine à glace: Le frigorigène Audiffren. *La Nature* 1828: 11–14.

Lami, E.O. 1885. *Dictionnaire encyclopédique et biographique de l'industrie et des arts industriels*, vol. 5, 481–482. Paris: Lami, Tharel et Cie.

Leslie, J. 1813. *A short account of experiments and instruments, depending on the relations of air to heat and moisture*, 176–178. London: Blackwood and Ballantyne.

Lhéritier, L. 1875. Appareil Carré pour frapper les carafes. *La Nature* 10849.

Lightfoot, Th.B. 1886. On refrigerating and ice-making machinery and appliances. *Proceedings of the Institution of Mechanical Engineers* 37: 201–260.

Malapert, P.P. 1836. Du mode opératoire pour obtenir la congélation artificielle de l'eau. Description de nouveaux appareils, et moyende transporter de la glace pendant les chaleurs. *Journal de pharmacie et des sciences accessoires* 5: 221–233.

Masters, T. 1844a. *An improved freezing, cooling, churning, and ice-preserving apparatus, the parts of which may be used separately or in combination*. British patent 9825, July 6; (1853). Improvements in apparatus for freezing, cooling, and churning, British patent 1089, 4 May.

Masters, T. 1844b. *The ice book*. London: Simpkin, Marshall and Co.

Masters, T. 1846. *Improvements in apparatus and means for cooling liquids and filtering and preventing liquids freezing*. British patent 11453, 17 November.

Masters, T. 1850. *A short treatise concerning some patent inventions and apparatus for the production of ice, and artificial cold, soda water, lemonade, nectar, and all aerated beverages*. London: Patent Office Library.

Masters, T. 1853. *Improvements in apparatus for freezing, cooling, and churning*. British patent 1089, 4 May.

Nairne, E. 1777. An account of some experiments made with an air-pump on Mr. Smeaton's principle; together with some experiments with a common air-pump. *Philosophical Transactions of the Royal Society of London* 67: 614–648.

Needham, J. 1976. *Science and civilization in China; Chemistry and chemical technology*, vol. 5, part 3, 226. Cambridge: Cambridge University Press.

Peclet, J.C.E. 1878. *Traité de la chaleur considérée dans ses applications*, vol. 3, 140–141. Paris: Mason.

Pfaundler, L. 1875. Sur l'inégale solubilité des diverses faces d'un même cristal et sur la relation existant entre cette propriété et certain principes généraux des sciences naturelles. *Bulletin de la Société Chimique de Paris* 23: 491–493.

Pouillet, C.S.M., L.B. Francœur, and J. Babinet. 1845. Rapport sur un appareil à faire de la glace, présenté par M. Villeneuve. *Comptes Rendus Hebdomadaires des Séances de l'Académie des Sciences* 20: 1706–1708.

Redenbacher, W. (n.d.). *Untersuchung der klein kühlmaschine "Autofrigor"*. http://www.eis-onki.de/pageID_7947164_1.html. Accessed 9 Dec 2011.

Reif-Acherman, S. 2009. The pre-history of cryoscopy: what was done before Raoult. *Quimica Nova* 32: 1677–1684.

Rudorff, F. 1873. Ueber die löslichkeit von salzgemischen. *Annalen der Physik* 148(456–475): 555–579.

Serbar, F. 1914. Une nouvelle machine domestique a glace. *La Nature* 2134: 348–350.

Siemens, C.W. 1855. *Improvements in cooling and in freezing water and other bodies*. British patent 1105, 16 May.

Siemens, C.W 1857. *Improvements in refrigerating and producing ice, and in apparatus or machinery for that purpose*. British patent 2064, 29 July.

Siemens, C.W. 1858. *Improvements in refrigerators, and in the treatment of the freezing or cooling materials used therewith*. British patent 2074, 13 September.

Silvestre, E. 1845. Exposé des motifs d'une proposition de prix pour la construction des glacières domestiques et la fabrication de la glace. *Bulletin de la Société d'Encouragement pour l'Industrie Nationale* 44: 309–310.

Stokes, D.H. 1920. *Refrigerating apparatus*. United States patent 1,362,757, 20 December.

Tancredi, L. 1607. *De fame et siti. Somasco*, Venice, Book I, chapter VII.

Thevenot, R. 1979. *A history of refrigeration throughout the world*, 24, 73–74, 172–174. Paris: International Institute of Refrigeration.

Toselli, J.B. 1866. Fabrication artificielle de la glace: glacière à vapeur d'eau. *Le Génie Industriel* 32: 11–13 (1968).

Toselli, J.B. 1868a. *Genre de glacière spécialement applicable à la production d'une petite quantité de glace*. Italian patent 79,895, 5 March; Apparatus for freezing. United States patent 78,159, 19 May.

Toselli, J.B. 1868b. *Mode de fabrication industrielle de la glace*. French patent 81,356, 15 June.

Toselli, J.B. 1873. *Método de fabricación industrial del hielo*. Spanish patent 4558, 17 September.

Toselli, J.B. 1876. *Appareil pour produire le froid artificial, applicable à la fabrication de la glace ou des glaces, ainsi qu'au refrodissment et à la conservation des liquids*. French patent 115,973, 11 December.

Vallance, J. 1824. Improved method or methods for freezing water by the rapid passage of rarified air. British patent 4884, January 1st; Abstracting or carrying off the caloric of fluidity from congealing water; producing intense cold; -applicable to other purposes. British patent 5001, August 28. *London Journal of Arts and Sciences* 8: 251–253.

Villafranca, B. 1553. *Methodus refrigerandi ex vocato salenitro vinum aquamque, ac potus quodvis aliud genus, cui accedunt varia naturalium rerum problemata, non minus jucunda lectu, quam necessaria cognitu.* Rome.

Villers, R. 1926. Une nouvelle machine frigorifique a absorption: La machine "Munten-Platers". *La Nature* 2730: 75–77.

von Lippmann, E.O. 1898. Zur geschichte der kältemischungen. *Zeitschrift für angewandte Chemie* 11: 739–745.

Walker, R. 1788. Experiments on the production of artificial cold. *Philosophical Transactions of the Royal Society of London* 78: 395–402.

Walker, R. 1789. Experiments on the congelation of quicksilver in England. *Philosophical Transactions of the Royal Society of London* 79: 199–211.

Walker, R. 1795. Observations on the best methods of producing artificial cold. *Philosophical Transactions of the Royal Society of London* 85: 270–289.

Walker, R. 1801. On the production of artificial cold by means of muriate of lime. *Philosophical Transactions of the Royal Society of London* 91: 120–138.

Walker, R. 1828. On the artificial production of cold. *Philosophical Magazine* 3: 401–406.

Wollaston, W.H. 1813. On a method of freezing at a distance. *Philosophical Transactions of the Royal Society of London* 103: 71–74.

Chapter 9
Carl Linde and His Relationship with Georges Claude: The Cooperation Between Two Independent Inventors in Cryogenics and Its Side Effects

Hans-Liudger Dienel

9.1 Introduction

Carl von Linde (1842–1934) and the 28-year-younger Georges Claude (1870–1960) were the two most important applied scientist-inventors and entrepreneurs in cryogenics of the twentieth century. Both can be seen as "independent inventors", as described by the eminent American technology historian Thomas Hughes. However, both would have preferred to be seen as scientists. Georges Claude, although he was referred to as the "French Edison" in the 1920s, sneered about the most famous of all independent inventors, Thomas Edison. He laughed about American industrialists, who handed large sums over to the inventor.

While Carl von Linde (born as Carl Linde, ennobled in 1897) is still honoured and praised as being one of the most important German engineers of the last two centuries, Georges Claude remains discredited in France. This is due to his strong support for Marshall Philippe Pétain and the Vichy regime during the German occupation and his antiparliamentary political convictions. He is no longer mentioned in the list of the Académie française (he was elected as member in 1924) and his important military decorations of WWI were confiscated. A staunch nationalist since 1900 and member of the Action Française in the 1930s, Claude notoriously endorsed Hitler's Germany during and after the invasion of France. Following the liberation of France in 1945, he was sentenced to life-long imprisonment. He was released in 1950 and died in 1960.

Claude and Linde not only accepted one another's scientific contributions. They liked and trusted one another as entrepreneurs. Following a fierce patent case in Britain in 1908, they decided to cooperate, rather than compete. They divided up the rapidly growing market for technical gases between their two companies, the Linde Company (Gesellschaft für Lindes Eismaschinen) and Air Liquide (Linde 1916;

H.-L. Dienel (✉)
Center for Technology and Society, Technische Universitat Berlin, Berlin, Germany
e-mail: dienel@ztg.tu-berlin.de

K. Gavroglu (ed.), *History of Artificial Cold, Scientific, Technological and Cultural Issues*, Boston Studies in the Philosophy and History of Science 299,
DOI 10.1007/978-94-007-7199-4_9, © Springer Science+Business Media Dordrecht 2014

Claude 1957). This close cooperation lasted for decades. Claude continued to cooperate closely, as part of a two-company syndicate, with Linde's sons during the 1920s, 1930s and early 1940s. Furthermore – what I am going to argue – Claude was absolutely convinced that France had to collaborate with Nazi Germany, not least because of his positive experiences with the reliable, long term cooperation with Linde.

Throughout this article, I will describe the relationship between and the different attitudes of Linde and Claude as scientists, inventors and entrepreneurs.[1] We begin with a look into mutual Franco-German perceptions. Then, we try to understand how Linde and Claude developed the new industry of cryogenics and their own companies, and why and how they collaborated. In the next chapter, I will highlight the political involvement of Claude and his exclusion from the French heritage after 1945. There is, aside from a remarkable recent niche publication (Baillot 2010),[2] nearly no historical scholarship on Claude. In the last chapter, I will summarise the friendship of and collaboration between Linde and Claude.

9.2 Mutual Franco-German Perceptions in Science and Industry

The relationship and collaboration of Linde and Claude in cryogenics has been embedded in the mutual perceptions of German and French science, technology and industry since 1870–1971 (Dienel 1999).

Not only science, but technology and economics as well, have always retained a notion of internationalism. However, this is due to very different reasons. Scientists and engineers have always been attracted to the notion that science is a body of absolute, universal truth, which owes nothing to personal prejudice or national identity. We have, however, many examples of the vulnerability of this internationalism of science. These examples prove that science and technology, in the era of nationalism, have been arenas for national competition and rivalry. Scientists have used national rivalries of the nineteenth and twentieth centuries to argue in favour of increasing support for and budgets of science and technology. There is no doubt that national governments have had national aims when supporting the sciences. Science and technology were and are dependent on governmental – and, thus, national – support. Still, even if scientific thought, in reality, cannot be completely de-contextualised, the ideal of international science is a persistent and potent one.

[1] The files of Georges Claude remain in the family. The recent biography on Claude made extensive use of these files. The files of Carl Linde can be found in the Linde Company archives in Munich. This article is based on my earlier work on the history of Linde and the Linde Company. Thus, I am more familiar with Linde's experimental work than with Claude's. For archival details on Linde, see Dienel (1995) and Dienel (2004).

[2] This book was written by a former head of communication of Air Liquide. There was no support of the company, which still has not found a way to integrate its founder's biography into its own history.

Economy and industry are also internationally biased, because the potential market is international. Krupp had to be interested in the French market demand for his canons, even after 1871. Therefore, Krupp remained active in France. However, there were strong links to the national system. There are many reasons for a close connection between country and company in a nationalistic world.[3]

Following the defeat of France in the Franco-German War of 1870–1971, the dominance of German science and industry was discussed openly in France (Fox 1990). Many French scientists emphasised the necessity to adopt a German style, at least in the funding of science and technology. The French picture of Germany described after 1871 was quite different from the Germany of romanticism and philosophical speculation, which had fascinated foreign observers during the early years of the century. Claude Digeion has sensitively approached the changing image of German intellectual life in France during the nineteenth century (Digeon 1959). According to most French scientists, the new Germany was distinguished for its rigour, which manifested itself in experimental science. At the same time, French scientists continued to view German science as uninspired, although efficient. In doing so, a consistent French perspective on Germany, and the Germans as such, could be maintained. The image of German education as being pedantic and inelegant and that of German science as being the handmaiden of industry characterised this perspective. Although the ideal of independent science was also present in Germany, the alliance of science and industry was considered by the French to be a 'German style'. The industrialisation of inventions, especially in the chemical dye industry, reinforced this French vision of German science. Once again, the atmosphere changed before the end of the century. French scientists began describing the productive connections in science, education, and industry in Germany. They did so to foster a vision that would serve their needs in a domestic debate. The German competition helped to reform the French system. In fact, the numerous institutes of applied sciences, founded in France after 1880, had a decidedly utilitarian orientation. These were as much application-oriented as any such institute in Germany.

French science continued to invest heavily in the world of congresses and committees. The German competitors, on the other hand, invested in the more costly, but more serious laboratories. The atmosphere of the scientific congresses changed in the last decade of the nineteenth century. It became less and less possible to control science via declarations. Robert Fox illustrates this change by describing the weak French position during the inauguration of the International Standing Committee on Electrical Standards in London in 1909. Germany, like the USA and Great Britain, had a major national laboratory. In the Physikalisch-Technische Reichsanstalt (PTR, now Physikalisch Technische Bundesanstalt PTB), the research required for tasks such as the standardisation of measures of resistance and current could be conducted. France could offer only the comparatively

[3] The structures and challenges of bi-national cooperation of technology companies have been described nicely by Hans Braun (1991).

ill-equipped Laboratoire Central d'Electricité. This was a sharp contrast to the PTR and the seriousness embodied by it.

We can say that, between 1871 and 1945, France and Germany were constantly competing in technological and scientific research and development. The necessary standardisation process was used as a weapon in this excited debate among hostile scientists and science managers of both countries. For some years following WWI, Germany was excluded from the international community and German scientists excluded from international scientific congresses. Ninety percent (90 %) of the international congresses, which occurred in the former Allied countries and their associates, did not accept the attendance of Germans during the early 1920s. However, this dropped to 50 % by 1925 and to 20 % in 1926 (Grundmann 1965: 806). Various political approaches – the Locarno Treaties and the admission of Germany into the League of Nations in 1926 – led the way. Still, the hostile tensions between French and German scientists continued. The investigations of Elisabeth Crawford in the Nobel Archives show that, during the interwar period, French and German scientists almost never nominated a possible candidate from the other country for the prize (Crawford 1988). Dominique Pestre was able to show that not only the scientific establishments, but young scholars as well, did not travel and work on the opposite side of the Rhine (Pestre 1984: 163). In 1930, there were 261 international scholars studying in German Kaiser-Wilhelm-Institutes. Only six of them were French. Furthermore, only one of the six was working in a hard science institute, while the remaining five studied law (Schroeder-Gudehus 1990: 111). There were only very few exceptions of eminent scholars trying to intensify cooperation with their French colleagues. Albert Einstein and Fritz Haber (1868–1934) are two examples. When the First World War broke out, Haber was appointed as a consultant to the German War Office. He organised the gas production and defences for the gas war. After the war, Haber made great efforts to re-establish the relationship between German science and that of other countries, especially France. In 1927, he travelled to Paris to give a speech, in French, for the occasion of the 100th birthday of the French chemist Marcellin Berthelot. After WW II, German-French contact reached a new level of intensity. Dominique Pestre describes this change as political détente, but also as an internal change within the sciences. This was a shift to a new type of pragmatic 'engineer-scientist', in both Germany and France, who adopted the behaviour of both entrepreneur and engineer. The more pragmatic approach enabled Linde and Claude to cooperate despite the hostile conditions.

Science, technology and industry are international and internationally-oriented. Yet, they are an aspect of national culture, prejudice and antagonism, and are dependent on the nation state. In most cases, science is more dependent on the state than is industry. This has changed only recently in Europe with the increase of European science funding. A valuable aspect of these funds is that they enable scientists to practice their internationalism.

The cooperation among companies of competing nations, such as France and Germany after 1871, was possible and realised. There were many economic reasons to cooperate. Furthermore, personal affiliation and friendship could emphasise and support these common interests.

9.3 Linde and Claude Enter Cryogenics

Cryogenics emerged as a high-tech industry after 1900. Based on the process of air liquefaction and separation, the first main products were oxygen, hydrogen, and inert gases. A growing number of technical gases followed soon after. The two leadings companies in the field were the Linde Company and Air Liquide. The Gesellschaft für Lindes Eismaschinen, founded by Carl von Linde in 1879, developed the cryogenics department after 1895/1902 and became the world market leader until the end of WWI. At this point, Linde Air Products, a former subsidiary, which was purchased by Union Carbide in 1917, took over. Second was Air Liquide, founded by Georges Claude and Paul Delorme. Third was British Oxygen, which was also co-founded by Linde before WWI, but operated independently of Linde after 1914. Between the 1970s and 2004, Air Liquide was the leader in the world of technical gases (Dienel 1995: 278–288). The Linde Company continued to be a technology leader in many fields. Since purchasing Swedish AGA and British Oxygen, the Linde Company has returned once again to the leading position.

The Linde Company, during its first 20 years of existence, did not produce any refrigeration machines itself. Rather, business was organised as a modern licencing system. Respected machine companies in Europe and the USA built all refrigeration machines and equipment on licence. However, contact to the customers remained in the hands of the Linde Company. This system enabled Linde to quickly expand internationally and become the world market leader in the refrigeration sector within only a few years and without significant investments. In the gas business, Linde and Air Liquide started production immediately. Additionally, the development of a worldwide gas production network of gas companies, cylinders and containers – a costly project – was begun.

Linde stepped down as CEO of the Linde Company in 1890, but continued to be an influential chairman of the board until 1930. He stepped down in order to go back into research and development. In 1890, he moved from Wiesbaden, which housed the headquarters of the Linde Company, to his summer home in Berchtesgaden. In 1892, he moved into the company-owned laboratory for refrigeration experiments in Munich. This then became the institutional research centre of the company. The successful development of air liquefaction techniques by Linde and his son, Fritz, in 1895, led to the establishment of an independent cryogenics department of the Linde Company. Linde arrived in Munich not as a retiree, but as a man with scientific and entrepreneurial ambitions. Both the experimental station itself and Linde's negotiations on the building of an aircraft factory demonstrate his desire to achieve a new technological breakthrough. The fact that he ended up with gas liquefaction, and fared so successfully in doing so, is largely due to lucky coincidences. The product could have been different. Gas liquefaction was situated in close proximity to earlier endeavours and, therefore, rendered itself an area to be developed. Even today, it allows the company to continue on as a development-oriented engineering company. Linde's son, Fritz, finished his doctoral thesis in 1895, before going into the experiment laboratory. It was on the electric resistance of liquefied gases, a topic that he and his father had surely agreed upon (Linde 1897).

Linde's experimental work on carbon dioxide refrigeration machines led him to gas liquefaction. In 1892, the Guinness brewery in Dublin ordered a carbon dioxide liquefaction installation. Linde accepted the contract, although the company had yet to construct such installations. This unleashed his interest in industrial gas liquefaction. In the following year, he held a lecture on the subject, in which he discussed the technical liquefaction of oxygen (Linde 1893). Linde had ordered a high-pressure compressor from the Maschinenfabrik Augsburg for the carbon dioxide machines. In the summer of 1894, he contemplated constructing an air liquefaction installation. In his letters, he called it an "oxygen machine", indicating the industrial use he had in mind.

Linde's two most important technical ideas were the continuous cooling in a counter-current and the use of internal work for cooling (Linde 1916: 83–89; Hausen 1935: 220–225). In contrast to the usual compression refrigeration process, Linde's idea was to cool gas through an expansion pass in a counter-current generator. The resulting circular process stabilised temperature levels on both sides of the counter-current generator: a warm and a cold side. On the warm side (i.e. the surrounding temperature), cooling occurs from the outside. Therefore, a continuous drop in temperature on the cold side becomes possible (Linde 1916: 84).[4] Linde's first idea was to achieve the drop in air temperature through the reduction of pressure by way of external work. His idea was to use a Ram cylinder similar to those used in air refrigeration machines. This would have faced major material problems in factory, since lubrication in a cylinder at −190 degrees is very difficult. Georges Claude ultimately pursued this idea. It was more efficient, but was also more complicated. Finally, in the 1930s, Linde built the first gas liquefier with external pressure release in turbines. However, before the work cylinder ordered from Sulzer arrived in Munich, Linde discovered that "alongside the equivalent of the external work, significant internal work can be applied in the refrigeration process, as soon as the temperature at the lower level of the counter-current generator approaches the lower critical temperature". This discovery of Linde's was, essentially, a development of an effect already described by the English physicists Joule and Thomson in 1857. The reduction of pressure in real gases (i.e. as opposed to the ideal gases used in thermodynamics) could result in a cooling process, even in the absence of external work below a particular temperature (i.e. the inversion temperature). The Joule-Thomson effect had become a standard in physics lectures. Aside from William Siemens in 1857 and Solvay, however, no one had yet considered its practical exploitation. The lack of practical use of the Joule-Thomson effect in gas liquefaction was presumably due to the fact that the effect was a minor one. The surrounding air temperature barely dropped as much as one quarter of one degree centigrade per atmospheric

[4] The regenerative method had a long theoretical tradition in refrigeration technology. It was not applied. Regenerativkaltluftmachine, *VDI-Zeitschrift* 2 (1858), 287.

company towards what he wanted it to be, a technological leader in a potentially unlimited field of technological development.

Georges Claude was born into a bourgeois Parisian family. He admired his father, an entrepreneur and engineer, who did not send his son to school, but educated him privately. In Claude's family, the democratic debates of the Third Republic were cursed with disrespect. Claude studied at the École de Physique et Chimie in Paris and graduated in 1886. He produced hundreds of inventions and patents. He started to work as an electrical engineer for the utilities service in Paris. After having heard of the liquefaction of air in 1898, he immediately began conducting his own experiments and writing articles and books on the topic. He immediately identified the great market potential. After successfully separating liquid air into oxygen and nitrogen in 1902, he founded Air Liquide, together with his business partner, Paul Delorme. In addition to air liquefaction and separation, he applied his knowledge of chemistry and began experimenting with noble gases. His aim was to find practical uses for them. Claude discovered that neon was sensitive to electricity and would glow if enough voltage were delivered through it. He worked to put neon in glass tubes with sufficient purity, allowing electrodes at each end to use electricity to illuminate the neon. After a successful display at the Paris Art Show in 1910, Claude conceived the idea of bending glass tubing in order to form letters. In doing so, he developed the first neon sign. In his high times, Claude was called 'the Edison of France'. He was a public figure and a respected inventor and entrepreneur in France. In 1922, the first neon signs were brought to Los Angeles from Paris, to advertise for the Packard Agency. During the 1920s, vivid neon signs became a trend in advertising. After his neon sign businesses had become profitable, the journal *L'espoir Française* published a special issue, within the series *Les élites françaises*, on Georges Claude. This was intended to educate the French youth. He wrote several technical books, which were translated into German and English. He continued to surprise the public with breathtaking and groundbreaking inventions and suggestions, such as the tidal power plant (Claude 1926).

9.4 Linde and Claude as Independent Inventors

According to Thomas Hughes, the era of independent inventors passed when large industrial systems were established. Most important was indeed their independence. The fact "that they were not limited by any organisation allowed them to devote themselves to problems, which once solved, became the core of new technological systems" (Hughes 1989: 30–33 and 62–64). Hughes claims that it was, however, precisely this radical creation of new industrial systems, which essentially left them high and dry. I do not agree to Hughes in his statement that their relevance in industry diminished after World War I, but I follow his description of the typical inventor's approach to the pursuit of an invention as being similar to that of a scientist seeking a discovery, this being characterised by the collection of scientific proposals and knowledge through outside consultation (Hughes 1989: 42, 73).[6] In general, inventors instructed mechanics to transform ideas and sketches into technical and electrical models. In most cases, inventors dreamed of living like monks, each with a small staff of disciples, in a well-equipped private laboratory work could be done in peace. The focus was on invention, not on conducting research. Some famous examples of inventors are the American Thomas Alva Edison and the cartoon figure of Gyro Gearloose, who was never at a loss of new technical solutions to various problems confronting the world of Donald Duck. Hughes describes the image of the inventor as that of a pragmatic being, regularly evaluating the competition's patents and purposefully seeking the weaknesses of contemporary technology. In the days before public research funding, it was an inventor's spectacular imagination and slick self-representation, which provided him the necessary support and attention to continue his work. According to Hughes, such independent inventors were common in the USA, especially. As a German example, he mentions Rudolf Diesel.

A number of the characteristics named by Thomas Hughes as being typical of the independent inventor can be found again in Carl Linde and Georges Claude. As a pioneer of machine technology, Linde was clearly focused on experimentation. He used theory as his guiding principle. Linde developed original methods of invention, surrounded himself with teams of mechanics and assistants, and became a partner in companies that benefited from his patents. The Munich laboratory served only occasionally as his primary workspace. More often, the primary workplace was the client's assembly site. His success as an inventor was a direct result of the close cooperation, which he cultivated with his clients. This is one important difference to the shy inventor, Linde, as portrayed by Thomas Hughes. Moreover, Linde's identity was manifold. He was not only an inventor, but also a scientist and an entrepreneur. His success can largely be contributed to the fact that he could distinguish among his scientific, technical and entrepreneurial ventures. This allowed him to move back and forth, walking a fine line between science, invention and business management. Finally, above all else, he was a brilliant entrepreneur.

Georges Claude also combined scientific interests with entrepreneurial spirit. He was, however, much more involved in public discussions than Linde was. He saw himself as a scientist, an entrepreneur and an independent, technical intellectual. Soon after Claude discovered the potentials of air liquefaction in 1898, he started to write and publish two books on the topic (Claude 1903). Claude not only acted, but also saw himself, in the French tradition of the more elegant, radical and inspired inventor. At the same time, he admired Linde for his German strength, rigor, experimental skills, continuity, steadiness and stability. He viewed Linde as an optimal German engineer. Claude was aware that he himself could rely on the French tradition, which would enable him to write, organise and shape scientific

[6] This was similar to the working relationships between Elmar Sperry and Prof. William Anthony (Cornell), Rudolf Diesel and Carl Linde, Charles Holl and Prof. Frank F. Jewett (Oberlin College), and Nicolas Tesla and a professor at the TH Graz. Edison employed 'a number of chemists who had received their doctorate degrees in Germany'.

discussion. He became prominent through his inventions and his writing. Two of his books on air liquefaction and technical gases were translated into German and English. By 1940, he had sent some 200 letters to the French Academy, informing it about his latest inventions. Linde, on the contrary, was not interested in writing and publishing. Rather, he tried to prevent a knowledge drain from occurring. His brightest employees, such as Helmuth Hausen, were not allowed to publish on the most important findings of multi-level rectification. Hausen began publishing extensively on the topic only after he became professor for thermodynamics (Hausen and Linde 1985; Kistenmacher 1987).[7] Linde was aware of the necessity of marketing and secured cleaned company publications, e.g. through the Institute International du Froid (Gesellschaft für Lindes Eismaschinen 1913).

Another difference between the two inventor-scientists was their approach to teamwork. Linde was clearly 'number one' in his company. His first employees were former students of his, such as Rudolf Diesel. Later, his sons, Fritz and Richard, and his son-in-law, Rudolf Wucherer, took over. Claude, from the start, had a close friend and business partner in Paul Delorme. Claude was the inventor, while Delorme acted as manager and, later, CEO of Air Liquide. Claude could only afford his radical lifestyle as an independent engineer and conservative public intellectual because he had such a reliable management partner.

9.5 Venture Capital of the Two Companies

On their way to becoming global leaders in the field of technical gases (i.e. cryogenics), both Linde and Claude needed financial investments in order to develop their companies. Georges Claude stemmed from a wealthier family than Linde, who was the son of a Lutheran minister, making him an ideal case for Max Webers discovery of the importance of protestant ethics for the spirit of capitalism (Weber 1934, orig. 1905). In 1879, when Linde, a young professor at the time, founded the Gesellschaft für Lindes Eismaschinen, he had no money at all. He was only able to contribute his patents on the ammonia refrigeration compression machine. The money had to come from his business partners, who were owners and directors of machine companies and breweries. Another important investor was the Jewish Franco-German railway entrepreneur Baron Moritz von Hirsch (Maurice de Hirsch (1831–1896)). He had become a millionaire with great railway projects in south-eastern Europe and Turkey. This success earned him the nickname of 'Türken-Hirsch' in Austria and Germany. Hirsch invested 100,000 Marks for 50 % of the shares of Linde's company. He was interested only in revenues and was the only shareholder without a seat in the board of the company. Without his investment, the company would not have been founded. After 2 years, the company started to become very profitable.

[7] An exception is Hausen 1937, because he had to counter American publications.

Annual revenues amounted to more than 100 %. Hirsch sold his shares to the other shareholders, thereby acting as an ideal early venture capitalist.[8]

When Linde developed air liquefaction and separation in 1895, he was already the head of a global technology company. Georges Claude, who was 28 years younger, had just started his first business. Claude began with money from his family, his business partner, Paul Delorme, and of 22 other friends. One of them, d'Arsonval, was a member of the Academie Française, and four others were engineers. Engineers held 85 % of the shares in 1902. This figure dropped to less than 20 % by 1913, as the number of shareholders increased from 24 to 1380 once Air Liquide was traded in the Paris stock market. We can identify two investors who signed 30 % of the new shares in 1903–1905, Babin and Lefebvre-Dibon. Air Liquide was, as was the Linde Company, an incorporated stock company from the beginning. The first boom phase for Air Liquide, between 1904 and 1913, was financed by external venture capital (Petit-Konczyk 2003: 25–58). In 1913, Claude and Delorme owned only 7 % of the company (Emptoz 2000: 677–692). In the case of Air Liquide, there was no single important venture capitalist (i.e. such as Maurice de Hirsch for the Linde Company).

Both the Linde Company and Air Liquide are good examples of the successes of early venture capitalists. Money was invested by individuals, such as Moritz von Hirsch, who were able to recognise the potential of the companies.

9.6 Political Orientations of Georges Claude and Carl von Linde

Claude saw himself not only as an inventor and entrepreneur, but also as a public intellectual and a technocratic political advisor. In 1904, he became politically active, publishing societal and political visions and concepts. In this regard, he is more similar to Rudolf Diesel (1858–1913) – Linde's most famous student and employee – than to Linde himself. Rudolf Diesel, who had a French mother and took many of his notes in French, was the best-paid engineer in Linde's Company between 1880 and 1893, 10 years of which he spent in Paris. In 1903, he published 10,000 copies of his book, "*Solidarism*". In this book, he proposed a cooperative societal structure to defeat poverty and to guarantee a good life for all members of society (Diesel 1903). Diesel was a brilliant communicator and used his abilities to build networks. This helped him, among other things, to secure and develop his Diesel engine (Dienel 2000). Diesel, however, had a cooperative and somewhat left-ist societal vision. Claude, in the antidemocratic tradition of his family, was an opponent of some core elements of democracy, such as the right to strike. He opposed Jean Jaurès and the socialists, and supported, in March 1914, a right-wing

[8]Towards the end of his life, Hirsch used his earnings to support early Jewish settlements in Palestine.

candidate for parliament. During the First World War, Claude was most frustrated about the weak preparation and unprofessional management of the French. He continued to admire the technical orientation of the Germans, which he could not recognise in French politics. He invented bombs to destroy barbed wire and was celebrated for his inventions and brave applications on the front. He was awarded, from Marshall Pétain, the *Croix des Guerre avec palme et la Lègion d'Honneur* in 1916. From then on, he was a faithful supporter of Pétain. After the war, in 1919, he published "Politiciens et Polytechniciens", with which he argued that France was in the hands of mere talkers who lacked strategic competence. In doing so, he was speaking for a strong French minority, which had come to similar conclusions. It is important to remember that the Third Republic was, at the time, far from being widely accepted and secured. From 1871 to 1940, the government struggled with 99 different cabinets, each having an average mandate of less than 8 months.

Claude was convinced that France had moved in the wrong direction. At the same time, he suffered from the continuing public rejection that he and his technocratic visions received. He saw himself as a martyr for technological progress. Furthermore, we must acknowledge that he saw himself as an important scientist and inventor. Thus, he was disappointed when the National Assembly of the French Parliament decided to use the German Haber process in large scale, rather than a process for ammonia synthesis, which he himself had developed.

Perhaps, this continued craving for recognition prevented Claude from acquiring a status as undisputed as that of Linde. Linde, in Germany, was member of several scientific academies, was ennobled, received honorary doctorate degrees and was president of various engineering associations. He was a shy and modest man, who gained trust and sympathy on account of his self-effacement and unpretentious diligence. Claude, on the other hand, was better-known by the public, because he was the more controversial figure. In 1924, he was finally appointed to be a member of the French Academy of Science, following three futile attempts. This was a big step in achieving public recognition. In 1928, however, he failed in an attempt to gain a seat in the National Assembly for the Union Nationale. He was not elected. From then on, he supported the ultra-conservative Action Française, which was founded in 1898 as a royalist and antidemocratic movement of the extreme right wing. He wrote a sarcastic book about his defeat in the 1928 election (Claude 1931). and iIn the early 1930s, he founded an independent movement called *Ordre et bon sens*. In the following years, his political publications and activities increased, coming to a boom after the occupation in 1940 (Claude 1938). He almost desperately toured France in an attempt to convince his fellow citizens to collaborate with Nazi Germany. He was convinced that this was the only chance for France to survive (Claude 1941, 1942, 1943, 1944). He gave many talks, even though he was not good at giving speeches.

Georges Claude's final step towards the Action Française, according to Rèmi Baillot, resulted from the lost election of 1928. Claude could no longer trust a democracy that handed power to irresponsible persons and demagogues. In 1940, Claude was 69 years old. From the beginning stages of the Vichy regime, he supported his old hero, Pétain. Further, he partly ran the propaganda division. Even after 1943, when the prospect of German defeat was becoming increasingly definite, he isolated himself and remained convinced that France had a future only as an ally of Nazi Germany. He went nearly deaf during those years. After liberation, he was jailed and convicted to a lifetime in prison (Claude 1946), but was released in 1950. During imprisonment, he was hailed as a martyr by some right-wing French.

The personal convictions of Georges Claude were obviously shaped by minor experiences. He felt the disapproval of the French democrats and science system. Simultaneously, he felt acknowledged and accepted by Marshall Pétain and the Vichy regime. The firm admiration of Pétain has a partly psychological explanation. The reliable partnership with the Lindes is a largely overlooked factor explaining his attitude towards collaboration.

Carl Linde considered himself to be politically indifferent, in the German tradition of the 'apolitical' man, as described by Thomas Mann in 1918 (Mann 1918). At the same time, he was prepared to act as president of the German Engineering Association in 1904, as co-director of the famous Deutsches Museum in Munich (the largest museum for science and technology worldwide at the time) and as member of a half dozen important associations for the funding and advancement of science and engineering. In 1904, he co-founded the first industry-funded institute, the Institute for Technical Physics, of his university, the Technische Hochschule Munich. The Institute for Technical Physics became one of the leading institutes for cryogenics and thermodynamics in the years to come. In the interwar period, the Institute for Technical Physics was larger than the German Physical Association. The two associations merged after 1945. Linde was an active member and supporter of the protestant church, which was a minority denomination in catholic Bavaria. In summary, Linde limited his political activities to the support of science and technology, such as aerial research. He subscribed to the Nationale Zeitung, a conservative-liberal daily newspaper. In 1933, when the university sent him the new forms for 'Ariernachweis' (EN: Aryan certificate), which were obligatory for higher civil servants, he refused to fill them out. This was tolerated, because of his age. He might have been the only professor at the TH Munich who did not fill out the form. Although he was 91 years old, he was still active as vice-president of the Linde Company board.

9.7 The Collaboration Between Linde and Claude

Carl von Linde, who was university professor at the Technische Hochschule Munich from 1868 to 1879 and from 1890 to 1934, considered himself to be a scientist more so than an entrepreneur. Claude and Linde respected one another as brilliant engineers and scientists. After the development of the basic inventions (e.g. air liquefaction in 1895, air separation through rectification in 1902, oxygen welding with the Fouché process in 1904), the two companies were entangled in a patent battle in different parts of the world. Linde and Claude both had worldwide ambitions in the marketing of oxygen and hydrogen. After 1906, both founded subsidiaries in different European countries and in the USA. However, in 1908, the two sides decided to collaborate internationally and divide the world market. Linde had tried previously to collaborate with other competitors in the fields. However, he soon decided only to collaborate only with the No. 2 company, which was Claude's, in order to defend the market from further competitors. Thus, the collaboration between Linde and Claude became important. After Linde won the patent battle against Claude in Great Britain, he used his good position and proposed a general agreement to Claude. In October of 1908, Claude visited the Linde works in Munich. Linde – who could speak French much better than he could English – and Claude became friends. When Linde published his autobiography in 1916, he wrote that Claude was the only innovative force in the field aside from the Linde works and declared this to be the reason for cooperation.

The two leading companies recognised a realistic chance of preventing the emergence of competitors internationally. In the USA and Great Britain, Linde controlled the market with his subsidiaries, Linde Air Products and British Oxygen Co. In November of 1909, Linde had to overcome resistance of the local executive committee of British Oxygen, in order to offer a share of the market to Air Liquide. This amounted to 33 % to begin and increased to 50 % over the next 2 years. In the USA, the situation was even more difficult. Linde's American partners resisted the integration of Air Liquide. Air Liquide became a partner only through the Internationale Sauerstoffgesellschaft, which was a Linde subsidiary co-owned by Air Liquide.[9] Additionally, Claude was an indirect partner, as patent holder, of Fouché's welding apparatus. The welding industry, until the 1930s, was the biggest consumer of oxygen. Thereafter and until today, steel production took over the lead on the demand side. In 1917, Linde Air Products was purchased by Union Carbide. Several months earlier, Linde had managed to beat sequestration by selling his shares to a New York bank. The shares were presumably shipped by the German submarine *Deutschland* through the British blockade and to New York.

By 1908, Claude and Linde had agreed to divide the market in Austria, Sweden, Belgium, Portugal, Greece, Turkey and Egypt, with each receiving 50 %. In 1909, an agreement on the markets of Switzerland, Russia, and Italy followed. On July 23rd, a week before the outbreak of WWI, the two companies jointly founded Indusgas for central Europe. During the war, both sides protected the interests of the different joint subsidiaries within their influence areas, resulting in mutual compensations after the war. British Oxygen and Linde Air Products had completely eliminated the German influence after 1914. The good collaboration with Air Liquide, however, lasted for decades. Air Liquide and the Linde Company even shared technical innovations with one another. Linde was able, in the 1920s, to start producing the cross-flow heat exchanger (DT: Kreuzstromgegenströmer),

[9] Correspondence between Linde, Brush and Javal. Seven letters from 20.11.1907 to 21.1.1908. In: The Charles F. Brush Collection. Case Western Reserve University, Special Collections. Box 3, Letterbook 1902–1910.

which had been first produced by Air Liquide. In exchange, Air Liquide received the Frankl process for oxygen production from Linde in 1930 (Linde 1962: 5).

Linde's sons and successors in the executive board, Friedrich and Richard, continued to seek a mutual, oligopolistic understanding until they resigned as directors in the early 1960s.

According to David Chandler, prime movers have a realistic opportunity to dominate the market forever (Chandler and Hikino 1990). Linde and Air Liquide did so. They were even able to overcome the American supremacy between WWI and the 1970s, a period which Thomas Hughes referred to as the "American Century". After WWI, Linde Air Products (i.e. the Linde division of Union Carbide) was the dominant world market leader for technical gases, while the leading technological remained in Europe with Linde and Air Liquide. Air Liquide managed to out-produce the American competition in the 1970s. Linde, meanwhile, became No. 1 for technical gases once more after purchasing Swedish AGA (founded by the Nobel Prize holder in Physics of 1912, Gustav Dalen) and British Oxygen (which was co-founded by Linde before WWI, but had been sequestrated during the war.).

When Linde published his autobiography in 1916, during WWI, he spoke of Claude as being "the only other scientist with significant contributions in the field of air liquefaction and separation technologies" (Linde 1916: 115). Additionally, he praised his parents' decision to hire only French-speaking nannies and recommended an early relationship to France, which was not typical in Germany in 1916. Linde spoke and wrote French fluently, while his English remained rather weak (Dienel 1999: 521). Claude, meanwhile, admired Linde's scientific and entrepreneurial capabilities. In his biography *My vie et mes inventions*, of 1957, Linde is referred to as being the leading company in the field. Rémi Baillot's 2010 biography of Claude presents more than 20 references to Carl von Linde and his company (Baillot 2010).

Following the November Pogrom in 1938, the Linde Company had to find refuge positions for two Jewish company directors. Fritz Linde asked British Oxygen to hire Philipp Borchardt, who returned to Linde in Munich following the war. Air Liquide was asked to hire Franz Pollitzer, the director of the chemical research lab of the company. Pollitzer was later imprisoned in Vichy's France and murdered in Auschwitz in 1942.

In conclusion, it is quite plausible that the positive collaboration of the two inventor-scientist-entrepreneurs and their companies was a foundation for Claude's positive assessment of Germany and his determination to come to terms with the German occupants and oppressors. He had positive and long-lasting personal experiences with a German partner, who treated his company in a correct and reliable way. In his book *La Seule Route*, published in 1943 on the necessity for collaboration with Nazi Germany, he emphasised the reliable and friendly contact, of over 40 years, to Carl von Linde and his sons. In his book, he referred to Linde as being one of the greatest scholars ever (Claude 1942: 95). This cannot, of course, justify or even completely explain the totalitarian and antidemocratic worldview of Georges Claude. It might, however, be a reason for Claude's conviction of collaboration with Nazi Germany. He confused Hitler with Linde.

References

Baillot, Rémi, and Claude Georges. 2010. *Georges Claude, le génie fourvoyé. Créateur d'Air liquide, du tube au néon, de l'énergie thermique des mers*. Paris: EDP Sciences.

Braun, Hans. 1991. *An uneasy relationship. Technology transfer between the Sulzer Co. in Winterthur, Switzerland and the Busch-Sulzer Brothers Diesel Engine Co.* (St. Louis 1911–1936), Universitat Hamburg

Chandler, A.D., and T. Hikino. 1990. *Scale and scope. The dynamics of industrial capitalism*. Cambridge, MA/London: Harvard University Press.

Claude, Georges. 1903. *L'air liquide, sa production, ses propriétés, ses applications*. Paris: Vve. C. Dunod.

Claude, Georges. 1926. *Sur l'utilisation de l'énergie thermique des mers*. Monaco: Institut Océonagraphique.

Claude, Georges. 1931. *Souvenirs et enseignements d'une expérience électorale*. Paris: Nouvelle librairie francaise.

Claude, Georges. 1938. *Plan de sauvetage économique du pays suivi d'un Appel aux ouvriers*. Paris: Fayard.

Claude, Georges. 1941. *Histoire d'une évolution. De l'hostilité à la collaboration*. Paris: Edition de France.

Claude, Georges. 1942. *La seule route*. Paris: Centre d'études de l'agence Inter-France.

Claude, Georges. 1943. *Francais, il faut comprendre!* Paris: Inter-France.

Claude, Georges. 1944. *Campagne d'Inter France*. Paris: Inter-France.

Claude, Georges. 1946. *Le procès de Georges Claude*. Paris: J. Vigneau.

Claude, Georges. 1957. *Ma vie et mes inventions*. Paris: Plon.

Crawford, E. 1988. Internationalism in science as a casualty of World War I: The relations between German and Allied Scientist as reflected in nominations for the Nobel Prizes in Physics and Chemistry. *Social Science Information* 27: 163–201.

Dienel, Hans-Liudger. 1995. *Zwischen Hochschule und Industrie. Kältetechnik und Thermodynamik in Deutschland und Amerika, 1870–1930*. Göttingen: Vandenhoeck.

Dienel, H.-L. 1999. Bilateral scientific and technical collaboration between hostile countries in Europe. France and Germany 1860–1950. *Innovation. European Journal for the Social Sciences* 12: 517–526.

Dienel, H.-L. 2000. Praktiker und Theoretiker in der technischen Thermodynamik. Die Unternehmeringenieure Rudolf Diesel und Carl von Linde. In *Oszillationen. Naturwissenschaftler und Ingenieure zwischen Forschung und Markt*, ed. I. Schneider, H. Trischler, and U. Wengenroth, 237–267. München: Oldenbourg.

Dienel, H.-L. 2004. *Linde. History of a technology corporation, 1879–2004*. London: Palgrave Macmillan.

Diesel, R. 1903. *Solidarismus. Natürliche, wirtschaftliche Erlösung des Menschen*. München: Maro-Verlag.

Digeon, Claude. 1959. *La crise allemande de la penseé francaise (1870–1914)*. Paris: Presses Universitaires de France.

Emptoz, G. 2000. La création de L'Air Liquide au début du XXe siècle. In *Créateurs et créations d'entreprises, de la révolution industrielle à nos jours*, ed. J. Marseille, 677–692. Paris: Association pour le développement de l'histoire économique.

Fox, R. 1990. The view Over the Rhine: Perceptions of German science and technology in France, 1860–1914. In *Frankreich und Deutschland. Forschung, Technologie und industrielle Entwicklung im 19. Und 20*, ed. Y. Cohen and K. Manfrass, 14–25. München: Jahrhundert.

Gesellschaft für Linde's Eismaschinen. (ed.) 1913. *Lowest temperatures in industry. Presented to the members of the III. International Congress of Refrigeration in Chigaco*. München.

Grundmann, S. 1965. Zum Boykott der deutschen Wissenschaft nach dem 1. Weltkrieg. *Wissenschaftliche Zeitschrift der Technischen Universität Dresden* 14: 800–810.

Hausbrand, E. 1903. *Die Wirkungsweise der Rektifizier- und Destillierapparate mit Hilfe einfacher mathematischer Betrachtungen dargestellt*, 5–14. Berlin: Springer.

Hausen, H. 1935. Gedanken und Erkenntnisse Carl von Lindes. *Zeitschrift für die gesamte Kälteindustrie* 42: 220–225.

Hausen, H. 1937. Austauschvorgänge bei der Zerlegung von Gasgemischen. *Zeitschrift für die gesamte Kälteindustrie* 43: 248–253.

Hausen, H., and H. Linde. 1985. *Tieftemperaturtechnik. Erzeugung sehr tiefer Temperaturen, Gasverflüssigung und Zerlegung von Gasgemischen*, 2nd ed. Berlin: Springer.

Hughes, Th. 1989. *American genesis. A century of innovation and technological enthusiasm, 1870–1970*. Chicago: University of Chicago Press.

Kistenmacher, H. 1987. *Grundlegende Methoden zur Berechnung verfahrenstechnischer Anlagen für die Stofftrenjhung im Wandel desletzten sechs Jahrzehnte*. Vortrag zum Festakt zum 90. Geburtstag von Professor Hausen. Höllriegelskreuth (Linde Berichte aus Wissenschaft und Technik, Heft 60).

Linde, C. 1893. Über die Verflüssigung der Gase. *Bayerisches Industrie- und Gewerbeblatt* 26: 116–119.

Linde, F. 1897. *Messung der Dielektrizitätskonstanten*. Diss. München.

Linde, R. 1962. *Meine Tätigkeit bei der Abteilung B der Gesellschaft Linde Typescript 1954/1955*, Munich.

Mann, Thomas. 1918. *Betrachtungen eines Unpolitischen*. Berlin: S. Fischer.

Pestre, D. 1984. *Physique et physiciens en France, 1918–1940*. Paris: Editions des Archives contemporaines.

Petit-Konczyk, M. 2003. Le financement d'une start-up. L'Air Liquide 1898–1913. *Finance Contrôle Stratégie* 6(4): 25–58.

Schroeder-Gudehus, B. 1990. Die Jahre der Entspannung. Deutsch-französische Wissenschaftsbeziehungen am Ende der Weimarer Republik. In *Frankreich und Deutschland. Forschung, Technologie und industrielle Entwicklung im 19. und 20. Jahrhundert*, ed. Y. Cohen and K. Manfrass, 105–116. München: C.H. Beck.

von Linde, Carl. 1916. *Aus meinem Leben und von meiner Arbeit. Aufzeichnungen für meine Kinder und meine Mitarbeiter*. München: Oldenbourg.

Weber, Max. 1904. Die protestantische Ethik und der 'Geist' des Kapitalismus. *Archiv für Sozialwissenschaft und Sozialpolitik* 20: 1–54.

Chapter 10
Meeting Artificial Cold: Expositions and Refrigeration, 1896–1937

Guillaume de Syon

"What a nice pavilion" wrote a columnist in summer 1896. "There are refrigerated wells for visitors interested in experiencing the sensation of –60 °C, or in studying its therapeutic effects." (Yung 1896) Though such descriptions could have best applied to a circus attraction, they were meant for a building at the Swiss National Exposition (SNE) held that year in Geneva. Four decades later, the Paris Universal Exposition of 1937 would feature a "… novel attraction" in the form of a snow tower announcing the achievements of artificial cold. These exhibits book end the presence of artificial cold as an end in itself at the international exhibits that marked Europe's celebrations of science and technology. Other shows featured exhibits pertaining to artificial cold, but none besides the two mentioned acknowledged the visual element of scientific discovery to such an extent. This article offers to consider the manifestations of artificial cold's appearance in public exhibits and how, though an important feature of the new developments in applied science, it did not show up front-and-center the way electricity did. By considering an overlooked aspect of expositions in the early twentieth century, I borrow from the notion of science as spectacle and its intersection with visual culture (Morus 2006) to clarify how artificial cold was "sold" in a way that enthused the general public.

10.1 Universal or National? Applied Science at Expositions

Starting with the 1851 Crystal Palace exhibit in London and continuing through to World War II, national and universal exhibits reflected a series of trends that included technological vision, encyclopedic display, and scientific cultism (Hilaire-Pérez 2012). Such tendencies implied tensions between industrial and political designs,

---

G. de Syon (✉)
History Department, Albright College, Reading, PA, USA
e-mail: gdesyon@alb.edu

K. Gavroglu (ed.), *History of Artificial Cold, Scientific, Technological and Cultural Issues*, Boston Studies in the Philosophy and History of Science 299,
DOI 10.1007/978-94-007-7199-4_10, © Springer Science+Business Media Dordrecht 2014

which the veneer of celebration often masked. More recent studies of exhibitions have teased out the many agendas hidden behind certain exhibits and thus enlightened the process by which such exhibits socialized the general public all the while emphasizing a positivist outlook (Ganz 2008; Geppert 2010). Even in the post World War I era when questions were raised about the true gifts of science and technology, the rhetoric of benefit continued, albeit in a more subdued way. These studies have extracted many lost elements and subtext, but the selling of science, with the possible exception of electricity, has often been left out. Contemporaries of the expositions that announced, and then accompanied the twentieth century are partly to blame for this. Whether such shows were regional, national, or universal, they shared in common a claim that substantial constructions and industrial progress should be part of their legacy. Consequently, "lesser" exhibits were removed, in line with the tradition of ephemeral display. Exceptions did occur, the most famous of which is arguably the Eiffel tower. Exhibits that included artificial cold thus disappeared, including the "bookend" cold exhibitions of 1896 and 1937. Yet to extract an archeology of artificial cold from such exhibitions is possible, but requires an acceptance of the fact that, unlike electricity, artificial cold was not well understood. To be able to see at night provided an instant benefit that few would have rejected. To enjoy cooling in any form required a change in life style that involved new attitudes. Though this aspect extends beyond the scope of the paper, the exhibitions of artificial cold point to what would become the main expectation of refrigeration in the 1930s: a consumer-oriented benefit. At its origins, though manifestations of artificial cold remained but novelties.

10.2 Preludes to Artificial Cold as a Twentieth-Century Show

Reporting on the 1867 Universal Exposition on Paris, the United States Commissioners who attended the event summarized the potential of artificial cold and marveled at the fact that the "old" Leslie experiment of 1810 was now carried out on a grander scale. Edmond and Ferdinand Carré, who had demonstrated an ice-making machine at the 1862 London universal show, now presented far bigger machines whose greater use of sulfuric acid helped them produce ice decanters (a popular item sold in Paris in the summer months) and big blocks of ice (Blake 1870). Whereas in London their exhibit had disappeared amidst the brouhaha of numerous displays, the show in Paris prompted more discussion. The American observers reporting on the show's scientific exhibits went on to summarize the advantages of various machines and what directions they may have, but they noted that the outcome thus far remained limited to ice in various shapes. This conclusion reflects an important facet: artificial cold was still in its infancy, both as an applied science and as a useful tool. Its applications were not yet clear. Subsequent displays of "ice" and its use would prove equally challenging.

Over the next two decades, Paris hosted another two exhibits. In 1878, the first saw Ferdinand Carré show anew his refrigeration process. Eleven years later another universal exposition was staged, intended to channel festivities associated with the centennial of the French Revolution. That exhibit became legendary because one of its artifacts was never torn down. The Eiffel tower, intended to display the value and strength of industrial steel, would stand, first as an eyesore, and eventually a national symbol. By then, artificial cold had become more common and some its applications, such as the practice of freezing plants for use in pharmaceuticals, or developing cold rooms for food storage was detailed in scientific exhibits (Reports 1891). The multitude of exhibitors present was also notable. Raoul Pictet, Swiss-French physicist, who the year before had successfully liquefied oxygen, showed his freezing process with the assistance of a company that bore his name. Out of the Swiss stand, a small German company named Linde used the services of the Sulzer industrial group to present its work (Dienel 2004). Whereas there was a common theme pertaining to electricity among other applied scientific developments, artificial cold did not benefit from such a link in the structure of the exhibit. The haphazard disposition of the machines and processes shown was confusing at best, and one finds little reference to cold per se in the catalog of the exposition. Instead, references to cold appeared in the surveys of the food industry at the exposition, and the technologies it used. This meant that a new applied science that should have drawn special attention at two exhibits did not in fact gain a foothold in the popular mind (Reports 1870).

To the general public, however, cold was not something that was easily graspable, partly for lack of a clear scientific agenda at such exhibits. As recent studies of nineteenth-century universal expositions have shown, industry dominated the realm of such shows, and did not yield to fully scientific displays until the early twentieth century (Raichvarg and Jacques 1991). When exceptions occurred, scientists needed the support of industrialists to show their discoveries. Such was the case of Raoul Pictet.

10.3 Raoul Pictet and the 1896 Swiss National Exposition (SNE)

Following his successful oxygen liquefaction method, Raoul Pictet accumulated success and honors in the late nineteenth century for further studying industrial cold (Pictet 1878). At the 1878 Paris Exposition, he had enjoyed popular success by demonstrating to the public gaseous liquefaction. He also used the opportunity to promote a new method of ice manufacturing that made use of sulphurous anhydride: “Thousands of hundred weights of ice were manufactured during the Exhibition in a special building in the grounds, erected by the Pictet Ice Company” noted an American observer (Reports 1880). That company which bore his name by agreement (known as *Société industrielle des procédés Raoul Pictet*) brought him a good measure of success but would also cause Pictet some difficulties.

Though successful as a scientist, Raoul Pictet also found himself in a difficult position as a businessman. His personality associated an idealistic sense of fair play with a strong ego. While that had served him reasonably well in the scientific realm, his sense of enterprise suffered from it (FAFP APS 1.5.1.22/3).

He was eager, for example, to affirm his credentials further and repair failed endeavors he had conducted in Egypt, failing to sell ice machines to the Sultan there (FAFP APS 1.1.1.4) He did not fare much better in France. A few years after the 1878 Paris exhibition, a series of arguments had opposed him to Théodor Turrettini, a Genevan engineer and investor who later sat on the board of the SNE. The dispute dragged on before culminating in summer 1890, into a protracted legal battle that opposed Pictet to the French company in charge of marketing his inventions. The court had found against Pictet, forcing him to relinquish to the company improvements he had made to his cold process (JdG 1885, 1890). Nevertheless, Pictet's fame and originality of thought was affirmed in court; it was his contractual obligations that he had failed to defend.

Thus, when the SNE organizers of the 1896 exhibit sought him out, Pictet eagerly accepted their offer to build a "Cold Pavilion" in Geneva for the duration of the exhibit (Herrmann 2000). To the authorities eager to find themes that would represent Swiss achievements (Exposition 1896), it was to attract the interest of the Switzerland's other cultures (primarily the German-speaking one). The Cold Pavilion was also a way of suggesting that scientific research could successfully happen in Western Switzerland rather than in Zurich alone, the site of the famed Federal Institute of Technology. The exposition would thus outline the paragons of an upcoming century that privileged science and industry, and made the Swiss nation part of this wave (Büchler 1970; Pauchard 1991).

As for Pictet, his motivation for constructing the Pavilion definitely included a wish to share knowledge with the public as will be seen further. However, his reasons likely went beyond selflessness and involved a face-saving measure following his legal troubles. Indeed, his former colleague Turrettini was involved in the organizing committee of the SNE, and though no specific correspondence exists concerning tensions between them, Pictet's actions in the preparatory stages are clear. He rejected any plans to place his machinery in a specific category of exhibits, and made sure he retained full control of the pavilion by signing on Paul Galopin, a former student of his from his years in Berlin, to manage the building (Herrmann 2000).

The Pictet Pavilion was a substantial undertaking, so big in fact that the 1894 outline of the project ran out of money. Conceived by architect Adrien Peyrot, it required a redesign to fall within budget. The pavilion was broken down into three sections (Exposition 1896 II). On one end, a 500-seat amphitheater was available and used for demonstrations as well as unrelated conferences. On the other end, a bar served refreshments cooled with ice made on the premises. However, beyond demonstrating the fast production of ice to an eager public, the building was also air conditioned and offered "frigotherapy." This attempt at using artificial cold to relieve pain represented a shot in the dark, yet its endeavor also reflected the scientific optimism that greeted large inventions, whereby the public became both spectator and actor in the exploration of scientific theory. For example, many commented on the hyperactive feeling subjects sensed when exiting exposure to cold. This was particularly intriguing to Pictet who viewed the stimulus as having great potential for treatments of various kinds, even tuberculosis (Pictet 1898).

The center of the Pavilion was where refrigerating machines were housed. Strangely, however, the ones producing ice, rather than coming from a Pictet design (these were built less than half-a-mile away in the industrial quarter of Geneva) were in fact contraptions built by competitors. Though Pictet explained this odd development away as a "courteous embrace" in the name of scientific enlightenment, it is possible that he viewed this as a way to avenge himself of his former associate Turrettini (Pictet 1896; Herrmann 2000).

The 1896 Exposition contributed to affirm Pictet's reputation as a "father of cold," a nickname the French press usually associated with Charles Tellier, but which the Genevan press quickly transferred to Pictet. Described as a patriotic scientist, he came across as a selfless man eager to share knowledge. His machines, however, were simply viewed as "able to lower temperatures." The brochure devoted exclusively to the Pavilion focussed on their inner workings, but nowhere was a purpose explained beyond the abundant production of ice, which was indeed welcome in the summer months (Exposition 1896 II).

However, prestige did not solve Pictet's challenges. The financial report of the SNE lists the Pavilion as one of the weakest moneymakers, despite its strategic location near the center of Geneva (FAFP AEP 2.19.2.5). The electrical bill alone was prohibitive, requiring additional fundraising to be paid off months after the exhibit was torn down. On the other hand, witnesses described the event as having demonstrated the importance of hands-on approaches to scientific and technical developments. At a time when optimistic attitudes towards science prevailed among European visitors to public expositions, seeing the public mind with a visual demonstration of cold's applications may have helped sway some regarding the complexities of science. In many ways, what historians have identified as both a turn and a split between scientists and the public in the early twentieth century due to the complexity of scientific endeavors did not occur as fast in this case. This was thanks to Pictet himself who, regardless of ulterior motive, viewed the notion of science as spectacle in a positive and necessary light.

10.4 Artificial Cold in Support of Exposition Shows

Later universal and national expositions saw little on artificial cold as a means in itself, but substantial manifestations of its application. At the 1900 exposition in Paris, United States arranged to pay for refrigeration aboard ships that carried fruit and vegetable across the Atlantic to be shown and served in France (Report 1901). By 1904 artificial cold was not only accepted, but deemed a necessity for any exhibit to run successfully. That year, for example, organizers of the St. Louis exposition combined form and function by asking the refrigeration installer to set up an ice production and cooling plant that would be visible to the public.

Though late in opening, the unit ran all summer and was described as "one of the interesting working exhibits" of the fair; this was likely due to the fact that the ice production allowed for the installation of an open air skating rink supplied with artificial ice made at the show (Francis 1913; Bertuca et al. 1996). The fact that skaters could enjoy themselves in a Midwestern summer resonated with most visitors. On the other hand, the notion of enjoying cold inside, other than as a novelty, did not catch on.

Pictet had invited spectators to experience coolness in artificial wells he set up inside his pavilion in 1896, and few spectators had enjoyed the experience beyond a few minutes. In fact, it is at United States exhibits that interior artificial cold received perhaps its biggest boost. The American Century of Progress exhibits in Chicago and later the New York each displayed air conditioning units for the public to experience. Where each exhibit departed from earlier shows, however, was by associating cold with a kind of technological utopianism. Such machinery was not for the immediate, but part of a projected future, in line with what American technological utopians envisioned (Segal 1987). This departure from earlier traditions in expositions meant that technologies that had yet to gain acceptance could be presented as simple "potentials" thus offending less, and possibly attracting more attention as non-threatening "gadgets."

As Gail Cooper and Marsha E. Ackerman have shown extensively, the acceptance of air conditioning was not a given, and the eventual success of the interior cooling trend in the 1960s came over all kinds of objections (Cooper 2002; Ackerman 2002). However, the presentation of a house of the future built around the concept intrigued visitors. Though air conditioning was placed at the center of the construction, what appeared to fascinate visitors to the Chicago fair was the fact that technology writ-wide would solve all housing problems. Air conditioning, however, was far from a concern, and summer heat was acceptable. The notion of *control* through refrigeration, however, seems to have begun registering and became eventually a central theme of advertisement around the show (Ackerman 2002). Though the country was reeling from the depression, the stress on future ways of living seemed to have gained acceptance, and would mature at an accelerated pace after World War II, when the era of great universal expositions ended. Across the ocean, this would not be the case.

10.5 Paris, 1937

The Paris exposition of 1937 represents the swan song of universal expositions in Europe. Planned as it was in the midst of socio-economic and political upheaval, its organizers nonetheless hoped to echo earlier events, notably the art deco show of 1925 and the 1889 and 1900 universal expositions that had emphasized the Eiffel tower and associated technologies. Thus, the theme of "art and technology in modern life" was chosen early on (Labbé 1935). As part of such professed modernity, a "Palais du froid" (Palace of cold) was planned. The goal of estheticizing artificial cold was a noble one, for it involved a triple challenge of finance, design and selling of the attraction to compete with other sets, notably the Trocadero building, which was to be opened to house expanded museums across from the Eiffel Tower. By 1937, attitudes had changed little in Europe regarding such things as air conditioning. Whereas a slow shift had become perceptible in United States through the commercialization of window units, the notion of summer heat was deemed part of life in Europe and air conditioning was not on the agenda.

Architect Maurice Fournier designed the "Palace" to attract as much attention as possible by stressing the progress made in artificial cold science. At a time when few households in France could avail themselves the use of a fridge, and household air conditioning was inexistent, the 300-ft long hall invited the public to see a replica of the cooling system for the *Normandie* oceanliner, then the crown jewell of the French commercial fleet. Architect André Hernant, drew some of the early interior designs (Hernant papers). Diagrams, photographs, truck models also complemented the exhibit designed to show that the importance of refrigeration went beyond the notion of luxury (Nature 1937). The main exhibit, associated with the building remained the "snow tower."

Echoing the nearby Eiffel tower, Fournier incorporated a 120-ft tall aluminum "snow tower" whose refrigeration system would produce ice while being lit at night. Its octagonal cross-section was divided into eight downward-pointing conical basins. Cooled through a solution of brined calcium chloride, they reached an average temperature of −10 to −20 °C, thus generating ice when air humidity made contact with the tower. The build-up of ice was then blown out into snow through ventilators. In darkness, four of the eight sides were lit up, thus affording the impression of a magical storm in the summer. To succeed, Fournier had incorporated another familiar trope from nineteenth century exhibitions: electricity. The message was nonetheless that the public was to *appreciate* cold, not just see it as a novelty.

Such magazines as *La Science et la vie*, while featuring the aesthetically pleasing snow tower on its cover (illustration), used the opportunity to deliver a full lesson on the peculiarities of low temperatures and their use (Science 1937) However, it seems that the very conception of the "Palace" seemed to draw as much attention than its purpose. The trade magazine *L'aluminium français* devoted a full article to the casting out of Alpax (a silicon-aluminum alloy) of the snow tower, lauding the quality of the metal used rather than the originality of the tower's purpose (Aluminium 1937).

The "Palace," however, ended up being more a stop on the way to other activities, such as an Argentinian restaurant opened near the "Palace" itself (Figaro 1937). The press did offer a few comments on the tower, but much as other exhibits received praise, little can be gleaned from reactions to the snow tower. It is true that the exposition took place in a time of rising tension while France was under a left-wing government that aroused suspicion among traditional elites. Far more media reports focused on the tensions between France and Germany at the exposition, and even expected discussion of the Hall of aviation set up to praise French flyers were passed over. The technological legacy of Paris 1937 did not last, whereas the art element of the exhibit's agenda continues to arouse passion and analysis.

10.6 Postscript and Conclusion

During the planning stages of the 1992 Universal Exposition in Barcelona to commemorate the five-hundredth anniversary of Columbus' sailing, the Swiss delegation considered the proposal of architect Vincent mangeat, entitled "Swice." Intended to make ice into an identity of Switzerland on par with stereotypical chocolate and banks, the entire structure would have stood through thanks to the uninterrupted production of a giant ice machine. Supporters of the architect's proposal suggested it was the perfect symbol of a rigidified society that could not change without melting away once machines were turned off. The project was not to be. Less so for the message it might send about Swiss identity, but because it was both ecologically unsound and energy-costly to operate (Schwarz 1991; Müller 2008). Cold as a novelty seemed finally to have run its course, much in the manner of its earlier iterations.

Just like the Cold Pavilion in 1896, the Palace of Cold was dismantled shortly after the Paris Universal Exposition closed in 1937. This was no surprise. Its cost, location, and lack of alternate function made it an unlikely candidate for preservation. The Hall of Aviation, too was shuttered, whereas the Palais de la découverte would house a variety of exhibits before returning to its initial function as a museum of science. Artificial cold may have its place there, but in historical terms, not as an applied scientific demonstration.

The presence of artificial cold exhibits at expositions in the late nineteenth and early twentieth centuries confirms that the need to seek acceptance by demonstrating the value of a new technology could go through demonstrative channels. Artificial cold was in fact part of the visual culture of science, but its presence raises other questions in turn: The assumption that the visual impact of a machine would sway viewers or attract them was not a given. Whether in 1896, or 1937, artificial cold machines impressed visitors, but did not revolutionize the public perception of cold. Early traces of consumerism hinted at its direction, however, as registered in the popular success of the air conditioned model homes shown at Chicago in 1933. Yet even there, the success of artificial cold was not instantaneous. It would take a slower evolution to gain public infatuation: Some two decades elapsed before such exhibits succeeded more widely. By then cold was hidden within the modernist lines of such consumer items as refrigerators and air conditioning units.

Bibliography

Ackerman, M. 2002. *Cool comfort. America's romance with air conditioning*. Washington, DC: Smithsonian.

Aluminium. 1937. *L'Aluminum français* 91(May):703–707.

Bertuca, D., D. Hartman, and S.N. Meumeister. 1996. *The world's Columbian exposition: A centennial bibliographic guide*. Norwich: Greenwood.

Blake, W.P. (ed.). 1870. *Reports of the United States Commissioners to the Paris Universal Exposition*, vol. III, 361–402. Washington, DC: GPO.

Büchler, H. 1970. *Drei Schweizerische Landesausstellungen Zürich 1883 Genf 1896 Bern 1914*. Juris Druck, Zurich.

Cooper, G. 2002. *Air-conditioning America: Engineers and the controlled environment, 1900–1960*, 110–139. Baltimore: The Johns Hopkins University Press.

Dienel, H.L. 2004. *Die Linde AG. Geschichte eine technologikonzerns, 1979–2004*. Munich: CH Beck.

Exposition. 1896. *L'exposition nationale suisse Genève 1896*. Geneva: Pfeffer.

Exposition. 1896 II. *Exposition nationale suisse Genève 1896*. Notice explicative sur le Pavillon Raoul Pictet. Geneva: Kündig.

FAFP (Fondation des archives de la famille Pictet, Geneva) AEP 2.19.2.5. *Undated report submitted to the shareholders of the Pavillon Pictet after the closure of the 1896 SNE.*

FAFP (Fondation des archives de la famille Pictet, Geneva) APS 1.1.1.4. *Egyptian correspondence of Raoul Pictet 1869–1875.*

FAFP (Fondation des archives de la famille Pictet, Geneva) APS 1.5.1.22/3. *Extrait des souvenirs de Raoul Pictet.* undated typed notice. pp. 3–4

Figaro. 1937. *Le Figaro*, 30 June, advertising.

Francis, D.R. 1913. *The universal exposition of 1904*. St Louis: Louisiana Purchase Exposition Company.

Ganz, C.R. 2008. *The 1933 Chicago world's fair. A century of progress*. Urbana: University of Illinois Press.

Geppert, A.C.T. 2010. *Fleeting cities: Imperial expositions in Fin-de-Siècle Europe*. New York: Palgrave.

Hernant papers Institut français d'architecture, André Hernant papers (1908–1978) 060Ifa, HERAN E-36-2 "*Exposition internationale de Paris 1937, stand de la république argentine au Palais du froid*."

Herrmann, P. 2000. Raoul Pictet, homme du froid à l'exposition natinale de 1896. In *Genève 1896. Regards sur une exposition nationale*, ed. Leïla el-Wakil and P. Vaisse, 80–91. Geneva: Georg.

Hilaire-Pérez, L. 2012. Les expositions universelles en France au XIXeme siècle. Lignes de tension et lignes d'horizon dans le champs technologique à l'ère de l'indutrialisation. In *Les expositions universelles en France au XIXe siècle*, ed. A.-L. Carré, M.-S. Corcy, C. Demeulenaere-Douyère, and L. Hilaire-Pérez, 13–34. Paris: Presses du CNRS.

JdG. 1885, 1890. *Journal de Genève*. 18 June 1885; 6 August 1890; 27 August 1890.

La Science et la vie 236 (February 1937): 141–146.

Labbé, E. 1935. *Conférence sur l'exposition internationale de Paris 1937*. Arts et techniques dans la vie moderne. N.p., Paris.

Morus. 2006. Seeing and believing science. *Isis* 97(1): 101–110.

Müller, C. 2008. Images of Switzerland at the world fairs – Ephemeral architecture as a symbol for national identity? In *Architecture and identity*, ed. P. Herrle and E. Wegerhoff, 113–122. Berlin: Lit Verlag.

Nature. 1937. *La Nature* 3000 (May 1):410.

Pauchard, P. 1991. *Les Suisses dans le miroir*. Lausanne: Payot.

Pictet, R. 1878. *Mémoire sur la liquéfaction de l'oxygène, la liquéfaction et la solidification de l'hydrogène; et sur les théories des changements des corps*. Geneva: Sandoz.

Pictet, R. 1896. Au pavillon Pictet. *Journal officiel de l'exposition nationale suisse*, 15.

Pictet, R. 1898. *La frigothérapie: ses origines, son but, les résultats déja obtenus*. Paris: Hugonis.

Raichvarg, D., and J. Jacques. 1991. *Savants et Ignorants. Une histoire de la vulgarisation des sciences*. Paris: Seuil.

Report. 1901. *Report of the General Commissioner for the United States to the International Universal Exposition in Paris, 1900*. Washington, DC: GPO.

Reports. 1870. *Reports of the United States Commissioners to the Paris Universal Exposition*, vol. IV, 649–658. Washington, DC: GPO.

Reports. 1880. *Reports of the United States Commissioners to the Paris Universal Exposition, 1878*, vol. IV, 25. Washington, DC: GPO.
Reports. 1891. *Reports of the United States Commissioners to the Universal Exposition of 1889*, vol. II, 631–632. Washington, DC: GPO.
Schwarz, H. (ed.). 1991. *Unter Null. Kunsteis, Kälte, Kultur*, 8. Munich: Beck.
Segal, H.P. 1987. The technological utopians. In *Imagining tomorrow: History, technology and the American future*, ed. J.J. Corn. Cambridge, MA: MIT.
Yung, E. 1896. L'exposition nationale suisse. *Journal de Genève*, 11 July.

Part III
Consuming Cold

Chapter 11
The Introduction of Frozen Foods in West Germany and Its Integration into the Daily Diet

Ulrike Thoms

Frozen foods are a matter of course today; they form part of our daily life and we hardly think about the long time span between the technical invention of freezers and the availability of frozen products in almost every supermarket and even corner grocery. From today's perspective frozen foods are an ongoing success story. Even though food expenditures in Germany somehow stagnate over the last decades and there is a sharp cut-throat competition in the food business, consumption figures and all-over turnover in frozen foods still increases and is expected to increase further on.[1] Nevertheless, this finding is a late development. Although the mere machinery for freezing was already available at the end of the nineteenth century, it took almost 50 years, that frozen foods became a widely used product of mass consumption.

Only recently Uwe Spiekermann (2009) has analysed product innovations in the German food industry. Using yoghurt and canned foods as examples he has shown, that it is not sufficient to limit the analysis of innovation to the perspective of production. Instead, it is important to pay tribute to the diversity in the appraisals of certain products by engineers, producers and consumers. Taking dietary habits of consumers, their economic and social situation, their knowledge about foods into account is as important as to take social and emotional aspects of the different involved actors seriously.

But so far, the existing studies in the history of frozen foods concentrate almost entirely on the development of freezing technology and its scientific foundations, the history of single enterprises in this field or the associated scientific and business associations. Many of these publications come from eye witnesses, can be regarded as commemorative literature or publications that serve the purpose of public relations.

[1] See 'Absatz-Umsatzentwicklung. Tiefkühlkostmarkt in Deutschland gesamt 1990–2010', http://www.tiefkuehlkost.de/tiefkuehlmarkt/statistiken/. Accessed 18 Feb. 2012.

U. Thoms (✉)
Institut für Geschichte der Medizin, Charité Universitätsmedizin Berlin, Berlin, Germany
e-mail: ulrike.thoms@charite.de

K. Gavroglu (ed.), *History of Artificial Cold, Scientific, Technological and Cultural Issues*, Boston Studies in the Philosophy and History of Science 299,
DOI 10.1007/978-94-007-7199-4_11, © Springer Science+Business Media Dordrecht 2014

Others concentrate almost exclusively on technical aspects (Hellmann 1990; Dienel 2004; Teuteberg 1991; Anderson 1953; Thevenot 1979; Hilck and auf dem Hövel 1979, Cooper 1997). Where the analysis has not been reduced to the very simple model of a trickle-down process, that relied on time and technical developments, the history of frozen foods does not appear as a straightforward success story, but has become more differentiated by taking consumers into account (Hamilton 2009; Oddy and Oddy 1998; Hietala and Leopist 1998).[2] The appropriation of the new technology was in fact a social process and the complicated and long cold chain is a large technical system which links many different fields and scientific and technical, legal and organizational problems. Problems had to be solved by negotiations between different agencies with different aims and purposes, such as the producers of foods, producers of freezers, the associations of transport and railroad companies, scientists, scientific institutions, the state, food control boards and last not least with consumers and consumer organisations.

In this context, Germany is a very special case: For long, it held a leading position in food research and engineering and especially in the production of cooling and freezing machines. On the other hand, many branches of the food production were not heavily industrialised before 1945, but remained smaller to medium-sized companies, whereas the influence of craftsmanship remained strong especially in meat and bread production (Ellerbrock 1993). The German State drove forward scientification by funding research in the field of food freezing already during WWI. During the National Socialist Era the interest for frozen foods increased against the background of the interest in healthy foods and vitamin preservation and the quest for food autarchy (Melzer 2003; Bächi 2009a; Thoms 2006). Consequently research and development in food freezing were granted massively, although the initiators were very well aware of the fact, that these expenses were long-term investments and as everything had to be built up from scratch. These efforts were made to serve the needs of the "Volksgemeinschaft", but not so much the needs of the individual person. Yet, after the break-down of the system it turned out that the frozen food business functioned only under the political and economic conditions of the National Socialist war economy. After the end of war many production sites were destroyed or damaged and so were transportation channels. Moreover the supply with raw materials was precarious in the after-war period with its major food shortages, so that many producers gave up their production, especially as the army, which had been the biggest consumer so far, was dissolved.

The producers had not only to cope with the damages of war, they had to realize, that the consumers were not as enthusiastic about the new foods as they were themselves. It turned out, that the technical invention was a prerogative, but formed only the very first step in the development of the frozen food business.[3] This chapter will

[2] Obviously even the Institute International du Froid has realized the importance of this perspective, as it recently published a history of refrigeration, that in its title points to the assumed needs of mankind, compare: International Institute for Refrigeration 2005.

[3] Classical for the role of innovations in economy Schumpeter 1912, in regard to innovation theory and the steps of innovations, see Rogers 2003.

ask why the reception of frozen foods took so long. Thus, it will concentrate on the role of the consumer in this process. In order to understand the development of the 1950s better, it will first give a short overview of the specific political and economical situation of the 1930s and the state-driven efforts to establish deep freezing as an affordable standard method of food preservation and a functioning frozen food industry. But its main part will concentrate on the after-war period and especially on the 1950s and 1960s, during which it turned out, that the consumers were rather reluctant to buy the new products. Consequently the market remained small until the 1960s, although frozen foods were regarded as to be the food of the future by industry and science, by politicians and consumer organisations alike. Finally it will question, why frozen foods finally became a matter of course in the nation's food budget.[4]

11.1 The Development of the Frozen Food Business Before 1945

The production of cooling and freezing machines has a long and strong tradition in Germany. It is connected to persons like prestigious Carl Linde, who had founded his firm for the production of cooling machines in 1879 and was selling them worldwide. Linde had realized the enormous role of research and development in this field from early on and had therefore founded the German Association for Cold (Deutscher Kälteverein) in 1878, a widely reknown and recognized scientific association (Plank 1960). He strongly influenced the foundation of the International Refrigeration Institute in Paris, which from its very first moment put food freezing and the construction of a cold chain on its agenda (Kuprianoff 1960: 25). Nevertheless, Linde's machines were mainly sold to breweries and ice factories, while a lively international trade with frozen meat had already developed. In 1909 165 ships were equipped with freezing or cooling facilities worldwide. They helped to balance out overproduction in South America and the demand of the Western World with their quickly growing populations (Critchel and Raymond 1912; Kallert 1967). Even though machines, equipment and transportation tripled transportation cost, the meat from South America still was 36 % cheaper than home-produced meat. Thus, it promised to improve the provision of industrial workers with their limited budgets (Fleischmann 1934). The number of ships in the international frozen meat business doubled until 1925, German imports boosted from some ten thousand tons in 1922 to over 123,000 t in 1927 (Kallert 1967: 129). The quality of this product was low resulting from the fact that it was not wrapped and thus not protected against drying out and freezer burn. But at the same time, the low temperatures did not reach today's norms. Both points contributed to the negative image of frozen meat's bad quality.

[4] Doing so, at least the second and main part of this study will not so much be the history of an innovation, but a history of consumption.

Government and traders were very well aware of the possibility, which lay within this trade. During the food crisis of WWI German government commissioned Rudolf Plank (1886–1973) with studies of the freezing of meat, beef and fish (Plank and Kallert 1915, 1916a, b). These works were meant to help to improve the critical supply of the German people with animal protein, but in fact the timely circumstances did not allow setting up large freezing plants.

The special importance of cold for the provision of large cities had been clearly recognized in the 1920s (Schipper 1925), during which Rudolf Plank had extended his studies from fish and meat to the freezing of fruit and vegetables and had continuously published on this topic (Plank and Schneider 1927–28; Plank 1927/28, 1929; Plank and Gerlach 1917). In 1925 a new Institute for Freezing was founded at the Technical University Karlsruhe and placed under Plank's direction. Its task was to drive forward scientific research in this field and to support German industry by the education of refrigeration engineers. Though being an engineer and a physician by training, Planck systematically extended the scope of his research. In 1926 he appealed to the Agricultural Ministry to use artificial cold for the preservation of perishable foods and to set up a large research program in order to reduce the health risk and the enormous economic losses resulting from spoilage of food (Plank 1926: 165).

Plank's ideas on the importance of food preservation were well heard by the Nazi Food and Agricultural Ministry, which like the Health Ministry adhered to the idea of natural foods and – at least in theory – opposed the addition of food colorants and preservatives to foods.[5] Moreover National Socialist physicians and physiologists were extremely interested in vitamins, which they thought would improve the physical performance of workers and soldiers (Thoms 2007; Bächi 2009a, b). The higher vitamin content of frozen vegetable and fruit in comparison to canned products paved not only way for their integration into the food scheme of the SS, but was its precondition.[6] Likewise, the director of the Food Department in the Administration for the People's Health (Hauptstelle Volksernährung im Hauptamt für Volksgesundheit) stressed the advantages of deep freezing in 1941 with the words: "A people can get healthier and physically more capable, as its lebensraum would otherwise allow, because it has possibilities to use high-quality and especially the most vitamin-rich fruit and vegetables from more favourable neighbour countries" (Wirz 1941). Thus, frozen foods held their place in National Socialist food policy, as surplus production during summer or in certain areas could in fact be stocked up for the winter months. Like many other state funded research project in the field of agriculture and breeding, this was conceptualized as a means to improve the German Reich's independence from food imports, which then would help to save precious foreign currency (Mosolff 1939).[7] Consequently, research projects

[5] Though chemical conservation methods were theoretically declined, many conservation methods were accepted, while justified by the needs of war, cf. Sperling 2011.

[6] See Letter Reichsführer SS, Persönlicher Stab to Chef des V.u.W. Hauptamtes, 21.2.1941, in: Bundesarchiv Berlin, BArch, NS 19/1323, Bl. 8.

[7] The limited production of fruit and vegetables had motivated the Nazi politicians to decline the general promotion of vegetarianism, although it basically fit the health goals of the party and Hitler himself was a vegetarian, cf. Thoms 2010.

from this sector reached far beyond the usual limits of food research. In fact they formed part of that body of projects that was considered as relevant for war and was massively funded under the 4-year plan.[8] Planck's institute profited from this policy, too.[9] In 1936 the Imperial Institute for Preservation of Food (Reichsinstitut für Lebensmittelfrischhaltung) was added as a separate department to the Institute for Refrigeration in Karlsruhe. It should help "to centralise the scientific guidance of research in this field" (Reichskuratorium 1936, 1937: 36) as the Minister of Food declared on the opening ceremony, which once more underlined the enormous role of deep freezing for the limitation of food losses by spoilage (Weber 1936).

The extensive research program brought together research in the construction of better freezing machines, freezers for shops and homes, transport vehicles, the different freezing methods, the choice of those fruit and vegetable varieties that were best suited for freezing, breeding programs to improve these varieties and packaging. Additional support in this field came from the Four Years Plan, which especially asked to double the fish production (Mosolff 1940). Research subsidies went to state institutions like Plank's Institute at Karlsruhe, the Research Association for Refrigeration of Vegetables and Fruit (Forschungsgemeinschaft für die Kühllagerung von Gemüse und Obst) in Calbe, the botanical Institute of the Technical University Karlsruhe, the Institute for Physics at the Veterinary University Berlin, the fishery administration at Wesermünde, the Society for Consumption Research in Nürnberg as well as to private freezing firms, like the firms Neuerburg and Joseph Pankofer and last not least to the IG Farben, which was involved in the development of packaging materials (Hilck and Auf dem Hövel 1979: 70–72). In other words: Research was carefully planned and covered almost every field involved. The state did not only further the foundation of firms in this business, it directly helped to found deep freezing firms itself, such as the "Solo-Feinfrost", which was mainly occupied with the freezing of food for the needs of the German Army (Hilck and Auf dem Hövel 1979: 34)

Especially the freezing of fish was massively supported, because fish is a cheap source for animal protein and protein provision needed to be improved for the sake of autarchy. German fish trading associations had already undertaken trials with fish freezing in 1924, but results and profitability had not been fully convincing so far. Now new trials were driven forward by Wilhelm Roloff, who held close contacts with Plank's institute in Karlsruhe. Roloff thought, that the whole development went too slow; therefore he took an informational trip to the USA, while Unilever had brought Clarence Birdseye to Germany in order to discuss the introduction of deep freezing with him. After Roloff's return he took up negotiations with Frosted

[8] See for example the projects of Bomskow from the Imperial Research Institute for Fishery, in: Bundesarchiv Berlin, BArch, R26III, no. 10 and the funding for the entrepreneur Pankofer, Bundesarchiv Berlin, BArch, R26III/20: 288–289. Plank himself was no Nazi at all. Like many others, he accepted the additional funding for political reasons, but he fought against his own pupil Rudolf Heiss, who became his serious competitor in this field and used his national socialist conviction to pursue his personal career. Moreover Plank helped Kuprianoff, who as a Russian had to leave the institute in 1933 to find a new job at the firm Bosch.

[9] Archiv der Universität Karlsruhe, Bestand Nr. 21011, Personalia, Nr. 238.

Foods, who meanwhile held the patents for the Birdseye process. Finally the NORDSEE took out an – expensive – licence, which meant that they had to pay a fee on every kg they produced (Hilck and Auf dem Hövel 1979: 35). This was an outstanding practice for the Nazi Regime, which according to common judgements preferred to develop new purely German methods to produce surrogates of low quality and from unsuited raw materials for high prices (Wengenroth 2002: 53). At least for the case of frozen foods the common judgement this was obviously not the case. In contrast, the argument of high quality products with their full nutrient content stood very much in the foreground and was pursued together with a thorough calculation of production cost. Though the long-term development seemed to be very clear and positive, the appraisals of the future prospects of the freezing industry remained cautious, the judgements were differentiated and cost calculations formed part of the plans in this field.[10] The involved parties tracked the expenses carefully. They saw them justified by the experimental character of the trials, but stated, that they would be an investment for the future (Heiss 1941). Building prototypes of floating freezing plants cost a lot of money. But the idea and the plans as such were by no means absurd, only the war conditions led them ad absurdity. Cost rose because the close-by fishery grounds got out of reach and ships had to undertake longer trips. Consequently expenses exploded and economic rationality was indeed questionable, especially as it turned out, that the prototypes of ships needed more repairs than assumed.[11] But still, the positive American development was seen as a major justification of all efforts and it was firmly believed, that this development would take place in Germany, too, although with a certain time lag.[12] Disregarding the many problems of the frozen food industry in the USA, the American development was seen as a success story and Clarence Birdseye was seen as the very hero of this development, while his bankruptcy in 1929 was not noticed at all (Mosolff 1939).

The efforts in the field of frozen foods concentrated almost exclusively on the production side. The very many upcoming problems document an immense, but overall naïve optimism of all parties involved. Not all producers had even storage houses, but had to rent storage room at slaughterhouses or refrigerating houses. Overall, the cold chain was the main problem at this point, especially in regard to the links between production and point of sale. Together with the Institute in Karlsruhe, the German railway company and some motorcar firms developed diverse types of transportation boxes, cars and cooling and freezing wagons (Wengenroth 2002: 53) ahead of time as well as diverse types of household and shop refrigerators. Nevertheless, it took until 1941 that the serial production of 6,000 freezers for wholesalers and shops was taken up at all. But even the very optimistic and propagandistic

[10] See only the many calculations for the fish freezings plants in Norway, for example in: Bundesearchiv Berlin, BArch Berlin, R17 II, Nr. 94–96.

[11] See the discussions and calculations in: Bundesarchiv Berlin, BArch Berlin, R 17 III, Nr. 04.

[12] Rudolf Plank did a lot to back up this view by his reports on another trip to the USA in 1937. From these we know, that the second travel was initiated by an invitation of the Massachusetts Institute of Technology in Cambridge. At these times, this was almost unusual for a German researcher as well as for the times of the National Socialist Regime.

official publications did not mention freezers for private households at this point (Mosolff 1941b: 18; Anonymous 1941; Bericht: 618). Though National Socialist food policy took psychological aspects into account, the consumer and her reaction was not considered at all in the freezing plans. Instead, she was only addressed as a passive receptor of the good message (Drews 2002). A massive propaganda was set up by the Food Ministry, which very much followed the general line of household instruction and nutritional advice, as it had been systematically extended by the National Socialist State as a part of its health prevention policy and of the girls' education for their job as housewives and mothers. Numerous more or less propagandistic articles on frozen foods were released, an informational Film *Blessing of the Cold* (Segen der Kälte) was produced together with the major film company UFA and the *Pocket Book on Freezing* was published in 1940 (Anonymous, 1940. Gefriertaschenbuch). Moreover a nicely illustrated brochure with the title *ABC of Deep Freezing* was brought out. It addressed not only the consumers, but another lay group, the shop owners, who were instructed how to handle the food products and the freezers (Mosolff 1941b). Covering any thinkable technical problem, the readers were informed about the complicated technique in a well-understandable and easy manner. Obviously this booklet intended to encourage the retailer's will to make major investment of a freezer. In order to lower the barrier even further the producers of frozen foods began to rent out freezing boxes to shop owners for little money or for free in order to ensure that their goods could be stored in the shops at all (Hilck and Auf dem Hövel 1979). Housewives were not specially targeted and trained beyond propaganda. Knowing that frozen foods were meant to close holes in the supply chain, the housewives surfaced only as "buyers", whose activities were foreseen as sure. It was expected that they would simply buy the food, take it home gladly and prepare it in due time (Mosolff 1941a, b).[13] This is not so much surprising because at this time, food was already rationed, so that foods were sought after. One scientist therefore went so far to conclude that there was no need to advertise frozen foods at all (Anonymous 1943. Gefrierkonserven). As we will see, this was a misconception that would cost the industry much money and delay the success of frozen foods enormously.

The mere production developed quite well. From 1939 until the end of the Nazi Regime 13 firms were producing frozen foods. Though larger experiments with food freezing had only started in 1939, the production reached 22,000 t in 1940, whereof 7,000–8,000 t were fish and 14,000 t were fruit and vegetables. During the following year the freezing capacity grew to 100,000 t and to 100,000–150,000 t in 1942. As 70,000 t of this amount went for fruit, vegetables and fish, the largest food group was meat, of which unknown quantities were delivered to the ever growing army (Anonymous 1941, Bericht: 618, Mosolff 1941a, b: 17–18; Anonymous 1944/45, Handbuch). But even the existing figures are remarkable. Once the war had ended, it took until the 1960s that these production figures were reached again for West Germany – although at this time the cold chain ended in the shops[14] and the

[13] This was seen as the normal case, see: Serger and Krause 1941, 173.

[14] See only the advert for the BBC frigidaires ibid., after p. 112.

propagandistic functions of frozen foods for the self-staging of the National Socialist State as a modern state that is able to solve its problems by techno-scientific solutions, can hardly be overlooked. However, the developments under the National Socialist system had long-term effects that reached well beyond mere propaganda. In fact a network of people, enterprises and research institutions developed during these years, which survived the defeat of the regime[15] and formed the basis from which the freezing business started anew after the end of war.

11.2 The Development After 1945

The end of war was a clear turning point for food freezing in Germany: Agriculture as well as production sites of the food industry and the transportation systems had been severely damaged during war. Especially the railroad system was severely damaged while rail cars as well as lorries fell short. Under the given circumstances the supply had to concentrate on bare necessities, whereas frozen foods as well as freezers were luxury goods (Anonymous 75 Jahre: 68–69). Consequently some producers gave up totally, some turned to other fields of business, especially as their most important client, the army, did not exist any more (Hilck and Auf dem Hövel 1979: 45–46).

By and large the future development was driven forward not by demand, but by production and industry and single entrepreneurs, which hoped for high profits. Producers, transport companies as well as political bodies were very well aware of the fact, that the cold chain was neither stable nor extensive enough to secure provision and to make frozen foods a rewarding business. So major efforts were undertaken to expand and stabilise the cold chain beyond the shops (Anonymous 1944/45, Handbuch).

Due to the enormous continuities in the food branch, the very same firms reappeared on the market after 1945 and the related research was continued by the very same set of people. Already in 1947 Rudolf Plank organized a first scientific conference on freezing in Ettlingen (Anonymous 1947. Kälte- und lebensmitteltechnische Tagung). Though under completely different political circumstances the actors followed up the very same vision that had formerly been supported by the state for the sake of autarchy and *Volksgesundheit*. This view was shared by international institutions, which now stressed the role of deep freezing as a means to solve the problem of hunger in Europe. As early as 1948 the United Nations had taken steps to develop the transportation system for perishable foods by establishing a group of technical experts. Even before the food situation improved after the West German currency reform of 1949 the European Productivity Agency (EPA) and the Organisation for European Economic Co-Operation (OEEC) supported the development of the cold chain massively in order to help up German agriculture and

[15] Food research in Germany showed in general a high grade of personal and institutional continuity before and after 1945, see Thoms 2006.

the food industry by modernization. The named agencies organised informational trips for food scientists, farmers and food technologists to different European countries. Doing so they wanted to further the exchange of ideas and to take up the experiences from the master example of the USA.[16]

Again, there was broad consensus, that a functioning cold chain was the precondition for the development and it was clear, that it depended on the coordination of the different actors in this field. In 1950 the *Transthermos* was founded in order to organize the Inter-European transportation of frozen food by rail, being owned by the German state at that time. In 1952 the *International Road Union* had formed a committee of experts, who worked out a statute for the organization of a street-bound transportation system on the supra-national level. The corresponding German association agreed upon their aim to extend the transport capacities, to improve the insulation of the trucks, the cooling machines as well as the whole construction of the refrigerated trucks. On March, 21st 1955 the German *Transfrigoroute* was founded, 7 days later the International *Transfrigoroute* followed with 140 member companies and their 350 special trucks. Given the limited number of member enterprises and a total of 350 refrigerated trucks, which had to supply entire Europe, their aim to make better use of the existing capacities can be well understood as their call to push the frozen food business beyond the regional limitations of production and trade (Anonymous 1960. 10 Jahre; Anonymous 2005. 50 Jahre; Anonymous 1962c. Die Zusammenarbeit).

Like before producers and research institutions firmly believed in the bright future of frozen foods. Most of the formerly active companies remained active in the field and continued to produce frozen food (Schöllgen 2008: 65–66). Arguing that the technical and organisational preconditions had already been worked out and were ready for use (Serger and Krause 1950: 265), they financed the necessary investments in the extension and stabilisation of the cold chain with income from other fields of their business.

When in 1955 the first West German food exhibition ANUGA (Allgemeine Nahrungs- und Genussmittel-Ausstellung, cf. Schüller 1999) took place, six firms presented their frozen foods, which were glorified in enthusiastic newspaper articles. The products were simply equated with modernity and the USA was seen as their very source. America was seen as the blueprint, so that industry always kept an eye on the developments going on there (Anonymous 1948/49. Produktionszahlen). One article read: "The Luxury of today is the consumer good of tomorrow. During the last years this was a frequently made experience, and it will, too, make a household machine popular, which is still to be found only seldom today, but which is a tried-out and approved household machine in the USA: the freezer. This bigger sister of the fridge is almost entirely a child of our time. Modern in any respect, it means a precondition of everyday kitchen practice for the American housewife.

[16] This conviction was further nourished on the trips of freezings experts to the USA, see only the extensive report: Anonymous 1951. The cold Chain. A whole series of trips to the USA and European countries is reported in the *Berichte über Studienreisen im Rahmen der Auslandshilfe der USA*, see for example: Kraut and Wirths 1951.

The revolution of the menu as well as the time saving rationalisation of shopping had not been possible without up-to-date aids. The "Freezer" almost replaces the larder and makes it possible to get along with only one or two shopping tours per week. Temperatures of −15 to 20 °C allow to store even perishable foods for any desired time and to have all that in the house, what we have always had to get freshly." (Anonymous 1959b. Tiefgekühltes). The American food industry presented itself on other exhibitions in a futuristic way, with completely electrified kitchens and tools, that would do all the work of the housewife automatically.[17] Such narratives resonated quite well with long-established ideas and phantasies of the mechanisation of the household in social theory and economic theory (Bebel 1979; Schmidt-Waldherr 1999).

But much to the disappointment of the producers of frozen foods, who had invested so much, the revolution did not take place. Though producers and researchers stressed the advantages of their products time and again, consumption remained below the war years. Brought up in a culture of mangle and deeply convinced of the enormous advantages of food freezing, producers and sellers did not understand these outcomes. They had neglected the consumer almost completely and had almost forgotten to take their economic situation, their wishes and demands, the structure of the whole food system and especially the likes and dislikes of the consumers and their dietary habits into account.

Michael Wildt has convincingly shown that the budgets of the German households were limited until the late 1950s. Despite all futuristic phantasies, austerity dominated the economy of most households. Very much in line with this the consumer representations did not so much discuss quality aspects, but price.[18] Moreover the home preservation of products from the own garden still played an important role. Introduced in 1900, the Weck system of preserving foods in glasses by sterilization was used widely. The home production of jams, pickles, fruit and vegetables played an important role not only in economic, but even in social terms, as German housewives took their pride in a larder filled with many colourful preserves (Spiekermann 1997). This system was firmly established, whereas freezing was new and made it not only necessary to buy an expensive freezer, but to acquire new skills and knowledge. In 1956 a home freezer would cost 1,000 to 1,300 Deutsche Mark, whereas the total household income during that year was only at 528,74 Deutsche Mark, of which only 39,60 Deutsche Mark went for household ware and almost 236 Deutsche Mark for food.[19] This small sum made it almost impossible to by a freezer. Consequently the distribution of freezers rose only slowly with the growth of wages and remained closely connected to income. In 1962/63 only 3 % of all German households had the opportunity to freeze food or to store frozen foods, among them mostly better-off households. In fact it took until the mid-1970s

[17] Female journalists were fascinated and attracted, but at the same time amused by these demonstrations of the possibilities of modern science and technology and their impact on food, see only the report: Schönfeldt, 1961.

[18] See only the articles in the journal *Verbraucherpolitische Korrespondenz*.

[19] Anonymous 1959a. Tiefgekühltes; Bundesministerium für Ernährung, Landwirtschaft und Forsten 1965, 531.

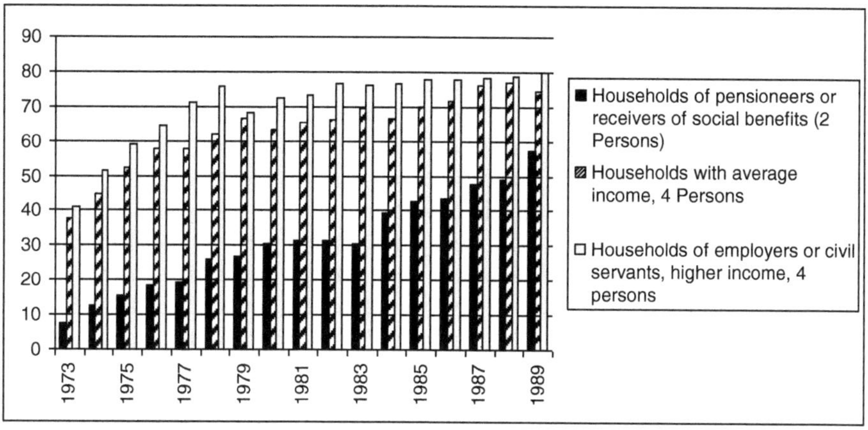

Fig. 11.1 Households with own freezers in West Germany 1973–1989 (per cent of all households in the respective income group) (Source: Statistisches Jahrbuch der Bundesrepublik Deutschland, 1977: 426, ibid 1980: 438; ibid. 1983: 454, ibid 1984: 467, ibid 1987: 488, ibid. 1990: 488)

that this number climbed over 50 % and reached its maximum of 68 % in 2003, while strong social differences remained (Andersen 1997: 108).[20] (See Fig. 11.1)

Until the late 1970s, the majority of consumers thus had to prepare their purchased items almost directly after the buy. But even shop owners were reluctant to buy freezers. This was at least one of the experiences the frozen food companies had to make during the mentioned ANUGA. Retailers would come by and have a look at the frozen products, but they would expect the producers to give them the freezers for free. When companies denied this, the small retailers would turn away because they were not willing to make the considerable investment on their own. Except that the small corner grocery would perhaps not have space enough to place the apparatus. A newspaper article mocked in 1956, that the number of freezers in shops was so small, that 12,500 people had to live out of one freezer (Anonymous 1956. Wachsender Appetit). At this time, the freezers in shops were relatively small. They were only able to store 25–30 packages, so that it would take very long to pay off the reasonable investment (Döring 1960).

At this time the German food industry was still very much production orientated. As explained above producers simply expected, that the consumers would just jump on their new products with their assumed advantages once they brought them to market and if not, advertising would make them do so. In this context it is very telling, that the companies which exhibited their products on the ANUGA did

[20] In recent times, figures are going down again, see: 'Ausstattung Kühlschränke/Gefrierschränke 2003–2008' on http://www.destatis.de. Last accessed 22 Feb 2012. The reasons may be the increasing number of singles who have no need to store large amounts of food or an increase in the number of fridges with a freezing compartment may have increased. These are not counted as freezers. But this may even be a statistical artefact, as such fridges are not counted in the freezer statistics.

not allow the targeted consumer to touch, see and smell their products, as the exhibition showed only empty packages.[21]

But as food had become more abundant and affordable, the consumers had become more critical in questions of choice and quality. Forcefully supported by the Americans, who saw the need to educate the German consumers in order to enable them to act economically and politically responsible, a system of consumer education was installed. Part of this education aimed at lifting the workload of mothers and wives in order to enable them to take part in political, social and cultural life. The Americans moreover expected, that women would increasingly work outside home. Therefore they had designated certain amounts of the Marshall Plan funds to rationalize the German households via household education or Home Economics. Not only did they use these resources to finance field trips of nutritionists, home economists as well as representatives from government and administration to the USA and other European countries to show them ways and means of extension services by practical demonstration and to encourage the institutionalization of home economics and consumer education. They promoted to build up an extension and information service. In 1950 the AID (Aufklärungs- und Informationsdienst) was founded as an institution of the Ministry of Agriculture (Anonymous 2010. Der AID; Büning-Fesel 2010). Very much against the silent resistance of many a male scientist the Americans insisted on a corresponding Federal Research Institute for Home Economics. When this institute was established in 1952, it drew back upon the traditions of German household education of the 1920s and addressed not so much the urban, but the rural households (Stübler 1985; Oltersdorf and Preuss 1996: 370ff). Thus the farmer's housewives were targeted as important members of agricultural enterprises and were consequently instructed how to facilitate their household work in order to save time, efforts and money by using the newest technology and scientific knowledge. The urban women though were told that they could lift their workload by showing them an idealized vision of modern life style and rationalized housekeeping, including time saving shopping and cooking such as ready cooked meals. The only problem was, that the housewives were reluctant in buying the new time-saving products, so that the sales figures for frozen foods remained low.

It turned out, that the producers in their strong belief in the future of deep freezing had built up huge overcapacities. Some of the veterans in the business like Joseph Pankofer from Munich were finally disappointed and frustrated, that after all the years of high investment, they were still not able to raise consumption. Even the mighty giant Unilever retreated from the German frozen food market in 1949/50, the newly founded firm Ahena followed in 1951 (Hilck and Auf dem Hövel 1979: 46), whereas others like Schöller retreated from the meat, fish, vegetable and fruit business and again concentrated on their former and more profitable core business in ice cream (Jones 2005: 26; Schöllgen 2008: 70–74; Winkler 2007: 372–373).

[21] See only the report of Holger Simonsen, who first was employed in his father's fish freezing plant, the "Hamburger Tiefkühlgesellschaft" and later on worked for Jopa, available from http://www.tiefkuehlkost.de/presse/meldungen/pm_2005_009.html. Last accessed 22 Feb 2012.

Finally it were the overcapacities of the fish industry, that brought a new beginning. As mentioned above, the idea of deep frosting surplus fish landings as to balance out seasonal differences has been discussed since the 1920s (Classen et al. 1924: 291–295; Schlienz 1930). Now, that the destroyed German fish fleet had recovered and was not able to sell its catches completely, this idea was revived. A couple of institutions became again interested in joining forces and in coordinating the building up of a functioning cold chain (Hilck and Auf dem Hövel 1979: 48–51). Finally they managed to make federal food policy agencies like the *Food Ministry* aware of the problem. A couple of conferences then helped to push things forwards.[22] The special impetus came from the *Fish Sales Consulting Service* under its director Wilhelm Kurjo and the entrepreneur Heinz Lohmann.[23] It was motivated by the overcapacities in fish landings, which they had frozen so far while expecting, that they would be able to sell the fish during periods with low fish landings. But the overcapacities were simply too large, the plan to sell frozen fish did not work and sales figures kept low, obviously because the reputation of frozen fish was low. The association of fruit and vegetable freezing companies faced the very same problems. This was the very moment that first steps were made to find out about the reasons of the missuccess and the chances of the business, which went beyond a self-critique of the members of the cold chain. Once more the state helped out: Advised by a sales consultant from the German Food Ministry representatives from fish and vegetable freezing companies the *Support Service for Fishery Sales* (Fischförderungsdienst), the *Association of the German Deep Freezing Industry in the Association of the Fruit and Vegetable Industry* (*Arbeitsgemeinschaft Tiefkühlunternehmen*), producers of frozen foods and several sales organisations joined and took steps to set up a test market in the area of Köln and Bonn. This market should cover 1.25 Million inhabitants. Over half of the 18,000 Deutsche Mark for this campaign came from the fish industry, whereas the rest of cost was divided almost equally between the vegetable industry and the consumer departments of the *Federal Ministry for Food, Agriculture and Wood* and the respective *Food Ministry of North Rhine-Westphalia*.[24] The test was prepared by a series of five adverts in the newspapers of this area and it was accompanied by several posters and by a press conference. Between 20 February 1956 and the end of August 1956 four discussion evenings took place, during which household teachers demonstrated the preparation of frozen foods to vendors who were provided with informational and recipe leaflets, which they should distribute to their clients. The local organisations of the *German Homemaker's Association* (*Deutscher Hausfrauenverband*) joined in as well as teachers of home economics

[22] See Kurjo, Hans Wilhelm. Veranlassung, Organisation, Durchführung [des Bonner Absatztestes]. In Landesarchiv Nordrhein-Westfalen (Hauptstaatsarchiv Düsseldorf) NW 123, no. 354, Bl. 53–54.

[23] Lohmann is a central figure in this development. Since 1932 he owned a fishmeal factory and a freezing plant since 1938. Already in 1954 he had set up the production of broilers on an industrial scale, which followed the American model, but lacked an efficient sales organisation, see Hilck und Auf dem Hövel 1979:52f; Anonymous 1962a. Brathendel.

[24] Sommer, Peter W. Verbraucheraufklärung und Werbung, Landesarchiv Nordrhein-Westfalen (Hauptstaatsarchiv Düsseldorf), NW 123–354, pp. 58–58,

with demonstrations to housewives, whereas the long-existing promotion agency of the fish industry organised demonstrations for the professional cooks in the catering business.[25]

For 6 months, the participating companies would offer their products from a total of 250 freezers. Once the test had begun, participation grew quickly to 400 retailers. The involved parties planned the reporting system systematically and involved several federal research institutes, such as the Federal Institute for Fishery in Hamburg, the Federal Institute for Food Preservation in Karlsruhe, the Federal Institute for Food Technology in Munich and the Federal Institute of Home Economics in Stuttgart. Representatives from the named Federal Institutes visited the area twice during the test period and drafted extensive reports on their observations, which were discussed during a conference of all participants in July 1956. One major finding was that the retailers had not fulfilled their tasks at all. Some had not even distributed the informational material, nor had they kept records or reported on their sales. Finally the wholesalers had to step in. The parties from the production side thus concluded, that "the producer would have to follow the way of his merchandise and the advertising very carefully up to the freezer. He instructs the retailers about the best opportunities for advertising in a very insistent way. The written word alone does not suffice. The retail trade has to be visited, in order to inform him thoroughly about good advertising for frozen foods."[26]

Although the financial means for this campaign had obviously been insufficient, the organizers were very content with the obtained results. For the very first time they obtained some insight into the composition of sales figures, the role of different products and the different types of shops and freezers. From their perspective, the campaign had forcefully demonstrated how important the cooperation between the diverse involved actors was, how much attention had to be paid to the shopkeepers and to advertising campaigns. Already while the test was still running and the results had not been evaluated, representatives from *Transfrigoroute* and *Transthermos*, the producers of freezers like *Linde, Escher-Wyss* and *BBC*, wholesalers like *REWE, SPAR* and *EDEKA*, and a couple of food companies from the meat, fish and vegetable industry joined in to found the *Working Association German Deep Freezing Chain* (Arbeitsgemeinschaft Deutsche Tiefkühlkette), which would later become the *Deep Freezing Institute* (Deutsches Tiefkühlinstitut) (Arbeitsgemeinschaft Deutsche Tiefkühlkette 1956).

From its very beginnings the *Tiefkühlinstitut* staged its foundation as a kind of virgin birth and an entirely new institution, which marked a new beginning. This remained unchanged until today. A recent image film does not even mention the earlier beginnings, but pretends that the German deep freezing industry began its life in the 1950s without any predecessors, although institutional, personal continuities were in fact as strong as the arguments for deep freezing. All three members of the founding management board had already been listed in the Who's Who of the

---

[25] ibid., p. 58.

[26] Ibid., p. 58.

Deep freezing Industry from the Third Reich (Anonymous 1944/45. Handbuch: 298) and after 1945 the former *Imperial Institute for Food Preservation* had become a Federal Institute under the very same director Rudolf Plank and kept most of its former personnel. Even the director of the former institute of the army, Rudolf Heiss, was still actively involved. And still, they justified their applications with the sake of frozen foods for "Volksernährung",[27] while the informative brochures from the 1950s and 1960s showed strong similarities with those from the 1930s and 1940s. As before, the development was imagined from the side of production and industry. And although the market test of 1956 took a look at consumption, the consumer was still looked upon from above as a recipient of messages and passive followers of advice. Thus the belief in the possibility to "make" markets by strong top-down advertising prevailed, which would be able to bring about almost every message. In contrast, the wishes and demands of the consumers were not mentioned at all, neither in the applications for financial support of the test nor in later reports on its results. Still in 1962, the agenda for the *Tiefkühlinstitut* listed the task to put through the guidelines for frozen foods, to inform and train the sales personnel, to increase the turnaround of frozen foods, to create new marketing opportunities, to promote research and to enable German production to stand the international competition. But the word "consumer" was not used a single time; the consumer was not recognized as a part of the business, which had to be taken seriously (Anonymous 1962b. Das Kältebewusstsein: 3). Reports from a conference on frozen food problems in 1968 show the ongoing tradition of this view. When a consumer representative demanded industry to not concentrate on the industry's view, but to widen it to the consumer's perspective, to see the consumer as a partner, as an active part of the market and not as a manipulable consumption slave, the representatives from industry simply laughed and called this a "provocative view" (Anonymous 1968. Provokatorische Verbraucher: 9–10).

Nevertheless: The development of the frozen food business in the 1960s is as much a child of modern marketing methods as a result of the economic development, which opened up new economic leeways. Though budgets may have been limited, the consumers may have been inexperienced with affluence and thus overcharged with choice, the growing budgets opened up options for choice. Finally the single company increasingly turned to market research to learn about the consumer's wishes. During the late 1950s almost every important market research institute in Germany studied the field of frozen foods in order to find out, who bought and consumed frozen foods, why he did so and which products were liked or disliked most. The J. W. Thompson GmbH published the results from an opinion survey on frozen foods among housewives in 1959, the important German Institute for Demoscopy

[27] "The introduction of the deep frozen goods requires a thorough information of the consumers. If this group of foods, which is very important for the people's food (Volksernährung) can be introduced at all or if the introduction is destined to fail depends on the right treatment of the frozen goods in the household." Letter Deutsche Fischwerbung e. V., Bremerhaven to Ministry of Food, Agriculture and Wood of North-Rhine-Westphalia, 11.1.1956, in: Landesarchiv Nordrhein-Westfalen (Hauptstaatsarchiv Düsseldorf), NW 123–354, p. 93.

and Infratest as well as the Nielsen Company followed in 1960 and the wholesaler GEG, the purchasing company of the consumer associations, produced a similar report in 1960.[28] Moreover, producers of frozen foods commissioned market surveys, for example by the renowned market research institute *Society of Consumption Research* (*Gesellschaft für Konsumforschung*).[29] Because of the obvious potential of the frozen food market for advertisers, the market research departments of publishers joined in with own inquiries as well[30] and the Institute Allensbach included frozen foods in their periodical surveys of the advertising market (Institut für Demoskopie Allenbach 1960ff). Since 1959 even single producers used test markets to reduce their risk of failure and misinvestment (Führer 1963: 49–55). Nevertheless, such surveys were expensive and meant reasonable investments, which were only accessible for large enterprises or institutions like the *Tiefkühlinstitut*. This explains, why the freezing industry underwent a concentration processes from early on.

But despite all these efforts and the enormous growth rates, the per-capita consumption remained low and did not fulfil the hopes at all. Surveys from 1959 and 1960 show, that 34–40 % of the interviewed had not even heard about frozen foods at all, while about a third had only vague ideas about frozen foods. Socio-economic data from the buyers and non-buyers which were selected to identify target groups for advertising and public relations show strong links between consumption, income and education. Moreover strong generational effects can be observed, as the typical frozen food buying housewife would be between 30 and 50 years, she or her husband would be a civil servant, a white-collar worker or run an own enterprise, have an available income of over 520 Deutsche Mark and live in a town with more than 30,000 inhabitants and in a household with four or more members. Housewives who had not heard about frozen foods, yet, would typically be over 60, pensioners, farmers or workers, have an available income of less than 320 Deutsche Mark, a lower educational level and live in a small household with one or two members (Infratest 1960. Verbraucherbefragung: 15; Anonymous 1959a. *Ergebnis*). From this viewpoint frozen foods were a modern product for young and better-off urban dwellers.

Surprisingly enough it turned out, that advertising in the shops was most effective for lifting sales, followed by newspaper advertising. About 60 % of the interviewed consumers said, that they would be willing to buy more frozen foods in the future. Those who did not, pointed to the high price, said, that enough other goods were available, that they had an own garden or that they would simply prefer to buy fresh food. Only 11 % of the non-buyers were willing to buy frozen foods in the future. Brands did play almost no role, as they were simply not known. The

[28] J. W. Thompson GmbH. 1959; Institut für Demoskopie. 1960. Nielsen A.C. Company GmbH. 1960; Arbeitsgemeinschaft deutsche Tiefkühlkost. 1960; Infratest GmbH. 1960. Tiefkühlkost-Marktforschungsergebnisse; Deutsches Tiefkühl-Institut. 1961.

[29] See for example the study from 1969: Die Verwendung von Tiefkühlkost (Findus GmbH, Frankfurt a.M.), Gesellschaft für Konsumforschung, Nürnberg, Historisches Archiv, S 1969–062.

[30] Their surveys were meant as a basis for the planning of advertising, see Axel Springer Verlag 1978; Verlag Axel Springer 1985.

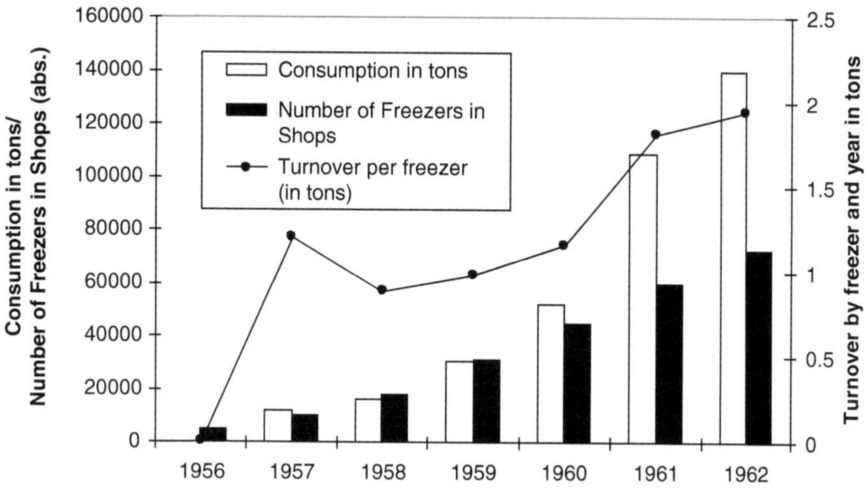

Fig. 11.2 The number of freezers in shops and the turnover of frozen foods per freezer in Germany 1956–1962 (Source: Calculated from Führer 1963: 29, 133)

promotion of the nutritional advantages of frozen foods had only limited effects. Though a third of the interviewed listed the aspect of freshness among the advantages, only 10 % named nutritional advantages explicitly, whereas half of the interviewed housewives stressed time and effort-saving effects as most positive. Price was an important, but not the decisive argument. Only a fifth criticized the prices for frozen foods as being too high, whereas the same share of consumers criticized the bad taste of products (Ibid.: 25). Overall, the picture was not too bad, and so was the development of absolute sales figures. Between 1957 and 1962 production tenfold up to 12,000 kg while producers, who had left the market after the end of war, now saw the potentials of this market and re-entered it in the 1960s, among them Solo-Feinfrost, Nestlé, Findus and Oetker (Führer 1963: 13). One important precondition for this development was the increasing number of shops that offered frozen foods on a regular basis and could rely on a corresponding network of wholesalers and carriers to provide them. The number of wholesalers indeed increased from 20 in 1955 to 600 in 1962 (Fig. 11.2). During the same period the number of freezers in shops grew from 5,000 to 80,000 (Führer 1963: 131, 133).

Nevertheless, per-capita consumption grew only slowly. In 1962 every inhabitant of West Germany consumed only 2.4 kg frozen foods on average, whereas Sweden had a per capita consumption of 4.4 and the USA of 24 kg (Führer 1963: 28). Still in 1964 only every eighth household bought frozen food on a regular basis (Michaels 1964) (Fig. 11.3).

This situation frustrated the producers. Further inquiries should find out, why the development went so slow. They were not only undertaken by the producers themselves, but to an increasing extent by the trading cooperatives REWE, SPAR

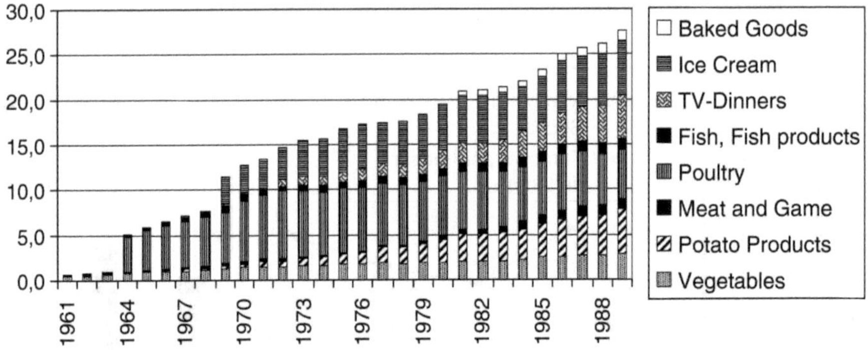

Fig. 11.3 Consumption of frozen foods in West Germany 1961–1989 (kg per capita and year) (Source: Statistisches Jahrbuch über Ernährung, Landwirtschaft und Forsten 1965: 149; ibid. 1970: 155; ibid. 1972: 159; ibid. 1977: 163; ibid. 1983: 165; ibid. 1989: 169; ibid. 1996: 196)

and EDEKA, which supplied their member retail shops with goods, market information and advertising material.

The freezing companies and the *Tiefkühlinstitut* identified the insecurity of consumers about what frozen foods were at all as one major problem. They stressed, that often, frozen foods were called "Gefrierkost", although the foods were not deep frozen, but only chilled (Führer 1963: S. 27). And in fact, very different terms for frozen foods were in use like chilled food, highly chilled foods, frozen and deep frozen foods. Before 1945, the term "Gefrierkost" had been used widely, but the *Tiefkühlinstitut* had decided to use "Tiefkühlkost" – probably to mark the new beginning – although the underlying German word "kühlen" means only "chilled". This opened up the way for misunderstandings, especially as the Food Ministry and retailers used different terms: Whereas the Ministry spoke of deep *frozen* foods (tiefgefrorene Lebensmittel), the Institute spoke of deeply *chilled* foods (Tiefkühlkost). (Bund für Lebensmitteluntersuchung und –kunde. 1962. In the institute's own perception, this created insecurity. Cartoons in the journal "Tiefkühl-Praxis" showed a housewife in front of products of different temperatures, being totally confused by the terms and the declarations on the packages (Fig. 11.4).

The German Association for Food Law and Science (Deutscher Bund für Lebensmittelrecht und –kunde) and the Ministry of Food and Agriculture worked out uniform quality standards, which were to become part of the food law and aimed at standardizing quality (Bund für Lebensmittelrecht- und kunde 1962). But despite all efforts made, this seems to have been a more or less academic discussion or a discussion between food law specialists.

Fig. 11.4 Housewife, confused by different temperatures of foods (Source: Anonymous 1962h: 7)

Fig. 11.5 The badly kept freezer (Source: Anonymous 1962j. In vielen Truhen: 18)

More important was the fact, that uninformed and unreliable shop owners harmed the trust in these foods. The journals from the freezing business are full of reports on retailers, who did not take proper care for their freezers (Fig. 11.5).

Consequently the goods were not kept at −18°, thawed and eventually showed freezer burn, so that the quality argument would not fit any more. Consumers who had trusted the quality promise and had bought frozen foods would disappointedly turn away. Another point, that was especially stressed by consumer representatives touched upon questions of trust, too: The packages of frozen foods had no uniform weights. Consumer representatives criticized, that the size of the packages would not serve the needs of consumers, as these did not fit to the different sizes of household nor to the size of usual meal portions. Moreover the different sizes did not allow comparing prices between different products easily (Zacharias 1959; Anonymous 1962d. Doch eine Irreführung; Anonymous 1962g. Krumme Gewichte).

In order to fight this situation, the Tiefkühlinstitut set up large informational campaigns for the shop-owners and the journals were full with instructive articles how to take care of freezers and how to arrange the packages best. Shop owners were recommended to determine a special department manager in every shop or – if the shop was too small to do so – to determine a sales assistant to take care of the freezers and look after them several times a day (Anonymous 1963a. Aus dem Alltag). While the first had to take care of the freezers, their functioning and cleanliness, the second would give advice to the housewives, how to shop, transport and prepare the new foods (Nebendahl 1964). Over time it turned out, that the freezers themselves were a problem, too: Some did not even reach the needed low temperature grades, in others the thermometer would not work properly (Brauneis 1970; Anonymous 1970b. Wer hat.) As the variety of frozen food products increased over time, the assortment of goods outgrew the dimension of the existing, rather small freezers. Larger freezers would not only be expensive, they would need more space as many a small corner shop had. This explains, why large food store chains like SPAR, REWE and EDEKA and modern self-service super markets were so active in this field. They had the financial means, could include space for the freezers in the building plans for new markets, build up own delivery, repair and advice services. On the other hand, frozen foods would help them to underline their modernity – apart from the fact, that they simply promised them a rewarding perspective (Anonymous 1963b. Sortimentsgestaltung). Believing, that much of the problem went back to missing information producers as well as state institutions poured out a steady stream of leaflets, brochures and articles for shop owners as well as for housewives. They organized numerous events, during which the right preparation of frozen foods was demonstrated and edited handbooks and cookbooks for frozen foods (Siemens 1969; Schönfeldt 1981; Wolter 1977; Berry 1976). One firm went so far not only to rent out freezers for little money, but to send out teams of two household teachers to the single households on request, where they would then demonstrate the proper handling and preparation of frozen foods (Anonymous 1963c. Jopa) (Fig. 11.6).

Only over time, industry realized that not missing information was the central problem, but the self-image of German housewives. An article in "Die ZEIT" from June 1964 for example cited a housewife from Swabia. Asked, why she did not buy frozen foods this woman stated: "I would like to try it, but when my neighbours see the frozen food package in my dustbin; they think I am a slut." (Michaels 1964).

Fig. 11.6 Demonstration of frozen foods to housewives (1962) (Source: Anonymous 1962f Hausfrauen-Vorträge)

Very similar, an interviewed woman stated, that she would not use frozen food because "my man is used to expect me to bring a meal on the table, which is cooked with love, when he comes home from his office. Of course this costs time, but I have it. Since the children are grown up and have left home, I am pleased to pamper him with excellent cuisine, self made, so to say." (Ingeborg 1970). These findings were similar to those of the motivational researcher Ernest Dichter. Based on surveys, he had undertaken at his Institute for Motivational Research Dichter concluded in 1964, that the introduction of frozen foods in the USA had been hindered by a vague feeling of guilt, which resulted from the aspect of the products' convenience. When over time women got convinced of the quality aspects of the foods, they would choose them for their superior quality, not for convenience. Even more, the growing variety of available frozen products would arouse their curiosity and would give them the feeling to act as discoverers. Going for a higher quality and the associated better taste, housewives would expect new, more and even better products. Once the aspect of novelty has gone, "women seek opportunities for expressing individual tastes", so that distinctive packaging and new, unusual flavours become decisive

(Dichter 1964: 44–45). On the other hand, the examples of spinach and poultry show that rational arguments could indeed outdo emotional aspects – if they were strong enough. This was especially the case with spinach and poultry. Consumption figures for poultry had never exceeded 2 kg per capita and year before 1955. Poultry was expensive and clearly classified as upper-class food. Especially as its preparation was time-consuming it was mostly served on festive occasions. Being easily digestible it was moreover the classical food for ill or weak people. But due to industrialized production methods, frozen poultry was getting cheaper now. As it came without feathers, no extensive preparations before frying or cooking were needed and it was easily possible to bake the broilers in the oven. During the 1960s and well into the 1980s, roasted chicken was a fashionable item at fast food outlets. Whereas overall per capita consumption of meat grew from 47 kg in 1955 to 82.5 kg in 1974, poultry consumption boosted from 1.7 to 8.7 kg that is by more than 500 %. And while overall meat consumption is on decline since the late 1990s, poultry consumption keeps growing. Already in 1991 it reached the amount of 18.5 kg (Teuteberg 1986: 240–241; Pro-Kopf-Verbrauch 2001–2009).

The other example is frozen spinach, which combined several advantages. Who has ever cooked fresh spinach, knows that its preparation is rather laborious, as a reasonable amount of spinach needs to be washed several times, but will shrink to almost nothing during cooking. In this case the time- and energy-saving aspect was rather obvious. Moreover spinach was and still is the prime example for the nutritional advantages of frozen foods. Many an article used spinach as an example to document the nutritional advantages of frozen spinach. It was argued, that the spinach on the family table, which came from the freezer had more vitamins than spinach, which had been prepared from fresh produce by the housewife (Zacharias 1962). Adverts as well as journal articles presented it as a modern, fresh and nutritious food that was produced by the most modern production methods (Anonymous 1962k. Spinat). In this case aspects of convenience, health and economic rationality came together and ruled out the aspects of guilt and laziness totally by the strong rational point of vitamin content and a price that was even below the price of the fresh produce. Indeed, during the early years of the frozen food business spinach made up for up to 70 % of the overall sales of vegetables, very much to the astonishment of the producers, who wondered, why other vegetables were so much harder to sell. Still in 1963 it held a share of all sales of about 40 % (Anonymous 1962e. Ein potentieller Markt; Anonymous 1962l. Wußten sie schon daß…).

As emotional arguments played their role, it might have had the negative effect that producers sought the advice of consumer organizations and of home economists. Consumer representatives pointed to the model of the rational consumer, too, and refused emotional appeals in advertising at all. According to them, the decisive arguments should be quality and – above all – fair pricing. Apart from that their brochures and the special issues of consumer journals stressed technical aspects and problems and so did the many books and booklets on deep freezing. Buying and keeping frozen foods the correct way thus was an art, which had to be acquired and learned during courses – and household economists had their own reasons to underline these aspects, as they strived for professionalization and signification. So, the dominance of information about technical aspects may have shied away the housewives instead of attracting them by pointing to new possibilities in an unagitated manner. Around 1970s the frozen food branch obviously realized this as a possible mistake: "It seems, as if frozen foods... had been depicted as too problematic." (Anonymous 1970a. Schwere Geschütze). And maybe, this is the reason why promotion in the shops and sampling by the retailers turned out to be most successful.

Producers were indeed in an unfavourable situation: They had extended their production and distribution apparatus and had built up overcapacities. Unfortunately their rational argumentation did not work. Especially ready meals did not sell very well to private households, where the housewife's work was not calculated in economic terms only. Things looked different, were cost for wages of kitchen personnel played a role. This was especially the case during the economic growth of the 1960s, when workforce began to run short and wages rose. In this situation frozen foods producers began to sell frozen meals to canteens, hospitals and the army. This tactic had the advantage, that they would sell large quantities with low efforts and distribution cost, whereas the institutions profited, too. Frozen foods and meals would allow them to save the money for the personnel and kitchen equipment, but still be able to serve many different pre-fabricated dishes. In fact, about a third of all frozen foods went into catering institutions already in the 1970s and these figures kept growing. Here, all the arguments, which were brought about by industry, met professionals, for whom cooking was no emotional affair at all, but linked to rational work organization and a sharp calculation of cost and efforts (Cf. Thoms 2005). Today, only 53 % of all frozen food are sold via supermarket and shops, whereas 47 % go to the catering business and most of the growth in sales is coming from the demand of caterers (Tiefkühlinstitut 2010: 10). This means, too, that the development of the frozen foods business cannot only be explained on the level of private households.

Over time, the structure of the consumption of frozen foods changed deeply: Poultry remained important, but lost its former strong position, whereas the proportion of ready meals and potato products increased from nothing to around 17 % in 1989. Entirely new product groups entered the business, like bakery products. In 1959 fish fingers were introduced with immense success (Töpper 2009; Josephson 2008; Heidbrink et al. 2008). Frozen pizza followed in 1970, which today holds about 15 % of the entire production, whereas potato products like fries make up for about 13 % and vegetables increased to almost 17 % (Tiefkühlinstitut 2010: 8).

Has frozen food once been bought mainly by the urban population, the situation changed until today. Freezers spread in the countryside from early on; first by the so-called locker plants, where one could rent a small freezer's department from a cooperative. These facilities had been established as larger corporate freezing facilities in central places. Lockers were used to store goods from own production, such as meats as well as vegetables and fruit from the garden (Gutschmidt 1955: 121; Anonymous 1955. Gemeinschaftsgefrieranlagen; Gutschmidt and Zacharias 1960; Zacharias and Thumm 1968; Anderson 1953: 287–302). As long as household freezers were expensive, this system seemed reasonable, though the ways to and from the facilities must have been rather time-consuming. But today, even the rural households do not grow their vegetables and fruit themselves any more and even

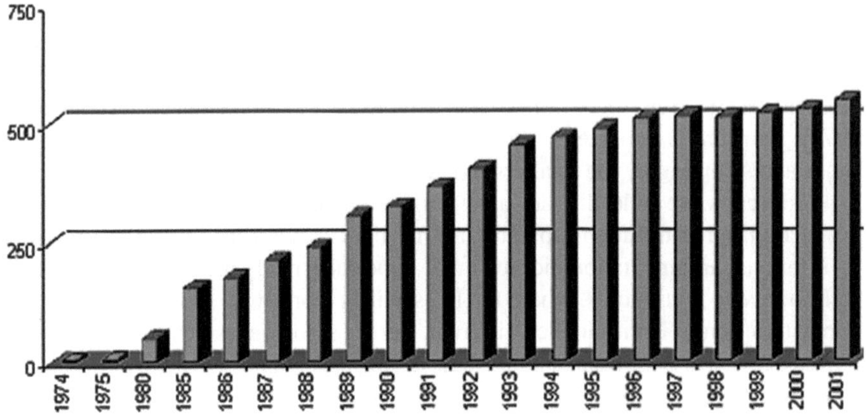

Fig. 11.7 Sales figures of the home delivery service for frozen food Eismann (1974–2001, in hundred thousand DM) (Source: Firma Eismann)

women in the countryside work outside home. For them, home delivery of frozen food is an ideal way to procure families in a time-saving and healthy manner, because shops are often far away and a car is needed to get there.

Had rural populations once hesitated to buy frozen foods, these now began to form a strong consumer segment. As many smaller shops in the countryside had to close, because they could not stand the competition of the cash and carry markets and supermarkets, many rural households now use home delivery services extensively to buy vegetables and other foods. Very much in contrast to the continuing crisis of the German food retail business, services like "Eismann", which was founded in 1974 developed very successful (see Fig. 11.7.). Today, the market leader in this segment, the company "bofrost" alone sells 20 % of all in Germany consumed frozen foods (Bofrost 2010).

Today, the social differences between consumers of frozen foods have become smaller: The highest rates of users are not from the highest income group any more but from the third lowest, while the highest groups makes up for only 10 per cent of the consumption. The larger the household, the lower the consumption and vice versa; highest consumption figures are to be found among single households. Classical frozen foods like spinach, fries, poultry and so on, are mainly bought by elderly consumers, whereas younger people prefer pizza, baguette and ready meals. These find many buyers in the group of men, who have no problem with their image anyway, if they use ready meals instead of cooking themselves (Axel Springer Verlag 1978, 59–70).

11.3 Conclusion

The example of frozen foods illustrates nicely, how production and trade of foods developed into large technical and transnational system. Nazi economists as well as food policy had realized this dependency from early on. The state promoted frozen food, while entrepreneurs made high investments. In fact the production of the late 1940s reached a relatively high production level, that was only achieved again late in the 1970s. But after the end of WWII, it came out that politicians as well as the company owners had concentrated too much on the technical aspects. Over this they had forgotten the consumer, his wishes and needs. Building up the infrastructure was of course importing. But consumers had another perspective. They did not want to know how to handle the existing problems, they wanted a problem-free product, that fitted into their food pattern and went along with their life styles. Caterers in contrast had different problems. For them economic arguments played a much greater role and this is why frozen foods were accepted in the catering industry much quicker. First conceived as a makeshift, respectively as a first entry into a problematic market, the major proportion of the frozen food production today goes to the catering business with its perspective to rationality and economy. From this point of view, frozen foods are a striking example for the dominance of the production perspective, which has been attributed to German economics by the business historians of marketing.

References

75 Jahre Eisfink. 1961. [Asperg/Württ.]: [Eisfink, C. Fink OHG].

Andersen, A. 1997. *Der Traum vom guten Leben. Alltags- und Konsumgeschichte vom Wirtschaftswunder bis heute*. Frankfurt am Main: Campus.

Anderson, O.E. 1953. *Refrigeration in America. A history of a new technology and its impact*. Princeton: Princeton University Press.

Anonymous. 1940. *Gefriertaschenbuch. Herstellung, Bewirtschaftung und Verbrauch schnell gefrorener Lebensmittel*. Berlin: VDI-Verlag.

Anonymous. 1941. Bericht über die VDI-Kältetagung. *Zeitschrift des VDI* 85: 618.

Anonymous. 1943. Gefrierkonserven im Haushalt. *Technik im Haushalt* 7.

Anonymous. 1944/45. *Handbuch der deutschen Kälteindustrie*, 5th ed. Mühlhausen: Markewitz.

Anonymous. 1947. Kälte- und lebensmitteltechnische Tagung in Ettlingen bei Karlsruhe am 6. und 7. Juni 1947. *Zeitschrift für Untersuchung der Nahrungs- und Genußmittel* 88: 199–200.

Anonymous. 1948/49. Produktionszahlen aus der US-Obst- und Gemüseverwertungs-Industrie. *Obst- und Gemüseverwertungsindustrie* 33/34: 299–300.

Anonymous. 1951. *The cold chain in the U.S.A. Report of a group of European experts*. Paris: OEEC.

Anonymous. 1955. Gemeinschaftsgefrieranlagen. *Kältetechnik* 7: 121.

Anonymous. 1956. Wachsender Appetit auf Tiefgekühltes. *Die ZEIT*, Nov 8.

Anonymous. 1959a. Ergebnis der Hausfrauenbefragung ueber Tiefkuehlkost in Muenster/ Westfalen, April.

Anonymous. 1959b. Tiefgekühltes. In *Die ZEIT*, 5 Sept, no. 41.

Anonymous. 1960. 10 Jahre Interfrigo. *Kältetechnik* 12: 328.

Anonymous. 1962a. Brathendl. Aus dem Wienerwald. *Der SPIEGEL* 28: 36–38, 11 July.

Anonymous. 1962b. Das Kältebewußtsein wecken. *Tiefkühl-Praxis* 3(7): 3.

Anonymous. 1962c. Die Zusammenarbeit in der europäischen Tiefkühlwirtschaft'. *Tiefkühl-Praxis* 3(10): 3–5.

Anonymous. 1962d. Doch eine Irreführung der Verbraucher! *Tiefkühl-Praxis* 3(10): 26.

Anonymous. 1962e. Ein potentieller Markt größten Ausmasses. Tiefkühlkost aus der Sicht der Verbraucher, des Handels und der Industrie. *Tiefkühl-Praxis* 4(3): 15–16.

Anonymous. 1962f. Hausfrauen-Vorträge. *Tiefkühl-Praxis* 3(10): cover.
Anonymous. 1962g. Krumme Gewichte bei Tiefkühlwaren. *Tiefkühl-Praxis* 3(10): 29.
Anonymous. 1962h. Richtlinien für tiefgefrorene Lebensmittel. *Tiefkühl-Praxis* 3(3): 4–7.
Anonymous. 1962i. In vielen Truhen herrscht Finsternis. Das Tiefkühlgeschäft aus der Praxis des Verkaufspraktikers. *Tiefkühl-Praxis* 3(5): 16–18.
Anonymous. 1962j. Spinat am laufenden Band. *Tiefkühl-Praxis* 3(12): 14–15.
Anonymous. 1962k. Wußten sie schon daß…. *Tiefkühl-Praxis* 3(1): 26.
Anonymous. 1963a. Aus dem Alltag des Tiefkühlchefs (Teil 4): Die schönste Aufgabe bleibt das Verkaufen! *Tiefkühl-Praxis* 3(4): 29.
Anonymous. 1963b. Die Sortimentsgestaltung bei Tiefkühlkost und ihr Einfluss auf den Absatz. *Tiefkühl-Praxis* 4(4): 8–9, 11.
Anonymous. 1963c. Jopa testet Haushaltsberatungsdienst. *Tiefkühl-Praxis* 4(4): 23.
Anonymous. 1968. Provokatorische Verbraucher. Antiquitierte Akzente auf der Tiefkühlkost-Tagung bei Duttweiler. *Verbraucherpolitische Korrespondenz* 15(33): 9–10, 25 Nov.
Anonymous. 1970a. Schwere Geschütze an die Tiefkühl-Front. *Food and Non-Food* 2(26): 41.
Anonymous. 1970b. Wer hat den schwarzen Peter. *Food und Nonfood* 2(8): 27.
Anonymous. 2010. *Der AID Infodienst. 60 Jahre gebündeltes Wissen*. Bonn: Ernährung, Landwirtschaft, Verbraucherschutz. Available from: http://www.aid.de/shop/pdf/0328_2010_60_jahre_x000.pdf?cb_content_name=0328,%2520Der+aid+infodienst+-+60+Jahre+geb%25FCndeltes+Wissen. Last accessed 26 Feb 2012.
Anonymous. undated [2005]. *50 Jahre Transfrigoroute Deutschland. Verband für temperaturgeführte Logistik,* undated [Frankfurt 2005].
Arbeitsgemeinschaft Deutsche Tiefkühlkette e.V. Niederschrift über die Gründungsversammlung der Arbeitsgemeinschaft Deutsche Tiefkühlkette am Mittwoch, dem 6. Juni 1956. Available from: http://www.tiefkuehlkost.de/publikationen/gruendungsprotokoll_dti/. Last accessed 2 Mar 2012.
Axel Springer Verlag. 1978. *Tiefkühlkost, Konserven, Fertiggerichte*. Berlin: Axel Springer Verlag AG.
Bächi, B. 2009a. 1933. Mikrochemie der Macht. Vitamin C und der Wandel von Körpermetaphern'. In *Arzneimittel des 20. Jahrhunderts. Historische Skizzen von Lebertran bis Contergan*, ed. N. Eschenbruch, V. Balz, U. Klöppel, and M. Hulverscheidt, 103–121. Bielefeld: Transcript.
Bächi, B. 2009b. *Vitamin C für alle!, pharmazeutische Produktion, Vermarktung und Gesundheitspolitik (1933–1953)*. Zürich: Chronos.
Bebel, August. 1979. *Die Frau und der Sozialismus*. Berlin: Dietz Verlag.
Berry, Mary. 1976. *Das Tiefkühl-Kochbuch*. München: Heyne.
Bofrost. 2010. … *with enjoyment and passion*. Bofrost: Straelen. Available from: http://www.bofrost.de/is-bin/INTERSHOP.enfinity/WFS/bofrost-shopDE-Site/de_DE/-/EUR/Onion-Literal?Id=o41783_23default_26language_3Ddec. Last accessed 20 July 2013.
Brauneis, Rudolf J. 1970. Mit dem Holzhammer. *Food und Nonfood* 2(8): 24.
Bund für Lebensmitteluntersuchung und –kunde. 1962. *Richtlinien für tiefgefrorene Lebensmittel. Gutachten vom 29. Nov. 1961*. Hamburg et al: Behr. (=Schriftenreihe des Bundes für Lebensmitteluntersuchung und –kunde, vol. 39).
Bundesministerium für Ernährung, Landwirtschaft und Forsten. 1957–1989. *Statistisches Jahrbuch über Ernährung, Landwirtschaft und Forsten der Bundesrepublik Deutschland*. Münster-Hiltrup: Landwirtschaftsverlag.
Büning-Fesel, M. 2010. 60 Jahre aid. Medienkompetenz in Sachen Ernährung, Landwirtschaft und Verbraucherschutz. *Ernährung im Fokus* 9: 186–189.
Classen, Th., et al. 1924. Eine Kühlhaus-AG für die Fisch-Industrie Deutschlands. In *Der Fisch. Mitteilungen über Fischerei, Fischindustrie, Fischhandel und allgemeine Fischverwertung*, vol. 2, ed. Hans Lengerich, 287–305. Lübeck: Verlag Der Fisch.
Cooper, A.J. 1997. *The world below zero. A history of refrigeration in the UK*. Aylesbury: ACR Today & Battlepress.
Critchel, J.T., and J. Raymond. 1912. *A history of the frozen meat trade*. London: Constable.
Deutsches Tiefkühl-Institut. 1961. *Sonderdruck: Die Ergebnisse einer Großhandelsbefragung*. Bremerhaven: dti.

Dichter, E. 1964. *Handbook of consumer motivations. The psychology of the world of objects*, 44–45. New York et al.: Mc-Graw Hill Book Company.
Dienel, H.-L. 2004. *Die Linde AG. Geschichte eines Technologiekonzerns. 1879-2004*. Beck: München.
Döring, G. 1960. Tiefkühlkostmarkt zwischen gestern und morgen. *Absatzwirtschaft* 3: 589–591.
Drews, J. 2002. Die Gleichschaltung im Stullenverzehr. Ernährungspsychologie im "Dritten Reich" – zwei Fundstücke. *Werkstatt Geschichte* 12: 82–93.
Ellerbrock, K.-P. 1993. *Geschichte der deutschen Nahrungs- und Genussmittelindustrie 1750-1914*. Stuttgart: Steiner.
Fleischmann, O. 1934. *Der Weltmarkt in Gefrier- und Kühlfleisch. Eine Untersuchung über Erzeugung, Handel und Verbrauch des Gefrier- und Kühlfleisches*. Ph.D. diss., University of Köln.
Führer, E. 1963. *Die Marketing-Konzeption dargestellt am Beispiel der Tiefkühlkost*, Ph.D. diss., University of Munich.
Gutschmidt, J. 1955. Gemeinschaftsgefrieranlagen in der deutschen Landwirtschaft'. *Kältetechnik* 7: 235–243.
Gutschmidt, J., and R. Zacharias. 1960. *Gefrierkonservierung. Untersuchungen an Gemeinschaftsgefrieranlagen*. Hiltrup: Landwirtschaftsverlag.
Hamilton, S. 2009. The economics and convenience of modern-day-living: Frozen foods and mass marketing, 1945–1965. *Business History Review* 77: 33–60.
Heidbrink, I., E. Hoops, and K. Jantzen. 2008. Fischstäbchen: Fisheries and fish processing as industrial heritage. In *Proceedings of the 10th conference of the North Atlantic Fisheries History Association Bremerhaven, 7–11 August 2006*. Bremerhaven: Deutsches Schiffahrtsmuseum.
Heiss, R. 1941. Die Aussichten der Gefrierdauerwaren. *Zeitschrift für Volksernährung* 16: 234–236.
Hellmann, U. 1990. *Künstliche Kälte. Die Geschichte der Kühlung im Haushalt*. Gießen: Anabas.
Hietala, M., and V. Leopist. 1998. Arctic Finland and the new technology of food preservation and refrigeration, 1850-1990. In *Food and material culture*, ed. Martin R. Schaerer and Alexander Fenton, 208–327. East Linton: Tuckwell Press.
Hilck, E., and R. auf dem Hövel. 1979. *Jenseits von Minus Null. Die Geschichte der deutschen Tiefkühlwirtschaft*. Köln: Tiefkühlinstitut.
Infratest gmbH. 1960. *Tiefkühlkost-Marktforschungsergebnisse, Teil I: Verbraucherbefragung über Tiefkühlkost*. Frankfurt: München: J. W. Thompson GmbH.
Ingeborg, H. 1970. Gespräch an der Truhe. *Food und Non-Food* 26: 46.
Institut für Demoskopie Allensbach. 1960ff. *AWA. Werbeträger-Analyse*.
Institut für Demoskopie: Aus der Gefriertruhe: Hausfrauenbefragung über tiefgefrorene Lebensmittel. Allensbach am Bodensee 1960: Institut für Demoskopie Allensbach.
International Institute for Refrigeration. 2005. *Serving the needs of mankind. A history of refrigeration/Le froid au service de l'homme*. Paris: International Institute of Refrigeration.
J. W. Thompson GmbH. 1959. *Ergebnisse der Hausfrauenbefragung über Tiefkühlkost in Münster/Westfalen*. Frankfurt: J. W. Thompson GmbH.
Jones, G. 2005. *Renewing Unilever: Transformation and tradition*. Oxford: Oxford University Press.
Josephson, P.R. 2008. The ocean's hot dog. The development of the fish stick. *Technology and Culture* 49(1): 41–61.
Kallert, E. 1967. Entwicklung und wirtschaftliche Bedeutung der Konservierung von Fleisch. In *Handbuch der Kältetechnik*, vol. 10, ed. Rudolf Plank, 127–207. Berlin/Göttingen/Heidelberg: Springer.
Kraut, H., and W. Wirths. 1951. *Mehr Wissen um Ernährung*. Frankfurt/M: Verlag Kommentator (=Berichte über Studienreisen im Rahmen der Auslandshilfe der USA, H. 74).
Kuprianoff, J. 1960. 50 Jahre Kälteanwendung auf dem Gebiet der Lebensmittelkonservierung und –verarbeitung. *Kältetechnik* 12: 24–30.
Melzer, J. 2003. *Vollwerternährung. Diätetik, Naturheilkunde, Nationalsozialismus, sozialer Anspruch*. Stuttgart: Steiner.

Michaels, H. 1964. 'Tiefgekühltes'. In *Die ZEIT*. 5 June, no. 23.
Mosolff, Hans. 1939. *Vorratswirtschaft durch Kälte*. Berlin: Haude und Spenersche Buchhandlung.
Mosolff, Hans. 1940. Einsatz der Gefrierwirtschaft für die Vorratswirtschaft. *Der Vierjahresplan* 4: 140–143.
Mosolff, Hans. 1941a. *Der Aufbau der deutschen Gefrierindustrie. Handbuch der Tiefkühlwirtschaft*. Hamburg: Hans A. Keune Verlag.
Mosolff, Hans (ed.). 1941b. *Tiefkühl-ABC*. Hamburg: Hans A. Keune Verlag.
Nebendahl, R. 1964. Aufgaben und Bedeutung des Fahrverkäufers. *Tiefkühl-Praxis* 5(9): 20–23.
Nielsen A.C. Company, GmbH. 1960. *1. Untersuchung Tiefkühleinrichtungen*. Frankfurt: Arbeitsgemeinschaft deutsche Tiefkühlkost.
Oddy, D.J., and J.R. Oddy. 1998. The iceman cometh. The effect of low-temperature technology and the British diet. In *Food and material culture*, ed. Martin R. Schaerer and Alexander Fenton, 287–307. East Linton: Tuckwell Press.
Oltersdorf, U., and Th. Preuss (eds.). 1996. *Haushalte an der Schwelle zum nächsten Jahrtausend. Aspekte hauswirtschaftlicher Forschung gestern –heute – morgen*. Frankfurt: Campus.
Plank, R. 1926. Das kältetechnische Institut der Karlsruher technischen Hochschule. *Zeitschrift für die gesamte Kälteindustrie* 33: 164–169, 181–193.
Plank, R. 1927/28. *Versuche über die Kaltlagerung von Obst und Gemüse*. Berlin: Springer.
Plank, R. 1928. *Haushalt-Kältemaschinen*. Berlin: Springer.
Plank, R. 1929. *Amerikanische Kältetechnik*. Berlin: VDI Verlag.
Plank, R. 1960. Zur Geschichte des deutschen Kältetechnischen Vereins. *Kältetechnik* 12: 1–8.
Plank, R., and V. Gerlach. 1917. *Über die Konservierung von frischem Beeren-, Kern- und Steinobst in Kühlräumen*. München/Berlin: Oldenbourg.
Plank, R., and E. Kallert. 1915. *Über die Behandlung und Verarbeitung von gefrorenem Schweinefleisch. Im Auftrag der ZEG durchgeführte Untersuchungen*. Berlin: Zentraleinkaufsgenossenschaft.
Plank, R., and E. Kallert. 1916a. *Über die Behandlung und Verarbeitung von gefrorenem Rindfleisch. Im Auftrag der ZEG durchgeführte Untersuchungen*. Berlin: Zentraleinkaufsgenossenschaft.
Plank, R., and E. Kallert. 1916b. *Die Konservierung von Fischen durch das Gefrierverfahren. Im Auftr. der Zentral-Einkaufsgesellschaft ausgeführte Untersuchungen*. Berlin: Verlag der Zentral-Einkaufsgesellschaft.
Plank, R., and E. Schneider. 1927–28. *Versuche über die Kaltlagerung von Obst und Gemüse*. Berlin: Gesellschaft für Kältewirtschaft.
Pro-Kopf-Verbrauch von Geflügelfleisch in Deutschland in den Jahren von 2001 bis 2009 (in Kilogramm). Available from: http://de.statista.com/statistik/daten/studie/186634/umfrage/pro-kopf-verbrauch-von-gefluegelfleisch-seit-2001/. Last accessed 11 Mar 2012.
Reichskuratorium für Technik in der Landwirtschaft/RKTL: Tätigkeitsbericht 1936, 1937.
Rogers, E. 2003. *Diffusion of innovations*, 5th ed. New York: Free Press.
Schipper, W. 1925. *Die Bedeutung der Kälteindustrie für die Lebensmittelversorgung der Großstadt*. München/Berlin: Oldenbourg.
Schlienz, W. 1930. *Die Kühlfisch-Aktiengesellschaft Wesermünde*. Berlin: Org. Verlags-Gesellschaft (Hirzel).
Schmidt-Waldherr, H. 1999. Emanzipation durch Küchenreform? Einküchenhaus versus Küchenlabor. *L'Homme. Zeitschrift für feministische Geschichtswissenschaft* 10: 57–76.
Schöllgen, G. 2008. *Der Eiskönig. Theo Schöller. Ein deutscher Unternehmer 1917-2004*. München: Beck.
Schönfeldt, Sybil Gräfin. 1961. Ausstellung in Hamburg: Beschossenes Essen und Dinner in Dosen. *DIE ZEIT*, 17 Nov, no. 47.
Schönfeldt, Sybil. 1981. *Das komplette Tiefkühlkochbuch*. Bonn: Hörnemann.
Schüller, M. 1999. *Wiederaufbau und Ausbau der Kölner Messe 1946-1956*. Stuttgart: Steiner.
Schumpeter, J. 1912. *Theorie der wirtschaftlichen Entwicklung*. Leipzig: Duncker & Humblot.
Serger, H. and H. Krause. 1950. *Konserventechnisches Handbuch (Serger-Hempel). Praxis und Wissenschaft der Konservenfabrikation*. 12th 3ed. Braunschweig: Serger-Hempel.
Siemens, A.G. 1969. *Die Eiszeit aus der Steckdose*. Berlin: Siemens AG.

Sperling, Frank. 2011. *"Kampf dem Verderb" mit allen Mitteln? Der Umgang mit ernährungsbezogenen Gesundheitsrisiken im "Dritten Reich" am Beispiel der chemischen Lebensmittelkonservierung*. Stuttgart: Deutscher Apotheker-Verlag.

Spiekermann, U. 1997. Zeitensprünge: Lebensmittelkonservierung zwischen Haushalt und Industrie 1880-1940. In *Ernährungskultur im Wandel der Zeiten*, ed. Katalyse e.V. and Buntstift e.V, 30–42. Köln: Katalyse e.V.

Spiekermann, U. 2009. Twentieth-century product innovations in the German food industry. *Business History Review* 83: 291–315.

Stübler, E[lfriede]. 1985. Bundesforschungsanstalt für Ernährung und Lebensmittel (BFEL). *Hauswirtschaft und Wissenschaft* 33: 123–132.

Teuteberg, H.J. 1986. DerVerzehr von Nahrungsmitteln in Deutschland pro Kopf und Jahr seit Beginn der Industrialisierung (1850-1975). Versuch einer qualitativen Langzeitanalyse. In *Unsere tägliche Kost. Geschichte und regionale Prägung*, ed. Hans J. Teuteberg and Günter Wiegelmann, 225–280. Münster: Coppenrath.

Teuteberg, H.J. 1991. Zur Geschichte der Kühlkost und des Tiefgefrierens. *Zeitschrift für Unternehmensgeschichte* 36: 139–155.

Thevenot, R. 1979. *A history of refrigeration*. Paris: Institute of Refrigeration.

Thoms, Ulrike. 2005. Industrialising catering. Technological developments and its effects in the twentieth century. In *Land, shops and kitchen technology and the food chain in twentieth century Europe*, ed. Peter Scholliers and Leen Van Molle, 278–295. Turnhout: Brepols.

Thoms, Ulrike. 2006. Einbruch, Aufbruch, Durchbruch? Strukturen und Netzwerke der deutschen Ernährungsforschung vor und nach 1945. In *Kontinuitäten und Diskontinuitäten in der Wissenschaftsgeschichte*, ed. Rüdiger vom Bruch and Uta Gerhardt, 111–130. Stuttgart: Steiner.

Thoms, Ulrike. 2007. "Vitaminfragen – kein Vitaminrummel?" Die deutsche Vitaminforschung in der ersten Hälfte des 20. Jahrhunderts und ihr Verhältnis zur Öffentlichkeit. In *Wissenschaft und Öffentlichkeit als Ressource füreinander*, ed. Sybilla Nikolow and Arne Schirrmacher, 75–96. Bielefeld: Transcript.

Thoms, Ulrike. 2010. Vegetarianism, meat and life reform in early twentieth century Germany. In *Meat, medicine and human health in the twentieth century*, ed. Christian Bonah, David Cantor, and Matthias Doerries, 145–149. London: Basingstoke.

Tiefkühlinstitut. 2010. *Cool facts 2010*. Köln: Tiefkühlinstitut.

Töpper, V. 2009. *Wie aus dem Fisch ein Fischstäbchen wurde*, welt-online 6 May 2009. Available from http://www.welt.de/wirtschaft/article3685045/Wie-aus-dem-Fisch-ein-Fischstaebchen-wurde.html. Last accessed 8 Mar 2012.

Verlag Axel Springer. 1985. *Tiefkühlkost, Fertiggerichte, Suppen*. Berlin: Springer.

Weber. 1936. Ansprache bei der Einweihung des Reichsinstituts für Lebensmittelfrischhaltung in Karlsruhe am 25. Mai 1936. *Zeitschrift für die gesamte Kälte-Industrie* 43: 185–186.

Wengenroth, U. 2002. Die Flucht in den Käfig. Wissenschafts- und Innovationskultur in Deutschland 1900-1960. In *Wissenschaften und Wissenschaftspolitik. Bestandsaufnahmen zu Formationen, Brüchen und Kontinuitäten im Deutschland des 20. Jahrhunderts*, ed. Rüdiger vom Bruch and Brigitte Kaderas, 52–59. Stuttgart: Steiner.

Winkler, R. 2007. Schöller, Theo. *Neue Deutsche Biographie* 23: 372–373. Available from http://www.deutsche-biographie.de/pnd129308501.html. Last accessed 8 Mar 2012.

Wirz, F. 1941. Foreword. In *Tiefkühl-ABC*, ed. Hans Mosolff. Hamburg: Hans A. Keune Verlag.

Wolter, Annette. 1977. *Das grosse Tiefkühl-Buch*. München: Gräfe und Unzer.

Zacharias, R. 1959. Was erwartet die Hausfrau von der Tiefkühlkost? *Verbraucherdienst* 25–27.

Zacharias, R. 1962. Nährstoffgehalt der Tiefkühlkost. *Tiefkühl-Praxis* 3(2): 4, ibid. 3(3): 8–10.

Zacharias, R. and G. Thumm. 1968. *Konservierung tischfertiger Speisen*. Hiltrup: Landwirtschaftsverlag.

Chapter 12
The Means of Modernization, Freezing Technologies and the Cultural Politics of Everyday Life, Norway 1940–1965

Terje Finstad

After World War II, the Norwegian government set out to modernize the country with the industrialization of agriculture and fisheries, the rationalization of retail and by heightening the standard of living in Norwegian households. The period between 1945 and 1960 was the formative years of the social democratic welfare state in Norway. Within this process, science was meant to guide the modernization and technology to serve as a tool (Sejersted 2011). Freezing technology became part of this process in many levels, and freezing plants were built in connection with the Norwegian fisheries, a fish freezing industry was established, frozen goods counters were installed in retail stores and ordinary households even obtained access to various freezing technologies such as freezer locker plants and home freezers. Within a period of 20 years, Norway became home to some of the leading producers of frozen fish in Europe, and had one of Europe's most developed cold chains. Moreover, it was one of the countries in Europe with the highest density of freezing technology per inhabitant, together with a high consumption of frozen foods.[1]

However, freezing technology represented something quite new in Norway at the end of the war, as there had been some attempts at freezing fish in the Norwegian fisheries before the war, although these had relied on natural ice to produce the needed cold temperature. As a result, freezing technology had to be appropriated, reshaped and woven into society. In order for freezing technology

[1] *Report on the meeting of experts on the marketing of deep-frozen products in Europe* (European Productivity Agency (EPA) of the Organisation for European Economic Co-operation OEEC), 1959); *Marketing and consumption of frozen fish in OEEC countries. Report presented at a meeting of experts on deep-frozen foods held in Verona from the 6th to the 12th of October 1959* (Verona: EPA of the OEEC, 1959).

T. Finstad (✉)
Department of Interdisciplinary Studies of Culture, Norwegian University of Science and Technology, Trondheim, Norway
e-mail: terje.finstad@ntnu.no

K. Gavroglu (ed.), *History of Artificial Cold, Scientific, Technological and Cultural Issues*, Boston Studies in the Philosophy and History of Science 299,
DOI 10.1007/978-94-007-7199-4_12,

to contribute to Norwegian fisheries, the retail sector and households, guidelines for its use and maintenance had to be developed, and workers, retailers and housewives had to learn how to use the technology as well. Additionally, the technology had to be given meanings that resonated with Norwegian cultural traditions and ideas. In short, the technology had to be domesticated into Norwegian society, but the society also had to be reshaped in the same process, thereby creating a two-way integration process (Berker et al. 2006; Latour 1991; Lie and Sørensen 1996).

This mutual shaping of technology and society means that technology cannot be considered as a neutral tool in our everyday lives. Even the most mundane of our everyday artifacts does something and is linked to what we could call a cultural politics of everyday life (Gjøen and Hård 2002; Sørensen 2004). Historians of technology have shown that kitchens and household appliances became linked to the politics of everyday life during the Cold War. The modern household became a sign used to show that the political system, whether capitalism or communism, was a success. This is perhaps best illustrated in the so-called "kitchen debate" between Nikita Khrushchev and Richard Nixon in 1959, in which the two superpower leaders argued as to whether the American kitchen symbolized the progress brought about by capitalism or not (Hamilton 2003; Oldenziel and Zachman 2009; Winner 1980). Thus, there can be no doubt that technology is both a reflection of social and cultural traditions, and involved in the politics of everyday life.

In order to study the cultural politics of technology, it is necessary to investigate technology in use (Edgerton 2008). It then becomes essential to see who promotes technologies, as well as their motivations for doing so. Furthermore, we must investigate how various actors or mediators interpret and inscribe technologies with programs and meanings that are thought to influence users' behaviors. These mediators can be the producers of the technologies, though they can also be governmental agencies or producers relying on the technology for their infrastructure and so on. By focusing on how they use the various technologies, it becomes possible to see how they utilize it to influence their everyday life. Their efforts may of course be more or less successful, as users can for instance develop anti-programs or simply ignore the technology, but by investigating the promotion of technology as embedded in a sociopolitical context it becomes possible to say something about how it was intended to affect the users, even if the users did not become interested in it (Akrich 1992; Oudshoorn and Pinch 2003).

In this chapter, I intend to investigate how freezing technologies were integrated into Norwegian society and how it was used to modernize important sectors of this society. I will look at three sectors, the Norwegian fisheries, the retail sector and households. As such, the chapter will demonstrate how in the period after WW II, freezing technology became both a technology for the conservation of foods, and became linked to various projects aimed at transforming the Norwegian way of life.

12.1 War, Protein and Freezing Technology

On the 9th of April 1940, Norway was occupied by Germany and the effort at introducing freezing technology into Norwegian fisheries was taken over by the Nazis. Hitler wanted his future empire to self-sufficient in terms of food, but for this to work he needed access to proteins. Fish is rich in protein and the Norwegian coast has excellent fishing grounds, so freezing therefore seemed like a technique that could secure enough fish for the planned German empire. Freezing had become the preferred choice for the Germans because the prospects for securing enough tin for the hermetic preservation of foods were poor. Hermann Göring had contacted his pilot colleague from WW I, Philip Reemtsma, and urged him to start building a German freezing industry back as far back in 1938 (Finstad 2004; Jacobsen 1996).

Reemtsma accepted the challenge and established the company, Hochseefischerei Andersen & Co. He bought a ship called the Hamburg, as well as some trawlers. The Hamburg was equipped with freezing machinery and filet machines, which made it the first floating factory for the mass production of frozen fish fillets. With the occupation of Norway, the Hamburg was to be stationed in the fishing harbors during the fishing season. In addition, the Germans built four land-based freezing plants in Trondheim, Bodø, Melbu and Hammerfest. The plant in Bodø had 14 Birdseye freezers and could freeze between 120 and 130 tons of fish per day, hence making it probably the largest freezing plant in the world at the time. An organization called the Fischeinkaufgemeinschaft was established to secure enough raw materials for the plants, and used ships to transport fish from local harbors into the central plants. After freezing, the fish were either transported to Germany by ship or rail, or distributed to Norwegian households (Andreassen 1995; Finstad 2004).

The German introduction of freezing technology during the war represented a turning point in Norwegian fisheries. Many went from producing dried or salted fish to selling fresh fish to the freezing plants, but it was also the first introduction of modern freezing technologies into the fisheries, as earlier attempts at freezing fish had relied on natural ice and salt mixtures. In many ways, the German occupation created excellent conditions for the building of large-scale freezing plants since the fisheries did not have to face economic risks, and the unstable market conditions from the 1930s were long gone. Under normal circumstances, the German production of frozen fish would have been an economic disaster, as the plants were too large to be profitable year round. However, the occupiers controlled the entire economic system, and did not have to balance the budget (Andreassen 1995; Finstad 2004).

In other ways though, one can question whether the Germans really established a modern freezing industry in connection to Norwegian fisheries. True, the plants were equipped with the best machinery of the time and could produce huge quantities of frozen fish, although they faced severe problems in supplying the plants with fish and lacked a proper supply chain. During a raid in Svolvær in 1941 the Hamburg was blown up, meaning of course that the ship could no longer transport fish from Norway. Consequently, the fish could lie for days before being frozen, and become exposed after freezing to variations in temperature due to the lack of refrigerated transport. As a result, it was thawed and rotten by the time it reached consumers (Andreassen 1995; Finstad 2004).

Even though the Germans experienced problems, there can be little doubt that their freezing plants marked the beginning of modern frozen fish production in Norwegian fisheries. For the Germans, freezing was a way to preserve a valuable resource that was needed for the future empire they were building. Thus, freezing became a means not only for preserving fish, but was also used in the geopolitical struggle of WW II. Technological systems such as railways, highways, electricity networks and so forth were essential for the vision of a future German empire, as these were infrastructures that were to link the Third Reich's various parts and make them governable. And as we can see, the establishment of a frozen food chain was part of an effort aimed at feeding the inhabitants of a future society, in addition to modernization (van der Vleuten and Kaijser 2006; van der Vleuten 2010).

Luckily, the Third Reich never materialized, though after the war the Norwegian government continued the German's attempts at introducing freezing technologies into the fisheries. Let us now turn our attention to how freezing technology was used to remake the northern parts of Norway after WW II.

12.2 The Arctic North and the Politics of Industrialization

After WW II, the Norwegian state accounts suffered. The amount of foreign exchange was close to none, and the government wanted to start increasing its exports. For many centuries, fish had been the main export commodity coming out of Norway, so after the war the government decided to get fish exports going again. If the Norwegian fisheries could produce a more processed and modern product than the dried and salted fish of the past, then foreign capital would again become readily available.[2]

For the Norwegian coalition government, industrialization became an important task after the war, and the industrialization of agriculture and fisheries was seen as an important step towards the modern welfare state, particularly in the northern parts of Norway. While approximately 12 % of the Norwegian population lived in the north, this part of the country only contributed 6 % of the gross domestic product and 2 % of the industrial production. Northern Norway's economy contributed to the exporting of fish, though it was scarcely industrialized (Fulsås 1996).

[2]The industrial committee – New York, *Freezing* (New York: The Royal Norwegian Ministry of Provisions and Reconstruction, 1945).

This meant that the traditional lifestyle in which people worked with a combination of fishing and farming, supplying themselves with what they needed and selling the surplus, needed to be transformed. The fisher-farmer was about to become a specialized industrial worker who was either employed in the fisheries or at the processing plants along the coast. The idea was that this would increase the region's contribution to the national economy, which was also connected to the belief that the northern parts of Norway were an "outpost" towards the communist Soviet Union. In order to build loyalty, one of the strategies was to ensure that people were employed and given a proper standard of living.[3]

The situation in Finnmark and Troms right after the war was excellent for modernization. Some cities, such as Hammerfest, were completely destroyed as the Germans withdrew and used their "scorched earth" strategy. This situation of utter despair for the local people meant that the rebuilding of the destroyed cities was not something to be discussed, but a bare necessity. The government set up plans for rebuilding the northern parts of the country that included economic support from the state, with the promise that the rebuilding would also include the modernization of fisheries, farms and thus, lifestyles. The traditional self-supporting fisher-farmer was to become a tax-paying industrial worker in the new welfare state, and his standard of life was to be heightened so that he became a trusted citizen of Norway, and would not be tempted by the promises of Soviet communism that even the former Socialist Labor Party had taken a harsh stance against. As such, there can be no doubt that the establishment of freezing plants was just as much a social and political project as an industrial and economic one.

As previously mentioned, the Germans had built four large-scale centralized freezing plants along the Norwegian coast. After the war, the experts at the Directorate of Fisheries Research Department (DFRD) decided that these plants could not be used to modernize Norwegian fisheries because they were simply too large to be profitable in peace time, thus more decentralized freezing plants were needed in order to maintain a healthy demographic structure in the north. Because of this, the German plants were disassembled and the machinery used for smaller and more decentralized plants built along the coast (Andreassen 1996; Finstad 1999, 2005). The decentralized plants simplified the distribution of fish between ocean and plant since they were located at the local harbors, but were also tied to cold war politics and the government's desire to maintain a decentralized settlement in the north. As such, we can say that this was a major factor in shaping how the freezing plants were built.[4]

The building of freezing plants however was not enough to establish a modern Norwegian freezing industry in connection with the fisheries. One problem that was soon encountered was that the people in the north were not used to working in an

[3]White Paper no. 10 (1947–1948), "The National Budget 1947".

[4]The buildup was very rapid. In the period between 1945 and 1949, more than 23 freezing plants were built and more were on their way. "Letter from Gustav Lorentzen on behalf of the department for construction and machine engineering to the director of fisheries", Sept. 1, 1949. Department of Fisheries Research Department, series 43.3, box 43.3/4, folder 82, the Regional State Archive in Bergen.

industrial regime, as they were not taught how to produce products of a standardized quality, how to deal with a rigid work hour regime, etc.[5] The same also applied for the fishermen who were working on the large trawl boats that were introduced to the fisheries in order to supply the plants with raw material. These boats represented something quite new for the Norwegian fisheries. Having relied before on small motorized vessels that were owned by the fishermen, these new trawlers were large and owned and run by sales organizations and companies such as Norwegian Frozen Fish and Findus, which were created to market frozen fish (Brox 1972; Jacobsen 1996; Tande 1957; Nordstrand 2000).

Before the war, there had been discussions about introducing trawlers in Norwegian fisheries, but the government rejected this idea because they feared that trawling would empty the sea and concentrate the economic power in Norwegian fisheries. Now, however, there were several reasons for introducing trawlers. First of all, more efficient fishing vessels were needed in order to supply the freezing plants with fish. Second, the government wanted to introduce trawling in order to lower the manpower needed in the fisheries. In doing this, it would be possible to transfer manpower from the fisheries to other sectors, which was an acute need. In short, fishermen were turned into industrial workers, and those who still wanted to be fishermen were to become workers on large industrial trawlers instead of being self-employed on small smacks (Brox 1972; Eriksen 1996; Fulsås 1996).

Hence, we can see that the building of freezing plants were part of an effort to industrialize the Arctic north, which was something that not only presupposed a transformation of the fisheries, but also the entire culture of fisher-farmers (Sagdahl 1975; Saxi et al. 1988; Slagstad 2001). In sum, the freezing plants were a technology not only for the preservation of fish, but also for the introduction of industrial society into the Arctic north. By industrializing the country, the social democratic government tried to diversify what had formerly been an economy that relied on primary industries, which was supposed to ensure that everybody had a higher standard of living through increasing the available capita for social security, health care and the like. Thus, freezing technology and industrialization were means used to both restructure the northern economy and create a modern welfare state. Therefore, what we see is the close link between technological modernization and the creation of the modern Norwegian welfare state.

12.3 Cold Chains and the Integration of Production and Retail

The Arctic north was not the only place where freezing technology was used to introduce transformations. In Oslo, the early 1950s brought with it what was called the fresh fish crisis. The problem was that the distribution of fresh fish to

[5] Charles Robertson, "The demand for quality in production and distribution", *Frionorbladet* (*Frionor magazine*), no. 2 (1949): 3–5; "What demands are posed to the phase of production and distribution in order to produce high quality frozen fillets?", *Frionorbladet* (*Frionor magazine*) (June 1960): 6.

Oslo and other eastern parts of the country was unstable and often too small in volume.[6] This caught the attention of the Directorate of Fisheries (DF), with the Minister of Fisheries, Reidar Carlsen, quickly claiming that the problem could only be solved by introducing frozen fish, which would therefore ensure a more stable distribution of fish. At this point in time, no frozen fish produced at the new freezing plants had yet to be sold on the Norwegian market. Instead, it was being exported to foreign countries, and the leader of the Norwegian Fish Dealers Association, Erling Moe, did not take kindly to Carlsen's suggestion. He viewed frozen fish as inferior to fresh fish, and stated that it would be a failure to introduce this product in Norway.[7] Why did two agents who were so central to the Norwegian fish trade disagree so heavily in regard to the issue of frozen fish?

Moe's skepticism was probably grounded in memories from the war. Some of the fish produced at the German freezing plants had been distributed to Norwegian consumers, but due to distribution problems, the fish were in a terrible state. For this reason, it is understandable that Moe did not have much faith in the quality of frozen fish. The fact that Carlsen was promoting the fish was no surprise since as the minister of fisheries, he was obliged to help further the new industry he had helped to create. By promoting frozen fish as a solution to the fresh fish crisis, he could reform the frozen fish from "wartime fish" into the fish of tomorrow, while rationalizing the distribution of fish at the same time. However, as a representative of the Norwegian authorities he also had other reasons for promoting frozen fish, though this was not to be found in the effort to industrialize the fisheries, but rather in the structure of the Norwegian retail sector.

At the end of the war, Norwegian retail consisted of small specialized desk-serviced stores. Fish had traditionally been sold by a fish dealer that specialized in fish products, while other kinds of food products were sold in other stores. While the production sector increasingly relied on the industrial mass production of standardized products, the retail sector was still on a small scale and not equipped to market and sell modern products. In reality, the fish dealers were one of the poorest parts of the Norwegian retail sector. This represented a problem for the frozen fish industry because its products demanded a technical infrastructure.[8]

[6] White Paper, no. 87 (1958), "Recommendations on the rationalization of the distribution of fish on the inland market, part II"; "The fresh fish distribution in Oslo discussed at meeting tomorrow", *Aftenposten* (*Newspaper*) (Oslo, November 5, 1951); "Will the fish distribution ever work?", *Aftenposten* (Oslo, May 17, 1953); "The evil spirit of the fresh fish distribution", *Verdens Gang* (*Newspaper*) (February 27, 1948); "The fresh fish distribution is no issue for the State", *Verdens Gang* (April 22, 1948); "It is snobbish not to like frozen fish", *Verdens Gang* (July 11, 1951); "The council of exports take responsibility for the inland fresh fish distribution", *Aftenposten* (Oslo, December 15, 1948); "Fresh fish all year if the fish is frozen and rinsed at the packing site", *Aftenposten* (Oslo, May 10, 1951).

[7] "It is snobbish not to like frozen fish".

[8] Nils L.S. Jacobsen, "The inland", *Norsk Fryserinæring* (*Norwegian Freezing Industry Trade Journal*), no. 6 (July 1949): 1–2.

In order to counter this problem, the companies, Norwegian Frozen Fish and Findus, bought frozen goods counters that were given to retailers,[9] and this effort was merged with a governmental effort to rationalize the retail sector. In 1946, the Norwegian Board of Trade (BT) had established the so-called Retail Committee (RC), whose job was to investigate the possibilities for rationalizing the retail sector. Later on, Norway received aid from the US Marshall Plan, and established a so-called Productivity Office (PO) to help promote rationalization in industry and retail. Both the RC and PO agreed that the Norwegian retail sector was in dire need of rationalization, and turned to American supermarkets for suggestions about what to do. Hence, Norwegian retail was to be modernized in accordance with American supermarkets.[10]

One part of the plan for modernizing the retail sector was through integration. As the RC delivered its proposition to the BT in 1957, it suggested that the small specialized stores should be phased out in order to give way to modern supermarkets where consumers could buy all types of food products in one place. The RC explained that the equipment in the retail sector had undergone large transformations, and that the sector was being integrated into cold chains stretching from producers' freezing plants to refrigerated transportation lines with the help of refrigerators and frozen goods counters. The committee further explained that "the technical developments that we have briefly mentioned have been so vigorous over the last years that a continuation will be judged by conditions in other countries ahead of Norway, which will not only save time and labor for the consumers, but be so socially radical that it will be proper to say that they are important means in a transformative process that will transform everyday life."[11]

This shows us how the introduction of freezing technologies was seen as something that could be used to transform the Norwegian retail sector. It was the freezing technologies that laid the groundwork for new grocery stores, where the consumer could buy everything he/she needed in one place. The RC promoted the

[9] J. Trondsen, "About the home market department of Norwegian Frozen Fish Inc.", *Frionorbladet* (*Frionor Magazine*), no. 1 (1951): 32; Per W. Bistrup, *Marketing of frozen fish on the home market* (Bergen: Norwegian School of Economics, 1952); "Frozen fish from frozen goods counters", *Husmorbladet* (*Housewives Magazine*) (May 16, 1951); "Window exhibitions", *Frionorbladet* (*Frionor Magazine*), no. 1 (1952): 38; "Regular visits – distribution of advertising materials", *Frionorbladet* (*Frionor Magazine*), no. 1 (1952): 39; Freia, *Annual report and accounts 1951* (Oslo, 1952); Freia, *Annual report and accounts 1952* (Oslo, 1953); Freia, *Annual report and accounts 1953* (Oslo, 1954); Leif Throne-Holst, *Fish and fisheries in Northern-Norway* (Oslo: Dreyer publishing, 1966).

[10] White Paper, no. 87 (1958), "Recommendations on the rationalization of the distribution of fish on the inland market, part II"; F. Wedervang, *The distribution of food in U.S.A. and Norway. Report from a study trip to the U.S.A.* (Oslo, 1957); *Retail trade in the U.S.A. Report from a Norwegian study group* (Oslo: Norwegian Productivity Agency, 1953); *Modern marketing. Report from a study trip to the U.S.A.* (Oslo: Norwegian Productivity Agency, 1954); Sverre Nilsen, *Self service stores and chains in the USA. Report from a study trip to the U.S.A.* (Oslo: Norwegian Productivity Agency, 1955).

[11] White Paper, no. 87 (1958), "Recommendations on the rationalization of the distribution of fish on the inland market, part II".

frozen goods counter in self-service supermarkets as the peak of rational retail. It was this technology that demonstrated that it was both possible and necessary to integrate the fish stores into supermarkets, and thus, into the infrastructure of the frozen fish industry (Finstad 2012). Production and retail were to become integrated into a rational system, with the consequence that everyday life would also become more rational.

For now, we can say that the frozen fish industry's attempt to build a market for their products in Norway was merged with the government's effort to rationalize Norwegian retail. In the long run, this was to seriously transform the conditions that the fish dealers worked under.[12]

12.4 The Disappearance of Fish Dealers

As has been mentioned, the freezing industry handed out frozen goods counters to retailers. The strategy changed a bit, but it was either sold or rented to the stores at a price lower than the purchase prize. Thus, the industry subsidized the frozen goods counters. If we look at who received frozen goods counters from the industry, we see that it was mostly general grocery stores and not fish dealers.[13] Also, the freezing industry took part in the greater modernization of the retail sector that was initiated by the government through a company called Merchants' Credit, which gave loans to merchants who wanted to modernize their stores.[14]

Why did the freezing industry want to transform Norwegian retail? To start, we can see that the fish dealers had put themselves in a difficult position due to their skepticism towards frozen fish. However, we can understand this better if we look at the product that the industry was to sell from the new frozen goods counters. As the freezing industry started to produce its frozen fish, the product was adapted to the standards of the American market. The industry had the idea that American consumers were not used to eating fish, and that the product therefore had to appeal to consumers who did not normally eat fish. Because of this, the product needed to be free from skin and bones.[15]

---

[12] In 1959 there were more than 6,500 frozen goods counters in Norway. Only Germany, the Netherlands, Sweden and England had more in Europe at the time: *Marketing and consumption of frozen fish in OEEC countries. Report presented at a meeting of experts on deep frozen foods held in Verona from the 6th to 12th of October 1959.*

[13] Bistrup (1952).

[14] Merchants Credit Inc, *5 years in the service of development* (Oslo: Merchants Credit Inc, 1958); Merchants Credit Inc, *10 years in the service of development* (Oslo: Merchants Credit Inc, 1963).

[15] Norwegian Frozen Fish, "Can't convince the kids to eat fish?", *Husmorbladet* (*Housewives Magazine*) (April 9, 1952); Norwegian Frozen Fish, "Good dinner with FRIONOR-filets", *Husmorbladet* (*Housewives Magazine*) (September 25, 1952); Freia, "Findus, a product of Freia", *Husmorbladet* (*Housewives Magazine*) (June 12, 1953); Arne L. Davidsen, "A sketch from the fish market in Copenhagen", *Frionorbladet* (*Frionor Magazine*), no. 1 (1951): 24–27; Anders Frihagen mfl., Norwegian frozen fish to the United States (Oslo: Department of commerce, 1946).

It was also important to adapt the product's packaging to the self-service American retail stores, as it was the packaging that would sell the products within such stores. Because the industry's aim was to sell its fish to both the average fish consumer and to consumers who did not normally eat a lot of fish, it was necessary to sell it in retail stores where a lot of consumers shopped. The fish dealer was not such a place. By installing frozen goods counters in ordinary retail stores, the industry could reach a larger group of consumers than they would by selling through the fish dealer.[16] This strategy was not unique to the American and Norwegian case, as the Norwegian freezing industry had already used the same tactics in both Switzerland and the Netherlands.[17]

At the same time as the freezing industry was allying itself with ordinary grocery stores, the government promoted frozen fish in its own way. After the war, the government had introduced measures in order to control the development of food prices in Norway. One such measure was to stipulate the gross profit that a retailer could take for various products. The government used this pricing policy in such a way that those who sold frozen fish gained advantages over those who were selling fresh fish. The allowed gross profit of frozen fish was set at twice that of fresh fish, which was interpreted as another step towards decreasing the amount of fish dealers, an interpretation that the Norwegian Fish Dealers Association shared.[18]

Of course, as in Switzerland and Netherlands, this did not sit well with the fish dealers who felt their livelihood was being threatened by the alliance between the freezing industry, the government's modernizing agents and the new supermarkets. In stark contrast to his earlier claim that frozen fish fillets were of poor quality, the leader of the Norwegian Fish Dealers Association, Erling Moe, now claimed that the fish dealers were the natural vendors for such products.[19] Later on, he and his organization even tried to transform Norwegian law so that special education and expertise were needed to sell fish.[20] Not surprisingly, the effort yielded no results, and during the 1950s and 1960s the number of fish dealers and specialized stores fell (Lavik and Brusdal 1996).

It would be wrong to say that the freezing technologies entering supermarkets was the cause of the disappearance of the fish dealer in Norwegian cities. The freezing technologies and frozen fish were only part of a new retailing system that came

[16] Ibid.; Robertson (1949); "What demands are posed to the phase of production and distribution in order to produce high quality frozen fillets?", *Frionorbladet* (*Frionor Magazine*) (June 1960): 6; "Give the market what the market demands", *Frionorbladet* (*Frionor Magazine*) no. 3 (1952): 1–2; Throne-Holst (1966); Freia, *Annual report and accounts* (1951–1954).

[17] Max Schmid, "Switzerland", *Frionorbladet* (*Frionor Magazine*), no. 1 (August 1949): 4; Frionor, *Frionor 1946–1971* (Oslo, 1976); Frionor, *Pioneering for 50 years* (Oslo, 1996).

[18] "Highest price of fish in Oslo og Bærum", *Verdens Gang* (March 11, 1952); "Fish dealers break with the price directorate. Refuse to negotiate today", *Verdens Gang* (January 11, 1952).

[19] "Highest price of fish in Oslo og Bærum"; "Fewer fish dealers in Oslo than before the war" *Verdens Gang* (June 10, 1954).

[20] "Highest price of fish in Oslo og Bærum"; "Fewer fish dealers in Oslo than before the war"; "Fish dealers wants qualifying exam and authorization", *Verdens Gang* (April 25, 1955); Bistrup (1952); "Give us back the fish store", *Verdens Gang* (June 12, 1962).

to the fore during these years. Still, there can be no doubt that the freezing industry's efforts at building a infrastructure for the sale of frozen fish significantly affected the everyday life of the fish dealers by promoting supermarkets and constructing a situation in which fish dealers and supermarkets became competitors. It was not the technologies that caused this situation, although they were used as a means for facilitating this process. We can say that it was enrolled in a network of agents who wanted to modernize retail along these lines. In the long run, this put many fish dealers out of business and moved the fish trade into supermarkets, which caused the death of the fish dealer as a specialized profession in Norway.

If this process of retail modernization greatly troubled the fish dealers, it would also mean a transformation of the everyday life of the Norwegian consumer. Let us therefore investigate how household freezer chests were used to link the consumer household to the frozen food chain, thereby altering the everyday life of Norwegian consumers.

12.5 The Socioeconomics of Freezers

It was by no means obvious that the individual household freezer chest was to become the most common freezing technology used for Norwegian households. The first freezing technology available to Norwegian households was the so-called freezer locker plants, which were plants comprised of a large refrigerated room that contained small lockers available for private rental. In some areas, the municipal authorities took responsibility for the construction of collective freezer facilities. Elsewhere, the construction was organized by individuals who formed cooperative societies dedicated to building small freezer-locker units. In many small towns in Norway, the Norwegian Dairy Cooperative was the most central actor in the development of freezer-locker facilities,[21] which were being built all over Norway in years from 1945 to well into the 1960s. They were promoted by the Norwegian expertise working at the Norwegian National Nutritional Council (NNNC) and the State Research Institute of Home Economics (SRIHE).[22]

---

[21] "The dairies build freezer locker plants in joint effort", *Norsk Fryserinæring* no. 4 (April 1950): 69–73; "Gran dairy's new cooling plant with 460 freezer lockers are now running", *Norsk Fryserinæring* (*Norwegian Freezing Industry Trade Journal*), no. 1 (January 1954): 9; "Hokksund dairy inaugurate large storehouse and freezer locker plant", *Norsk Fryserinæring* (*Norwegian Freezing Industry Trade Journal*), no. 3 (March 1953): 55.

[22] "Report from a meeting in the committee of home economics, Norwegian National Nutritional Council (NNNC)", (Oct. 19, 1951), Box 2, Norwegian Institute for Consumer Research (NICR); "Report from a meeting in the committee of home economics, NNNC", (Nov. 13, 1951), Box 2, NICR; "Report from a meeting in the committee of home economics, NNNC", (Nov. 20, 1952), Box 2, NICR; Bergliot Qviller Werenskiold, "Cooling and freezing of food", *Tidsskrift for husstelllærerinner* (*Journal of home economics teachers*) no. 10 (Oct. 1942): 81–90; "Deep freezing" (Oslo: Norwegian National Broadcasting (NNB), July 7, 1954), Box 18.1: NRK etc., NICR; "The newest news about the freezing of food", *Hus og heim* (Oslo: NNB, Sept. 4, 1950), Box 18.11,

These facilities were seen as technologies that could improve the Norwegian diet by making the riches of summer and fall available during the cold winter months. To improve the Norwegian diet had been an important part of Norwegian social policy since the 1930s when the medical establishment started to integrate social questions into their medical practice, forming a kind of social medical school, and joined ranks with nutritional experts in order to improve the public health. The result was that you had state-led nutritional campaigns and programs that aimed at modernizing the Norwegian diet (Lyngø 2003; Elvebakken and Stenvoll 2008).

The locker plants were tightly linked to the nutritional policies of the 1930s. At the end of the war, Norway was still a rural country with low levels of urbanization, and during a period of sparsely stocked shop shelves, the ideal of self-sufficiency enjoyed a strong position. People went fishing and berry picking and grew their own crops, the so-called "kitchen garden movement," to obtain enough food for themselves and their families during the food shortage that lasted well into the 1950s (Avdem 2001; Avdem and Melby 1985). At the same time, the NNNC and the SRIHE pointed out that freezing could give nutritional benefits since it was the storage method with the smallest impact on the food's nutritional value.[23] Consequently, we can say that the freezer locker plants were domesticated into an already existing food culture in Norway, where self-support in regards to food was not uncommon. In short, it was a technology that was easily adapted to the customs of the old rural country.

Nonetheless, the freezer units must also be seen in connection with the economic situation in Norway at the time. Centralized freezer-locker units reduced the need for imported parts, and the building of locker plants for private use could be coordinated with the construction of facilities for the refrigeration and freezing of dairy products and so on. Moreover, the electricity supply was rather unstable during this period, and generators could be used to protect collective freezer units against the effects of the quite regular power cuts experienced during this period. Obtaining the same protection for private households would be both difficult and costly, thus collective technological solutions seemed more rational than individual solutions (Kvaal and Wale 2000).

This must be seen in light of the economic situation, while also bearing witness to a specific ideology of collective consumption. In hard times, it was important to make the most of society's resources. In that way, everybody could take part in the technological and economic progress. This was an idea that had followed the Social Democrats from their socialist past in the early twentieth century, even if they had distanced themselves quite a bit from their former socialist legacy. During the 1950s the situation changed, as industrial modernization started to have its effect, urbanization made its mark and the economy improved. More products such as frozen fish started to enter grocery stores and the rationing of foodstuffs was loosened.

NICR; *The freezing of food* (Oslo: State Research Institute in Home Economics (SRIHE), 1951); *The freezing of food* (Oslo: SRIHE, 1953).

[23] Bergliot Qviller Werenskiold, "Freezing as conservation for vegetables", (1948), Box 18.5 NICR; "The newest news about the freezing of food".

In addition, household technologies that were supposed to rationalize housework started to appear from the many industrial ventures that had been promoted, and the electricity situation greatly improved (Avdem 2001; Myrvang 2009).

The economic improvement also meant that the restrictions on the import of refrigeration machinery could be lifted in 1956, which actually marked a transformation in Norwegian economic policy. Whereas the years right after the war were marked by a strong will to regulate economic life on behalf of the politicians, the 1950s witnessed a gradual relaxation of the import restrictions and pricing policies. The Social Democrats were transforming Norway into a social democratic consumer society. In many ways, the transfer from freezer locker plants to individual freezers symbolizes the transfer from an ideology of collective consumption to an ideology in which the individual household became the central consuming unit of society (Sejersted 2011).

Thus, the introduction of individual freezer chests can be said to be linked to a transformation of the Norwegian political economy and to symbolize the coming of a consumer society. As such, there can be no doubt that the freezer chest belonged in another socioeconomic context than the freezer locker plants. Even so, the freezer chests were not only a reflection of the transformation in a larger socioeconomic context, they also became a node in the frozen food chain built by the freezing industry.

12.6 The Technology of Consumer Society

The household freezer chest that was placed within the individual household quickly became a popular technology among consumers and the household expertise at the SRIHE. The freezing industry also perceived this technology as being a useful tool, though it had run into a big problem as it started to promote frozen foods in the Norwegian market in the 1950s. The memory of the rotten frozen fish from the war made many consumers skeptical of their product. In short, consumers viewed frozen fish as a low quality food.[24] We can see that this was a problem for the freezing industry in a quote from Lillemor Erken, a famous cookbook author hired by the freezing industry to promote frozen fish: "It will be a Sisyphean task to convince housewives who still have the taste of frozen fish from the war in their mouths."[25]

The freezer chest became one of the technologies used to change the consumer's minds in regard to frozen fish. The secretary of the Norwegian Association of Freezing Plants claimed that: "Home freezing contributes more than anything to breaking the traditional resistance against a new way of preserving (food)."[26] In a report from a study trip to the US where he had examined the American cold

[24] "No hope for bigger food rations", *Verdens Gang* (June 22, 1948); "It is snobbish not to like frozen fish".

[25] "Mrs. Erken demonstrates frozen fish", *Frionorbladet* (*Frionor Magazine*), no. 3 (Sept. 1951): 38–39.

[26] Jacobsen, "The inland".

chain, the idea of home freezing as something that would eliminate resistance to the new industrially frozen food was repeated by Nils W. Pettersen-Hagh, a manager at Norwegian Frozen Fish. The general manager of Norwegian Frozen Fish, Otto Hanssen, also claimed that the lack of home freezers could explain the underdevelopment of the frozen food industry. Consumers were not "frozen food minded" because they were unacquainted with freezing as a way of preserving food. He said that the home freezer would make the advantages of frozen foods visible to consumers.[27]

Here, we have several claims made by central players in the freezing industry which illustrate that the home freezer was seen as an important technology by the industry. While it was true that the home freezer made it possible for consumers to produce their own frozen fish, this would not necessarily lead to a decrease in the sale of frozen fillets. Using home freezers and freezing food on their own would change the distrust held by consumers towards frozen foods. When consumers found that freezing worked for preserving their home produce, they would also start believing that industrially frozen products were of high quality. The home freezer was therefore thought to produce new associations within the minds of consumers that did not link frozen fillets to experiences from the war, but rather to the daily activities of the post-war kitchen.

If we are to believe the actors from the freezing industry, the freezer chest became an even more important part of the frozen food chain as the 1950s turned into the 1960s. At the National Refrigeration Congress in 1960, the leader of the national Deep Freezing Office, Rolf Kirkvaag, claimed that "Home freezing is the same for the freezing industry as the highway for the car."[28] At a meeting of the Norwegian Refrigeration Engineering Association 2 years later, he explained that the experiences from the US and Sweden had shown that an increased sale of home freezers led to an increased sale of industrially frozen foods; this again led to an increase in the sale of freezing equipment, which then led to a higher frozen food consumption. In the end, the freezer would become more a place for the storage of food from the freezing industry than from the household's own production.[29]

We can see that the freezer chests were seen as the last link in the frozen food chain and as an important link because it familiarized people with the use of artificial cold for the preservation of foods. The home freezer further represented a completion of a technical infrastructure stretching from ocean to household, and once this

---

[27] Nils W. Pettersen-Hagh, *Freezing and refrigerated storage in the USA*. (Oslo: Norwegian Association of Freezing Plants/Norwegian Productivity Agency, 1956); *Marketing and consumption of frozen fish in OEEC countries. Report presented at a meeting of experts on deep frozen foods held in Verona from the 6th to 12th of October 1959.*

[28] Rolf Kirkvaag, "Deep freezing in Norway", *Norsk Fryserinæring* (*Norwegian Freezing Industry Trade Journal*) no. 1 (Feb. 1960): 3–13. The Deep Freezing Office was an organization established by actors from the freezing industry and producers of freezing technology in order to promote the sale and consumption of frozen foods and freezing technologies. In short, it was the freezing industry's propaganda office.

[29] Rolf Kirkvaag, "The future of deep freezing", *Kjøleteknikk og Fryserinæring* (*Refrigeration and Freezing Industry Trade Journal*) no. 6 (Dec 1962): 150–154.

system was completed, the consumers would start viewing frozen foods as their natural choice over fresh or self-produced foods. The home freezer and frozen foods were symbiotic products that presupposed each other. The promotion of freezer chests was part of a strategy to create a "frozen food minded" Norwegian consumer household that was thoroughly linked to the freezing industry's frozen food chain.

At this point, the freezing industry and the household expertise at the SRIHE seem to have promoted freezer chests in the same way. While the freezing industry promoted freezers and frozen food as effective technologies designed to create more free time for overworked housewives, the household expertise promoted freezer chests as technologies that would give housewives more ease in the preparation of food, while allowing them to take work outside the home or spend more time with their children. Even if these actors did not cooperate in a direct way, they promoted freezing technologies as rationalization technologies that would be central in the creation of a new and easier way of life, which would be closely linked to larger social visions about the household of tomorrow.

12.7 Freezing Technology and Visions of Tomorrow

For both the freezing industry and the public, there was a well-developed belief that the frozen food chain would transform food distribution, as the future household would not have a kitchen. People would simply buy prepared frozen meals, and heat them in a small oven placed in a corner of their living room next to their home freezer, which was filled with frozen dishes. At the same time, they would eat from cardboard plates and use plastic cutlery. The food would be delivered directly from centralized warehouses so the retail store would in fact be superfluous, and if this did not happen, the store would be automated so that consumers just punched in their order and picked up their shopping bag. In this way, frozen foods and freezer chests would be the basis of a new food system.[30]

This kind of vision about the food system and household of the future does not seem to have been unusual. Norwegian newspapers and magazines reported from the household masses, and claimed that the household of the future would not contain kitchens, but instead small ovens and freezer chests. Food would be delivered from a central kitchen and stored in the freezer chest until preparation. The new frozen food and freezing technologies would diminish the need for food

[30] Rolf Kirkvaag, "Frozen TV-dinners in the Norway of tomorrow", *Kjøleteknikk og Fryserinæring* (*Refrigeration and Freezing Industry Trade Journal*) no. 2 (April 1962): 41–44; "The food of the future", *Norsk Fryserinæring* (*Norwegian Freezing Industry Trade Journal*) no. 11/12 (1959): 14; Rolf Kirkvaag, "Home freezers – some statistics", *Kjøleteknikk og Fryserinæring* (*Refrigeration and Freezing Industry Trade Journal*) no. 5 (Oct. 1963): 101–102; "Revolution in the retail trade. A glimpse into the store of the future", *Frionorposten* (*Frionor Magazine*) (Oct. 1965): 18–19.

preparation, thereby easing the consumer's everyday life.[31] For this reason, the food system of tomorrow would not only entail the end of drudgery in the kitchen, but also transform gender relations so that women could work outside the home without having to worry about neglecting their household duties. Moreover, these visions can be understood as envisioning a society in which rational systems would be able to free humans from labor, and be a society where leisure and consumption were the main activities of the population (Cowan 1983; Hayden 1981).

Even if these dreams of technological emancipation must be seen as Utopian, they say something about how the direction of contemporary societal development was interpreted by historical actors. These actors saw this technological development as bringing more efficient systems for production, retail and consumption, hence envisioning a future society in which technological systems freed humans from old and time consuming practices. Of course, we know that there was another side to the technological coin. Even if the mundane tasks of everyday life were delegated to technological systems, there would be no emancipation, as the population would still have been turned into consumers and thus become embedded in a consumer society, while the society of leisure was not realized because of heightened standards of life. Nevertheless, there can be no doubt that the freezing technologies were utilized for transforming everyday life.

12.8 The Cultural Politics of Artificial Cold

In this article, we have seen how the introduction of freezing technologies into Norwegian fisheries, retail and households can be understood in light of larger socioeconomic transformations. The introduction of freezing plants, frozen goods counters and freezer chests are all episodes in the creation of the frozen food chain, which not only marked the beginning of a new era in the history of food in Norway, but also in the way one thought about society. Society was to be understood as a system with interlinked parts. The historian Thomas P. Hughes has argued that in the post WW II years, we witnessed a proliferation of technological systems, but also that the system became a metaphor for understanding both culture and nature (Hughes 2004). As such, we can say that the introduction of freezing technologies and the creation of a frozen food chain not only marked the coming of a new food system, but also a new way of thinking about social organization.

The period after WW II can be called the breakthrough of technocratic planning in Norway (Slagstad 2001), as every aspect of everyday life was to be adapted to new knowledge created by scientists, planners and industry. The new society was to

[31] "Tomorrows home presented at Danish exhibition" *Aftenposten* (Oslo, Dec. 12, 1966); "The food of the future"; "When the central kitchen takes over" *Husmorbladet* (*Housewives Magazine*) (Jan. 20, 1955); "Household schools has lost their purpose", *Aftenposten* (Oslo, Jan. 29, 1963); "Frozen, prepared foods", *Aftenposten* (Oslo, Sept. 14, 1955); "Deep frozen pre-cooked dinner and deep frozen school lunch are no longer just a dream", *Aftenposten* (Oslo, Apr. 2, 1959).

be ordered, effective and well planned, and technology became one of the means through which this society could be achieved. Even the everyday life in private homes was to become part of the ideology of efficiency, and new technologies such as freezing plants, frozen goods counters and freezer chests were parts of the social machinery. What this perhaps most of all shows us is that the freezing technologies that were introduced during the period from 1940 to 1965 were part of a larger process of modernization. And not least, that this process of modernization not only brought new technological artifacts, but also furthered sociocultural transformations and ways of thinking about everyday life.

Thus, what we have here is an example of what we can call the cultural politics of technology, which shows us that new technology is never simply imported into a society and adapted to this, but that technology and society are mutually adapted. Technological integration is a two-way process in which both the new and the old are somehow adapted to each other. Technology does not determine development in a simplistic way. What we see however is that several actors thought that freezing technologies would transform everyday life and society, and that they acted on this belief. Freezing technologies were not only products of modernity, but also a means of modernization, while at the same time artificial cold was means for conservation and transformation.

Literature

Akrich, M. 1992. The description of technical objects. In *Shaping technology/building society. Studies in sociotechnical change*, ed. W.E. Bijker and J. Law, 205–224. London: MIT Press.

Andreassen, D.K. 1995. *Kjøle- og fryseteknologi fra planer til industri: Den norsk-tyske filetfabrikk på Melbu 1940–45*. [University of Bergen]. (Cooling and freezing technology from plan to industry: The Norwegian-German filet factory in Melbu).

Andreassen, D.K. 1996. Firkantet fisk: Kjøle- og fryseteknologi blir fiskeindustri. *Volund: Årbok for selskapet Norsk teknisk museum*, 55–81. Oslo: Norsk teknisk museum. (Square fish: Cooling and freezing technology becomes fishing industry).

Avdem, A.J. 2001. *Husmorparadiset*. Oslo: Det norske samlaget. (The housewives paradise).

Avdem, A.J. and K. Melby. 1985.*Oppe først og sist i seng. Husarbeid i Norge fra 1850 til i dag*. Oslo: Universitetsforlaget. (Up first and last to bed. Housework in Norway from 1850 until today).

Berker, Th., M. Hartmann, Y. Punie, and K. Ward (eds.). 2006. *Domestication of media and technology*. Maidenhead: Open University Press.

Bistrup, Per W. 1952. *Marketing of frozen fish on the home market*. Bergen: Norwegian School of Economics.

Brox, O. 1972. *Hva skjer i Nord Norge? En studie i norsk utkantspolitikk*. Oslo: Pax Forlag. (What is going on in the North of Norway? A study in Norwegian district policy).

Cowan, R.S. 1983. *More work for mother. The ironies of household technology from the open hearth to the microwave*. New York: Basic Books.

Edgerton, D. 2008. *The shock of the old. Technology and global history since 1900*. London: Profile books LTD.

Elvebakken, K. T. and D. Stenvoll (eds.). 2008. *Reisen til helseland. Propaganda i folkehelsens tjeneste*. Bergen: Fagbokforlaget. (The travel to health land. Propaganda in the service of public health).

Eriksen, Erik Oddvar. 1996. Nord-Norge som laboratorium. In *Det nye Nord-Norge. Avhengighet og modernisering i nord*, ed. E.O. Eriksen, 11–20. Bergen: Fagbokforlaget.

Finstad, B.P. 1999. Freezing technology in the Norwegian fish processing industry, 1930–1960. In *Technological change in the north Atlantic fisheries*, vol. 3, ed. P. Holm and D.J. Starkey, 89–113. Reykjavik: Esbjerg: Fiskeri- og Sjøfartsmuseet.

Finstad, B.P. 2004. The Norwegian fisheries during the German occupation: Change and continuity. In *Fish, war and politics in the North Atlantic fisheries, 1300–2003*, ed. D.J. Starkey, F.R. Loomeijer, and R. Robinson, 113–119. Den Haag: Institut voor Maritieme Historie.

Finstad, B.P. 2005. *Finotro: Statseid fiskeindustri i Finnmark og Nord-Troms: Fra plan til avvikling*. Institutt for historie, Det samfunnsvitenskapelige fakultet, Universitetet i Tromsø. (Finotro: State-owned fishing industry in Finnmark and Northern Troms: From plan to wind up).

Finstad, T. 2012. Cool alliances: Freezers, frozen fish and the shaping of industry-retail relations in Norway, 1950–1960. In *Transformations of retailing in Europe after 1945*, ed. R. Jessen and L. Langer, 195–210. London: Ashgate.

Fulsås, N. 1996. Kvifor fekk ikkje industrikapitalismen fotfeste i Nord-Noreg? Utsyn over nordnorsk økonomi frå merkantilisme til velferdsstat. In *Det nye Nord-Norge. Avhengighet og modernisering i nord*, ed. E.O. Eriksen, 39–69. Bergen: Fagbokforlaget. (Why did industrial capitalism not get a footing in northern Norway? Perspective on northern Norwegian economics from mercantilism to the welfare state).

Gjøen, H., and M. Hård. 2002. Cultural politics in action: Developing user scripts in relation to the electric vehicle. *Science, Technology and Human Values* 27(2): 262–281.

Hamilton, S. 2003. Cold capitalism. The political ecology of frozen concentrated orange juice. *Agricultural History* 77(4): 557–581.

Hayden, D. 1981. *The grand domestic revolution: A history of feminist designs for American homes, neighborhoods, and cities*. Cambridge, MA: The MIT Press.

Hughes, Th.P. 2004. *Human-built world. How to think about technology and culture*. Chicago: University of Chicago Press.

Jacobsen, A.R. 1996. *Fra Brent Jord til Klondyke. Historien om Findus i Hammerfest og norsk fiskeripolitikks elendighet*. Oslo: Universitetsforlaget. (From scorched earth to Klondyke. The history of Findus in Hammerfest and the misery of Norwegian fishery politics).

Kvaal, Stig, and Astrid Wale. 2000. *En spenningshistorie. Trondheim energiverk gjennom et århundre*. Trondheim: Trondheim Energiverk.

Latour, B. 1991. Technology is society made durable. In *A sociology of monsters. Essays on power, technology and domination*, ed. J. Law, 103–132. London: Routledge.

Lavik, R. and R. Brusdal. 1996. *Varehandelens utvikling. Kjøp ute og hjemme*. Oslo: Cappelen Akademisk Forlag. (The development of the retail trade. Purchases out and at home).

Lie, M., and K.H. Sørensen (eds.). 1996. *Making technology our own?: Domesticating technology into everyday life*. Oslo: Scandinavian University Press.

Lyngø, I.J. 2003. *Vitaminer!: Kultur og vitenskap i mellomkrigstidens kostholdspropaganda*. Dr.art., Det humanistiske fakultet, Universitetet i Oslo. (Vitamins!: Culture and science in the interwar period nutritional propaganda).

Myrvang, Christine. 2009. *Forbruksagentene. Slik vekket de kjøpelysten*. Oslo: Pax.

Nordstrand, L. 2000. *Fiskeridirektøren melder. Fiskeridirektoratet 1900–1975*. Bergen. (The director of fisheries reports. The directorate of fisheries 1900–1970).

Oldenziel, R., and K. Zachman (eds.). 2009. *Cold war kitchen. Americanization, technology and European users*. London: MIT Press.

Oudshoorn, N., and T. Pinch (eds.). 2003. *How users matter: The co-construction of users and technologies*. Cambridge, MA: MIT Press.

Robertson, Charles. 1949. The demand for quality in production and distribution. *Frionorbladet* (Frionor magazine), 2: 3–5 (See footnote no 5).

Sagdahl, B. 1975. *Trålfisket i norsk fiskeripolitikk: et konfliktskapende fiske*? Bodø: Høgskolen. (Trawl fishing in Norwegian fisheries politics: Confrontational fisheries).

Saxi, H.P., K.N. Onsager, and A. Bogen.1988. *Innovasjon og politisk motstand: Konfliktlinjer i spørsmålet om tråldrift*. Vol. 10/88. NF-rapport. Bodø: Nordlandsforskning. (Innovation and political opposition: Conflict lines in the question of trawling).
Sejersted, F. 2011. *The age of social democracy: Norway and Sweden in the twentieth century*. Princeton: Princeton University Press.
Slagstad, R. 2001. *De nasjonale strateger*. Oslo: Pax. (The national strategists).
Sørensen, K.H. 2004. Cultural politics of technology: Combining critical and constructive interventions. *Science, Technology and Human Values* 29(2): 184–190.
Tande, T. 1957. *Norsk Fiskeripolitikk. En analyse av fiskerinæringens utvikling siden 1920*. Oslo: Studieselskapet Samfunn og Næringsliv. (Norwegian fisheries politics. An analysis of the development of the fisheries since 1920).
Throne-Holst, Leif. 1966. *Fiske og industri i Nord-Norge* (Fish and fisheries in Northern-Norway). Oslo: Dreyer (See footnote no 9).
van der Vleuten, E. 2010. Feeding the peoples of Europe. Transnational food transport infrastructure in the early cold war, 1947–1960. In *Materializing Europe*, ed. A. Badenoch and A. Fickers, 148–177. London: Palgrave Macmillan.
van der Vleuten, E., and A. Kaijser (eds.). 2006. *Networking Europe. Transnational infrastructures and the shaping of Europe, 1850–2000*. Sagamore Beach: Watson Publishing.
Winner, L. 1980. Do artifacts have politics? *Daedalus* 109(11): 121–136.

Chapter 13
The Invention of Refrigerated Transport and the Development of the International Dressed Meat Trade

Jonathan Rees

> *I now feel that the time is not far distant, when the various portions of the earth will each give forth their products for the use of each and of all; that the overabundance of one country will make up for the deficiency of another; the superabundance of the year of plenty serving for the scant harvest of its successor; for cold arrests all change.*
>
> – Thomas Sutcliffe Mort, September 2, 1875 (Mort, 255, 1875).

In 1898, a terrible scandal occurred involving the beef fed to U.S. soldiers during the Spanish American War. The commander of the Army, General Nelson A. Miles charged that the military had provided condemned meat to its fighting men. Some of the beef that Miles described in testimony before Congress came from cans. Most of it had been preserved by refrigeration. According to Miles, 337 tons of this refrigerated meat was preserved by "embalming." That meant injecting it with dangerous chemicals. In response to these charges, President William McKinley appointed a commission headed by Major-General James F. Wade to investigate. He found no evidence that any of the beef fed to troops was tainted. Instead, Wade's investigation blamed the way the meat had been handled between the Chicago packing plants where it originated and the soldiers in the field. Poor sanitation, slow delivery, lack of cold storage warehouses in Cuba, lack of ice to preserve the meat en route to the battlefield; these were the causes of the problems which led to the complaints (Young 1989: 135–37).

The same was not true of American dressed beef shipments to England, which had begun over two decades earlier. Dressed meat came from animals that were killed in one place and shipped to another in pieces to save money through economies of scale in production and less waste in transportation. Even in the mid-1870s,

---

J. Rees (✉)
Colorado State University, Pueblo, CO, USA
e-mail: drjonathanrees@gmail.com

K. Gavroglu (ed.), *History of Artificial Cold, Scientific, Technological and Cultural Issues*, Boston Studies in the Philosophy and History of Science 299,
DOI 10.1007/978-94-007-7199-4_13,

American dressed beef was fresh enough and cheap enough to appeal to consumers more than beef from England itself. "Even when beef is at 10d. per pound there is not much profit in it," wrote one farmer to the *York Herald* in 1877, "but if we have to compete with beef at 6d. per pound we shall be beaten out of the market, as no farmer can afford to feed beef at that price…[I]f much of this American beef comes into our English markets I can foresee that England will no longer be a beef-producing country (York Herald 1877)." Lamb from Australia, on the other hand, did not find a niche in the British market for years after its introduction because of its inferior quality. Like that meat shipped to Cuba, the root of the Australian dressed meat industry's problem was its poor transportation and preservation infrastructure. During the early history of the dressed meat trade, logistics mattered more than the quality of the product because without a proper cold chain the quality of even the best product degraded by the time it reached consumers. The effectiveness of any country's transportation infrastructure depended upon the technology that their equipment used in order to produce cold in the first place. Ineffective machinery meant ineffective preservation both at the point of production and at every point from there to the point of consumption. The effectiveness of that machinery depended to a surprising extent upon the basic decision of what refrigerant the manufacturers of that equipment chose to use.

To this day, refrigerating engineers refer to the infrastructure that makes the transportation and storage of perishable food possible as cold chains. Specifically, cold chains are a linked set of refrigeration technologies needed to preserve and transport perishable food from the point of production to the point of consumption.[1] The most visible manifestation of the cold chain is the modern household refrigerator, but that is just its end point. Storage, transport –even the display cases in grocery stores– have to be reliably refrigerated for a cold chain to work well, and the development of all these technologies took time and enormous amount of technological innovation.

Long cold chains, the kinds that stretched between continents, had to wait for a technology that could preserve perishable foods over long journeys. That technology came with the development of mechanical refrigeration during the late nineteenth century. Even before large ice factories and cold storage warehouses became common in the United States, entrepreneurs from around the world experimented with long distance refrigerated shipping in order to transport meat. Meat was a highly-valued commodity on world markets. More importantly, a fundamental imbalance existed between the supply of cows and sheep in the places that raised them and the demand for them in the European countries that wanted to eat them. Shipping live animals by boat was not economical since too much space was taken up by parts of animals that most people would not or could not eat. Slaughtering and shipping their dressed carcasses, on the other hand, was profitable as long as a means existed to preserve those carcasses en route. That means was mechanical refrigeration.

[1] I have only seen the term "cold chain" used in refrigeration circles. Nonetheless, I think it has great value for historical analysis too.

Thanks to the development of refrigerated transport, many trade routes for dressed meat between continents existed during the late nineteenth and early twentieth centuries. This essay considers two of them in depth in order to suggest the importance of infrastructure for making a route successful. American beef quickly captured the English market during the mid-1870s. It was fresh, cheap and well-preserved because the cold chain between Chicago and London was so effective. Australian mutton, on the other hand, took years to find a market in England (and even then it was a product consumed mostly by poor people) because Australia's cold chain was so ineffective. British customers faced the same kind of trouble with Australian meat that American soldiers in the Spanish-American war did with the Chicago product. Technological decisions made by the machinery manufacturers in both countries affected the course of the dressed meat industries for decades afterward.

Where a continuous cold chain existed, like in the Chicago to London trade, dressed meat exports thrived. Where it did not, like between Chicago and Cuba or between Sydney to London, this kind of trade faced significant difficulties. The success of a country's dressed meat trade greatly impacted the success of that country's exports in perishable foods of all kinds. The advanced state of American mechanical refrigeration technology greatly contributed to their success vis-à-vis the rest of the world in exports of other perishable products since the same technology could keep other foods cold besides beef or mutton. The development of refrigerated transport for dressed meat, in turn, gave an impetus to the development of other aspects of mechanical refrigeration. While Australia has long since adopted the refrigerated transport techniques that once served the American dressed meat industry so well, America has kept its lead in mechanical refrigeration innovation ever since.

13.1 The Long Wait for Mechanical Refrigeration

Mechanical refrigeration was a goal of inventors all over the world long before it was finally perfected during the late nineteenth century because of its obvious value for preserving perishable food. The ability to preserve surplus food this way greatly expanded the market for the sale of many products and eliminated a substantial amount of food waste. The final development of commercially viable mechanical refrigeration systems, however, required many innovations during the late-eighteenth and nineteenth centuries. Jacob Perkins, an American working in London, created the first complete working refrigerating system in 1834, utilizing ether as the refrigerant. The system included all the principle parts of modern refrigeration machinery: the compressor, the condenser, the expansion valve and the evaporator. The machine could actually make small quantities of ice, but with no demand for ice in Great Britain at the time the invention attracted no attention (Frick Company 1944: 37). Professor Alexander Twining of Yale University, in a series of experiments begun in 1847, improved on this ether compression technology and made it commercially available (Twining 1857: 3). He patented his system in England in 1850 and America in 1853. A machine Twining installed in Cleveland produced 1,600 lb of ice per day. Unfortunately, since Cleveland (and, in fact, much of America) could be easily supplied with cheap natural ice, Twining's machine could not compete with the local product. The same system was used more often in Europe, particularly France between 1857 and 1864, but Twining made no money from these efforts as they were not done with his authorization (Thévenot 1979: 40; Twining 1870: 1–4).

While these inventors worked in relative obscurity, new kinds of mechanical refrigeration attracted more public attention after 1850. That year, a Florida doctor named John Gorrie patented an "improved process for the artificial production of ice" which used air as the refrigerant (United States Patent Office 1851: 1). He discovered it accidentally in the course of inventing an ancestor of modern air conditioning for his yellow fever patients when his cool air machine "iced up (Hutchins 1958: 122)." Gorrie developed his ice machine in New Orleans and Cincinnati and licensed it overseas, but his efforts failed commercially (Donaldson and Nagengast 1994: 119–20, 124). In 1869, the German engineer Franz Windhausen patented an improved cold air machine along the same lines as Gorrie. The English firm of J. & E. Hall made further improvements in 1880. This technology proved particularly useful for early experiments in refrigerated shipping, but carbon dioxide systems eventually surpassed them (*Ice and Refrigeration* 1892: 270).

Considering the relative position of the two industries at the end of the century, it's ironic that the first inventor/entrepreneurs to experiment with refrigerated transport in the entire world were Australian, not American. James Harrison built Australia's first ice plant at Rocky Point in 1854. It used ether as the refrigerant. Machines that used ether as a refrigerant were highly inefficient compared to the machines that used other refrigerants that were developed by the end of the nineteenth century. A compressor big enough to a create cold from ether needed to be 17 times greater than that of an ammonia compressor (the most efficient refrigerant available by the end of the century) (Siebel 1906: 159). Ether is also highly flammable. Finding a safer refrigerant proved absolutely essential to the development of the industry. Developing refrigerated transport was much harder than developing mechanical refrigeration itself for many reasons that are outlined below. Harrison recognized the great value that this technology held for himself and his country if he could make it mobile. While he could freeze meat and keep it fresh for months on end, his one attempt at using this technology on board ship failed in 1873 (Bruce-Wallace 2011).

In 1861, the Australian industrialist Thomas Mort and his French-immigrant engineer Eugene Nicolle built the first plant in the world designed for freezing meat (Morris 1958: 47). What separated Mort from Harrison and other failed inventors in this field was his clear vision of the future. He saw where the Australian dressed meat industry needed to go in order to exploit its natural advantages, but unfortunately he could not perfect the technology to get there. Mort had a long career in other aspects of Australian commerce, but had expressed an interest in preserving meat for shipment to England as early as 1846. In 1866, he attended a meeting convened by the pastoralist Augustus Morris. Morris had invited extraordinary ridicule by suggesting that a 100,000 lb prize be raised to reward whomever would develop a way to transport freshly-killed meat from Australia to England. Starting with this meeting, Mort's efforts to revolutionize the international dressed meat trade began in earnest. For the rest of his life, Mort showed more passion for the development of mechanical refrigeration than he did for all his other endeavors (Barnard 1961: 177–78, 194).

At a 17 July 1875 speech before the Council of the Agricultural Society of New South Wales in Sydney, Mort explained his assessment of the supply and demand for meat between Australia and Great Britain. First, Mort noted the abundance of livestock in four Australian states (New South Wales, Victoria, South Australia and Queensland) relative to the people there. In 1874, there were 1,760,855 people in these four colonies compared to 5,648,709 cattle and 47,824,299 sheep. According to Mort's calculations, which allowed for each person to consume 350 lb of meat per year, there was an excess of nearly 300,000 tons. Mort went on to suggest other problems Australia faced because of its meat surplus, overfeeding of its population and, in a particularly modern section of the speech, the destruction of forests due to the careless stewardship of impoverished shepherds and cattle ranchers. Britain, on the other hand, Mort explained, faced a terrible shortage of fresh meat for the masses. Many people there had expressed a great concern as to whether the meat supply there would be able to keep up with the growing population. As the *Times* of London explained the problem (in an article that Mort cited in his speech), "There are parts of the world adapted for maintaining vast herds of cattle and sheep, and adapted for no other purpose, and from these the more densely populated countries might be permanently supplied with sustenance," assuming, of course, somebody could develop a means to get fresh meat across the world without it spoiling (Mort, 247–49, 1875; Perren 81, 2006). Fresh was the operative word here. Australia exported canned meat to Britain as early as 1843, but the poor quality of the product hindered sales (Farrer 2005: 30). The willingness of English consumers to purchase 22 million cans of Australian meat, mostly mutton, was a result of it costing less than half that of the fresh product rather than taste (Tannahill 1988: 312). There was much room for growth in trade here if fresh meat could be shipped across the world without having to can it.

But Mort had to raise more money in order to make shipboard refrigeration technologically viable. Noting the amount of money he had spent already, the minutes of the meeting where he gave this speech summarize Mort as saying that he:

> might be disposed [to send a shipment of meat on a long voyage] at some future time if he had the means. He considered that he had spent enough money in this matter, and it would be inconvenient for him to send away a large shipment of meat which might involve a large outlay. A small shipment he would not send. It would not be out of place to mention that several gentlemen had applied to him to know in what way the objects of his work might be furthered. Messrs. Hay, Holt, Watt, Knox, and Alger had sent him letters recognising the importance of the work in hand, and asking him to communicate to them some plan by which the aid which they think ought to be given to him, should be given (cheers). He received the application with a great deal of satisfaction, because it was some little omen that people were beginning to believe in what he was doing (Mort 255).

To help underscore the feasibility of the technology, he passed out samples of frozen meat. One attendee noted that it was "just as good and fresh as when it was first killed (Holt, 259)." Mort's well-received speech and demonstration raised enough money to try an experimental shipment of lamb to London, but all the money in the world would not necessarily allow Mort to overcome the technological obstacles this venture faced.

13.2 The Experimental Shipment

By the time of Mort's speech before the Agricultural Society of New South Wales, he and Nicolle had clearly demonstrated that they could create a refrigerating system that worked on land. Putting the same system shipboard to keep those carcasses cold presented many additional problems such as the need for reliability on a months-long ocean voyage (since replacement parts would be unavailable) and the need to create machinery that could withstand the salty sea air. Another problem was the size of the system, particularly all those extra brine pipes. The more space the refrigerating equipment took up on the ship, the less space there was for meat. In fact, while the refrigerating equipment was essential for bringing chilled meat from Australia to England, on the way back to Australia it just took up space that might have gone to something else that could have been sold in order to make the entire trip profitable (*Ice and Refrigeration* 1901: 95).

Overcoming these obstacles required more money than Mort had ever expected. Therefore, in November 1875, with no meat shipment yet made, Mort's primary investors addressed a letter to him that was printed up as a flier for general distribution to all other investors and other interested parties. Mort's response made it clear that he did indeed need more money. He asked for between 20,000 and 25,000 lb from additional investors ("Shipment of Meat under the Freezing Process," Mort Papers, Reel CY2091). If Mort's investors really had that much faith in Mort and Nicolle's technology, one would think they might have offered the rest of the money Mort needed themselves. Perhaps this was the first sign that Mort knew that his plan might not work. It could not have helped calm Mort or his investors that Harrison's refrigerated transport system failed while Mort was trying to raise additional money.[2]

Finally, on 8 July 1877, the company packed its initial experimental shipment on a converted iron barge, the *Northam*, to take it to London. To install the refrigerating machinery required permanent structural changes to a vessel. The ship's owners resisted these changes being made for just one client. The owners also feared the high pressure that refrigerating machinery had to keep its refrigerant

---

[2] Harrison stayed in London where he continued to make improvements to his ether-based system. He died in 1893 without having succeeded in the goal of a single successful shipment. For more on Harrison see W.R. Lang, *James Harrison: Pioneering Genius* (Geelong, AU: James Harrison Museum, 2003).

under for fear the pipes would rupture and the vessel would sink as a result. According to Nicolle, "[O]wing to the refusal of the [ship's] agents to delay the time of sailing, the [refrigerating] apparatus was not fully complete when shipped (*Sydney Mail*, 28 September 1910 in E.D. Nicolle Papers Reel A2751)." Nicolle probably accepted this arrangement because he planned to sail on the *Northam* to London to fix whatever went wrong with the system along the way. Unfortunately for Mort, this experiment failed before it ever left Sydney.

The exact reason for this failure is unclear. The report of the Board of Directors of the Sydney Fresh Food and Ice Company to stockholders explained the situation this way: "On the 9th, the refrigerating apparatus was to all appearances working satisfactorily, having reduced the meat chamber temperature from 52(degrees) to 35(degrees) in 60 hours, but in the afternoon of the same day an accident occurred through the action of the nitrate of ammonia on the coils…which caused immediate stoppage of operation ("Report of Directors submitted to the General Meetings of the Shareholders of the New South Wales Fresh Food & Ice Co.," 30 August 1878, Mort Papers, Reel CY 2901)." Ammonia corrodes steel and therefore could have led to that result. In 1904, C.C. Lance, the commercial agent for the New South Wales government in Britain, offered another explanation. "When all was ready," he recounted, "the accidental opening of a cock – said to have been done by Nicolle himself – led to an explosion of the pipes (Lance 119)."[3] No matter what the ultimate reason for the failure of Mort's experimental shipment, the result was the same. The company removed the meat from the ship before it sailed. Ironically, Mort himself recognized the importance of safety in his speech before the Agricultural Society's Council. By accepting an incomplete installation of the machinery to please the ship's owners, Mort disregarded his own advice in his haste to please his investors. Mort would not survive to see his vision become a reality. Indeed, the failure of his experimental shipment affected his health and thereby accelerated his death from pneumonia in 1878 (Cole 1926: 132).

13.3 The Triumph of American Dressed Meat

While Mort and Harrison struggled, the demand for meat in England that Mort intended to fulfill was being met through other means. The direct shipment of American beef to England using natural ice cut from lakes and streams as the preservative began in 1875 (*New York Tribune*). Since ice took up far more space than refrigeration equipment, there was plenty of incentive for American meat exporters to experiment with new technologies. By 1878, seven other businesses had joined the transatlantic meat trade using both ice and mechanical refrigeration. By the end of 1880, 400 carcasses of frozen lamb had arrived in England from Australia, shipped by other Australian dressed meat pioneers. By comparison,

[3] This explanation is compatible with the one given to stockholders. It is easy to imagine the detail being omitted in the earlier explanation in order to help Nicolle save face.

120,000 tons of fresh beef had been imported from America (Critchell and Raymond 1912: 13). American beef captured 94 % of the British market between 1880 and 1889, its most successful period (Higgins and Gangee 2010: 209). This kind of domination depended both upon the technology that the Americans employed, as well as the infrastructure they built in order to make use of it.

The infrastructure that constituted America's dressed beef cold chain began at the point of production, inside Chicago's many slaughterhouses and storage rooms. As one refrigerating engineer later explained it, demand from meat producers and brewers to keep their products cold drove the growth of the mechanical refrigeration industry throughout the United States during the late-nineteenth century. "These industries were at that time in a very flourishing state," he wrote in 1916, "which in turn gave a great impetus to the art of mechanical refrigeration, rendering the business of building refrigerating machinery relatively much more prosperous than it is today (Doelling 1916: 159)." America's ample supply of ice refrigerated the chilled meat leaving those mechanized facilities along the routes from Chicago to the port cities of the east coast. [Nobody perfected a mechanically refrigerated railway car until the 1950s.] By the 1890s, Smithfield Market, London's main meat hub, developed cold storage facilities in order to keep the products it imported fresh while it awaited purchase by hungry British consumers.

The advanced state of its cold chain gave American meat an advantage over its competition based not just on price, but on quality. In 1877, the *New York Tribune* reported that, "The beef-loving people of Great Britain, attracted at first, it is said, by the novelty of this new American enterprise, found upon trial that the imported meat was sweeter and more juicy and in every respect superior to the beef to which they had been accustomed (*New York Tribune* 1877)." This was a function of the technology that preserved it. Since their meat was chilled rather than frozen solid it was widely acknowledged to have as much flavor as that which was freshly killed (Perren 2006: 48). English observers claimed that butchers often substituted American beef for British beef in order to gain a price premium and avoid a stigma that some consumers had against buying anything from the United States (Higgins and Gangee 2010: 210). Most domestic English beef, on the other hand, seemed scraggily by comparison. Salted or canned beef, another alternative, simply could not hold a candle to fresh meat and quickly became the exclusive province of those who could not afford the fresher kinds. The beef and mutton from Australasia or South America, which was always frozen until the years immediately preceding World War I, was also a lower quality product because of the way that it was preserved.

Freezing affects the physical structure of meat, making it less moist. It can also rupture small blood vessels in the muscle fiber, making it look less appetizing (Perren 1978: 125). *Ice and Refrigeration* described the pitfalls of trans-oceanic frozen meat transportation this way in 1901: "The chief danger is in freezing the meat too rapidly, so that a coat or wall of frozen material (virtually insulating material) is formed around the warmer interior part of the meat, driving the heat further and further toward the center. Thus gases are imprisoned, decay of bone and marrow sets in, and when the meat is thawed the interior foulness becomes evident (*Ice and Refrigeration* 1902: 99)." Frozen meat also created a problem with damage caused by melt water en route, which could leave the meat discolored. Such problems greatly hindered the development of the Australian dressed meat industry, which depended upon freezing its product because its trip to Britain was so long. Because so many shipments from Australia proved unmarketable upon reaching the Britain, Australian exporters had to pay huge insurance premiums for each voyage. This drove up their costs, which in turn eventually priced their product out of the market as they faced greater competition (Berges 1913: 328). Distance also constituted a real barrier to delivering high quality meat around the world. Various Australian firms tried five times to keep meat chilled rather than frozen on its way to England starting in 1895, only to succeed for the first time in 1909. [Even then the meat had to be sprayed with formaldehyde in order for it to keep all the way through the long voyage, which undoubtedly affected its taste in a different way (Critchell and Raymond, 248, 252)]. Shipments of meat from Australia to other parts of the world remained miniscule until the industry there was finally able to abandon freezing.

Another important reason for the poor quality of Australian dressed lamb was lack of cold storage in and around Sydney, the place from where these shipments departed. That meant the product might be damaged before its global voyage even began. "A proper chilled room to protect the meat should it not be sold on arrival is a *sine qua non*," wrote the American consul there in 1891:

> It was the want of this that was one of the principle causes of the failure of the Orange Company to establish a fresh-meat trade, for as the company had no other means of protecting the meat when not sold, they had to take the price offered by the retail butchers or allow it to stink. They were, in fact, largely at the mercy of the trade; whereas with a proper chill-room the salesman could put the meat in and wait a fortnight if necessary for better prices.

There were not even enough cold storage facilities near the abattoirs in Australia, thereby increasing the chances that the meat would be damaged before departure (U.S. Department of State 1891: 76–77). By 1896, total Australian meat exports had grown to 1,565,360 tons, but the meat that passed inspection was still of poor quality (*Ice and Cold Storage* May 1898: 5). As the London-based trade journal *Ice and Cold Storage* reported in 1898, "It is undoubtedly the case that, although the numerous steamers engaged in the Australia trade all deliver their cargo in much the same manner to the various London stores, some of these cargos are found to be "off-colour," out of shape, batten-marked, discoloured and blood-stained, and even mouldy." Out of 46 ships that arrived in London in the first half of 1897, 21 contained at least some meat that had to be condemned upon arrival (*Ice and Cold Storage* April 1898: 7).

To make matters worse, the Australian meat export industry also faced other problems not directly related to the freezing of its product. The physical separation of freezing works from slaughtering houses in Australia gave the meat more opportunity to spoil en route traveling over dry land. In New Zealand, a country where the distance between the slaughtering and the freezing was much shorter, this problem did not exist and the mutton shipped was therefore of consistently higher quality (*Ice and Cold Storage* February 1902: 51). Drought in Queensland also posed problems. In 1897, the *Australian Pastoralists' Review* cited an English market report. "[C]aused by drought," it explained, "many shipments having been so unsuited to the requirements of the market that buyers have turned their attention to other kinds of beef," the shortage of food having left Queensland cattle too skinny for the British consumer (*Australian Pastoralists' Review* 1897: 26). These other kinds of beef came from other growing markets which would curb the growth of Australian meat exports into the long run. The amount of American meat arriving in England topped that of both beef and mutton from Australia and New Zealand every year from when it first arrived in 1876, through 1901 when it started trailing off towards zero in 1914, largely because of a shortage of crops to feed cattle in the United States (*Ice and Cold Storage* 5, June 1902: 167; *Ice and Cold Storage* 6 February 1903: 33; Perren 1978: 147)."[4] Even in the shipment of lamb and mutton, Australian firms faced stiff competition from New Zealand, which shipped over 2.2 million carcasses that same year compared to a shade less than 1.6 million that same year (*Ice and Cold Storage* 1 April 1898: 5). In 1901, the increase in beef received in the United Kingdom from South America, mostly the valley of the Rio de Plata in Argentina, over the previous year was greater than the entire quantity shipped there from Australia (*Ice and Cold Storage* 5 February 1902: 38). "Australia," explains Richard Perren, "the pioneer of the frozen meat exporters, did not retain the lead given by her early start and her other two competitors forged well ahead (Perren 1978: 184)." By 1922, the industry was in crisis as demand for beef in Great Britain had cratered and meat from Argentina could significantly undersell the Australian product. "A sea voyage of 15,000 miles constitutes a real barrier," reported the United States Department of Agriculture (U.S. Department of Agriculture 1922: 61).

13.4 Different Refrigerants for Different Tasks

The technological superiority that allowed America to ship its beef chilled while Australia froze its mutton hinged on the refrigerant that each country's industry primarily used. While this distinction may seem insignificant in retrospect, such details determined the strengths and weaknesses of the entire refrigeration infrastructures. Technically, cold is the absence of heat, not a force in its own right. Therefore, the process of refrigeration or cold storage requires that a substance be used to draw heat away from the area where one wishes to lower the temperature. Substances able to perform this operation are called refrigerants. Many substances can be used as a refrigerant. When heated, a refrigerant rises steadily in temperature to its boiling point where it can exist either as a liquid or a gas. When further heat is applied, the temperature of the water remains constant and the extra energy is used

[4] Revelations about the American beef industry in Upton Sinclair's 1906 muckraking classic *The Jungle* undoubtedly didn't help either.

by the conversion of the refrigerant to a gaseous state. Only when all the liquid is converted will the temperature rise above its boiling point. Once gaseous, the vapor only releases the extra energy when condensed back into a liquid. The energy given off during condensation is always equal to that absorbed during conversion from the liquid into a gas. This constant is known as latent heat and it varies from substance to substance. Substances with the highest latent heats can absorb the most energy from surrounding areas, and hence make the most effective refrigerants.

Because the relationship between the temperature, volume and pressure of gasses is fixed, the volume and pressure variables could easily be altered to manipulate the refrigerant's temperature or potential energy to the point that it changes its state. To do this, most refrigerating machinery puts the refrigerant through three stages: compression, evaporation and expansion. In the first step of artificial cooling, the refrigerant in its gaseous state must be compressed into a smaller space. During compression, heat is generated in proportion to the amount of pressure put upon the gas. The hot, compressed gas is then forced through a condenser of coiled pipes submerged in cold water until enough heat is drawn off to liquefy the refrigerant. At this point, the now-liquefied refrigerant is drawn through an expansion valve. The expansion valve lowers the pressure to a point at which the refrigerant returns to a gaseous state. This liquid-to-gas transformation requires heat. Since the condenser absorbed the heat from the last state of change, the liquid refrigerant draws the heat from its retransformation into a gas from the surrounding area. The resulting removal of heat in the surrounding area lowers the temperature there. Owing to the expense of most refrigerants, all refrigerating machines operated as closed systems recycling the gas and liquid so that they could undergo compression and expansion once again.

"It has not taken long to give you the outline of the history of my travels in search of the agent required to accomplish the object with which I set out," Mort told the Agricultural Society, "but believe me, the road has been a weary one, for invention is a plant of slow growth (Mort 246, 1875)." Nicolle tested seven different refrigeration systems up until to the removal of the experimental shipment from the *Northam* in 1877 (Barnard 1961: 189). Reviewing Nicolle's career in refrigeration, Michael Organ and Fred Turnidge write:

> With the completion of construction of each new machine he would immediately move on to the next, continuing to update his method or even taking an entirely new direction. A system using ammonia would be dropped for one utilising compressed air, which in turn may be rejected; different metals and alloys would be experimented with for carrying coolant and maintaining pressures; and he would constantly be on the lookout for new and more efficient insulation materials (Organ and Turnidge 2009).

The system that Nicolle finally settled upon was an absorption system using a solution of ammonium nitrate as the refrigerant (Selfe 1895).[5] In his *Pioneers of Australian Industry*, Percival Cole suggests that Mort and Nicolle's freezing

[5] Not all sources agree on the exact details of the refrigerating system used for the experimental shipment. However, Selfe, an engineer and an employee of Mort's at the time of the venture, was in the best position of all writers on this subject to understand the technological details of the refrigerating machinery on the *Northam*.

system was "a device new to the scientists of Europe (Cole 1926: 130)." He was right. Indeed, today ammonium nitrate is best known as a key ingredient for fertilizers and explosives rather than as the basis of a refrigerant. To the best of my knowledge, no inventor except Nicolle has ever used an ammonium nitrate solution for refrigeration. The problem that led to the failure of the experimental system may explain why.

While an ammonium nitrate solution was a novel refrigerant, Nicolle's absorption system clearly derived from the absorption refrigerating process invented by the Frenchman Ferdinand Carré in 1859. Carré's was the first refrigerating machine to use ammonia as a refrigerant. It was the dominant system of mechanical refrigeration around the world during the 1860s and 1870s. This was in large part due to improvements in that system pioneered by Nicolle (Selfe 1896: XIX, VII). In the common absorption refrigeration system, a mixture of ammonia and water was heated until it passed through a condenser under pressure. In this manner, the ammonia was liquefied, after which it expanded in the coils of the refrigerator. During that process, it drew heat from the area to be chilled. Afterwards, the ammonia was reabsorbed into the water and the cycle began again. Over time, the ammonia absorption system would prove to be a dead branch on the evolutionary tree of mechanical refrigeration. Other systems better met the needs of companies that purchased such machinery. Nevertheless, aspects of this system like principle of temperature exchange between refrigerating agents greatly aided the development of mechanical refrigeration for meat on shipboard and for refrigeration in general (Barnard 1961: 207).

Another system that would prove to be more popular –ammonia compression refrigeration technology –was in its infancy during the 1870s. Lacking that choice, ship owners around the world turned to carbon dioxide as the refrigerant for their vessels. It would prove to be the best choice possible for many reasons. Franz Windhausen of Brunswick, Germany built the first refrigeration machine using carbon dioxide as the refrigerant in 1860. These machines subjected carbon dioxide to the same kinds of processes that ammonia experienced in ammonia compression machines. However, carbon dioxide had to be placed under much higher pressure than ammonia in order to create refrigeration. In the early years of the industry, it was difficult to solder pipe well enough to stand the strain from the gas. This meant that it took time for CO_2 compression machinery to develop into a viable alternative to the ammonia compression system. However, initial problems with carbon dioxide compression systems were worked out through various technological improvements. Because ammonia has a chemical affinity for copper, copper pipes could not be used for any kind of ammonia refrigeration installations. This was important because copper is resistant to seawater. Steel, even galvanized steel, would eventually corrode when used for refrigeration machinery on an ocean-going vessel (*Ice and Cold Storage* 4 April 1901: 99). With the technological issues surrounding carbon dioxide refrigeration solved, the efficiency of these systems matched that of machines that used ammonia as the refrigerant. After it licensed the Windhausen patent, J. and E. Hall, Ltd. of London conducted pioneering experiments with carbon dioxide compression in the late 1880s. Its improvements made this technology economically viable and the firm the largest manufacturer of refrigeration equipment in that country (Miller 1895: 71–72). These improvements spurred the spread of CO_2 compression technology. Other manufacturers, especially British manufacturers, who had advocated machines based on other refrigerants, gradually began to make their own CO_2 machines, especially for oceangoing vessels (*Ice and Cold Storage* 16 September 1913: 219). This, in turn, spurred the overseas meat trade around the world.

Two other considerations drove the use of carbon dioxide refrigeration on the high seas. First, CO_2 is not poisonous like ammonia and, second, it could not harm edible cargo. Safety was an important consideration on ship. Since leaking ammonia can explode under the right conditions, this worked to its disadvantage. Even if leaking ammonia did not explode, it could still damage equipment (like it might have done on the *Northam*) or taint food, especially something as susceptible to contamination as meat. Either of these possible accidents were potentially disastrous to a meat export venture. As a result, vessel owners flocked to carbon dioxide refrigeration systems. By 1906, 90 % of refrigerating machines installed on ocean vessels were of this type, despite some disadvantages associated with CO_2 compression machinery (Goosman 1906: 152). Even as ammonia compression systems became more popular on land, carbon dioxide systems developed a stranglehold on this niche in the refrigerating equipment industry.

In the British Empire, the transcontinental meat shipments that carbon dioxide refrigeration made possible captured the imagination of a generation of entrepreneurs, but still did not lead to the development of a complete cold chain to bring fresh meat from producer to consumer until well into the Twentieth Century. Butchers in Newcastle-Upon-Tyne, for example, "relied entirely on the natural temperature for the preservation of their stocks." In Falmouth, meat that was not sold in the limited window during which it was fresh would be salted and sold to departing ships (United States Department of State 1891: 199, 177). There was almost no retail distribution system for ice to preserve meat in the country apart from that which could be bought through fishmongers. Writing in the American context, Roger Horowitz explains, "It is difficult to turn a living thing into a meal for human beings…Developing the apparatus for killing animals and then preserving and disseminating their meat entailed massive capital investment by business organizations, the labor of hundreds of thousands of workers, and the creation of machinery to speed production and distribution (Horowitz 2006: 2)." In every meat-eating country around the world, refrigeration has played a vital role in the development of a larger cold chain. Even comparatively small meat shipments have traditionally been the gateway to a greater use of refrigeration of food of all kinds.

Australia is an excellent example of this phenomenon. Successful shipments of meat from the Antipodes launched the use of refrigeration in other areas entirely. Apples, butter and rabbits being just a few other commodities that Australia exported by the turn of the century (Lance 1904: 119–120). Refrigeration had an even greater effect on industry in Britain. "The initial application of refrigeration was devoted to the importation of frozen and chilled meat from all parts of the globe, to meet the constantly-increasing demand of our ever-growing population," explained *Ice and Cold Storage* in 1898:

> From this commencement the freezing industry has grown apace. Refrigerating machines of various kinds are now extensively used for preserving all kinds of dairy produce; for brewery purposes; fruit importation; bacon curing; India-rubber manufacture; natural-ice skating rinks; preserving fish, poultry and game; chocolate cooling; gunpowder works; smokeless powder factories; private mansions, hotels and asylums; and last, but by no means least, on mercantile ships of all nations and men-of-war of our own and other Powers (*Ice and Cold Storage* 1 July 1898: 77).

These successes, of course, were the direct precursor to the multinational, globalized food chains of today. That, in turn, has had a direct effect upon the diet of every person in the industrialized world.

References

1875. 'The Hon. THOMAS HOLT, M.L.C....'. *Journal of the Agricultural Society of New South Wales* 2: 259.

1897. *Australian Pastoralists' Review* 7, March 15.

Barnard, A. 1961. *Visions and profits: Studies in the business career of Thomas Sutcliffe Mort.* Melbourne: Melbourne University Press.

Berges, P. 1913. Methods of transporting by water meat under refrigeration. Development from 1868 to 1913. In Third International Congress of Refrigeration, *Proceedings,* vol. 3, 328–343. Chicago.

Bruce-Wallace, L.G. 2011. 'Harrison, James (1816–1893)'. *Australian Dictionary of Biography.* http://adb.anu.edu.au/biography/harrison-james-2165. Accessed 17 Oct 2011.

Cole, P. 1926. *Pioneers of Australian industry: Australian industries and their founders*, 127–132. Sydney: George B. Philip & Son.

Critchell, J.T., and J. Raymond. 1912. *A history of the frozen meat trade*. London: Constable and Company.

Doeling, L.K. 1916. Twenty-five years evolution of refrigeration. *Industrial Refrigeration* 51: 159–161.

Donaldson, B., and B. Nagengast. 1994. *Heat & cold: Mastering the great indoors*. Atlanta: American Society of Heating, Refrigerating and Air-Conditioning Engineers.

Farrer, K. 2005. *To feed a nation: A history of Australian food science and technology*. Collingswood: CSIRO Publishing.

Frick Company. 1944. *Seventy-five years of progress*. Hagerstown: Hagerstown Bookbinding and Print Company. (Reprint of the 1928 edition.)

Goosmann, J.C. 1906. *The carbonic acid industry*. Chicago: Nickerson and Collins.

Higgins, D.M., and D. Gangjee. 2010. "Trick or treat?": The misrepresentation of American beef exports in Britain during the late nineteenth century. *Enterprise and Society* 11: 203–241.

Horowitz, R. 2006. *Putting meat on the American table: Taste, technology, transformation*. Baltimore: The Johns Hopkins University Press.

Hutchins, R.E. 1958. Apalachicola: Birthplace of mechanical refrigeration. *Popular Mechanics* 110: 122–123.

Ice and Cold Storage 1. 1898. April, May, July.

Ice and Cold Storage 4. 1901. April.

Ice and Cold Storage 5. 1902. February, June.

Ice and Cold Storage 6. 1903. February.

Ice and Cold Storage 16. 1913. September.
Ice and Refrigeration 3. 1892. October.
Ice and Refrigeration 21. 1901. September.
Ice and Refrigeration 23. 1902. September.
Lance, C.C. 1904. The growth and influence of refrigeration in Australia. *Ice and Cold Storage* 7: 119.
Miller, H. 1895. *Halls of Dartford 1785–1985*. London: Hutchinson Benham.
Morris, T.N. 1958. Management and preservation of food. In *A history of technology*, vol. 5, ed. C. Singer et al., 26–52. Oxford: Clarendon Press.
Mort, T. 1875, July 17. On the preservation of food by freezing, and the bearing it will have on the pastoral and agricultural interests of Australia. *Journal of the Agricultural Society of New South Wales* 15(2): 244–260.
Mort, T. S. Thomas Mort Papers. Mitchell Library, State Library of New South Wales, Sydney, Australia.
New York Tribune. 1877. April 21, p. 5.
Nicolle, E. E.D. Nicolle Papers. Mitchell Library, State Library of New South Wales, Sydney, Australia.
Organ, M., and F. Turnidge. 2009. *E.D. Nicolle – Australian pioneer of refrigeration*. http://www.michaelorgan.org.au/nicol1.htm. Accessed 11 May 2009.
Perren, R. 1978. *The meat trade in Britain, 1840–1914*. London: Routledge & Kegan Paul.
Perren, R. 2006. *Taste, trade and technology: The development of the international meat industry since 1840*. Aldershot: Ashgate.
Selfe, N. 1895. Refrigeration in the Antipodes. *Ice and Refrigeration* 8: 93–95.
Selfe, N. 1896. *The machinery employed for artificial refrigeration and ice making*. Paper read before the Engineering Section of the Royal Society of New South Wales, Mitchell Library.
Siebel, J.E. 1906. *Compend of mechanical refrigeration and engineering*, 7th ed. Chicago: Nickerson and Collins.
Tannahill, R. 1988. *Food in history*. New York: Three Rivers Press.
Thévenot, R. 1979. *A history of refrigeration throughout the world.* Trans. J.C Fidler. Paris: International Institute of Refrigeration.
Twining, A.C. 1857. *The manufacture of ice on a commercial scale....* New Haven: Thomas J. Stafford.
Twining, A.C. 1870. *The fundamental ice-making invention'. United States house of representatives, committee on patents*. Washington, DC: H. Polkinhorn and Company.
United States Department of Agriculture. 1922. *Weather, Crops and Markets* 1 January, 21.
United States Department of State. 1891. *Refrigerators and food preservation in foreign countries*. Washington, DC: U.S. Government Printing Office.
United States Patent Office 1851. *Improved process for the mechanical artificial production of ice*. Patent # 8,080, May 6.
Young, J.H. 1989. *Pure food: Securing the federal food and drugs act of 1906*. Princeton: Princeton University Press.
York Herald. 1877. January 18

Chapter 14
'Fresher than Fresh'. Consumer Attitudes Towards the Development of the Cold Chain in Post-2WW Greece

Faidra Papanelopoulou

In studying the development and role of home appliances, either from a historical or anthropological or a sociological perspective, there emerges a kind of symmetry: what they do is, in effect, what humans can do. Appliances just do it better, they do it faster, they are more thorough, they do it more safely. Electric stoves, gas ovens, vacuum cleaners, washing machines even electric irons have their 'counterparts' in human activities and, at times, in human skilfulness. They can, in effect, be regarded as concentrated labour. There is, however, an intriguing exception to this rule: home refrigerators and to a certain extent, air conditioning. In other words, appliances producing artificial cold do not appear to be instantiations of concentrated labour. And if one is to search any kind of symmetry analogous to the appliances mentioned above, one will have to search it not with respect to human labour and skills, but with respect to natural conditions. Refrigerators '(re)produce' what happens in cold winter days or what exists in mountain tops; refrigerators are regarded as appliances which harness and control 'something' found in nature. And, thus, refrigerators and air conditioners, as opposed to all other home appliances, produce something for consumption. They produce cold and, hence, they regulate its consumption. Furthermore, refrigerators not only produce material forms of cold, that is ice cubes (there is no analogue for heat), but they also produce a quality that creates an added value to a vast number of products. The very possibility that foodstuff can be preserved in cold, gives to these products an extra value. To have consumer goods, whose rate of perishing can be diminished with the aid of a specific appliance, makes these goods more valuable.

It is these characteristics of home refrigerators that generate a number of questions concerning changes in food habits and daily life. Long-distance transportation of perishable produce, long-term preservation, new ways of packaging

F. Papanelopoulou (✉)
Department of History and Philosophy of Science, Athens University, Athens, Greece
e-mail: fpapanel@phs.uoa.gr

K. Gavroglu (ed.), *History of Artificial Cold, Scientific, Technological and Cultural Issues*, Boston Studies in the Philosophy and History of Science 299,
DOI 10.1007/978-94-007-7199-4_14, © Springer Science+Business Media Dordrecht 2014

and new approaches to management and distribution of foodstuff have all led to a reconceptualization of notions such as 'fresh' and 'freshness'.[1] It is, perhaps, not surprising that the motto 'fresher than fresh' is used in many advertisements for frozen foodstuff around the world (Freidberg 2009).[2] However, the story of home refrigerators has not been an uninterrupted success story. Bad practices in the early days of refrigeration, as well as fear among producers that refrigerated and frozen foodstuff would reduce the price of fresh products, created public hostility to the cold-storage industry that was hard to battle – even though with more contemporary standards, it is especially difficult to perceive how refrigeration had serious difficulties to gain social legitimacy (Petrick 2006). In this paper I would like to examine consumers' attitudes towards refrigeration, and in particular towards the electric domestic refrigerator that was introduced around the 1950s in Greece. Although today the electric refrigerator is the home appliance found in almost every kitchen in western societies, in the 1950s not only the electric refrigeration was considered to be a luxury product, but consumers seemed to have had doubts about its practical benefits. Even when prices came down to reasonable levels, it took a great deal of news-items, advertisements, popular science articles, and practical advice on how to use them in order to gain consumers' trust. I will argue that most of these articles, found in the daily and periodical press, cookbooks, books on nutrition, official reports on the development of the Cold Chain in Greece and some textbooks on refrigeration, came as a response to consumers' reservations about the nutritional value of refrigerated or frozen food, and their fear for food poisoning –something that did, indeed, happen to many Greeks.

14.1 The Cold Chain in Greece. A Brief Overview

Despite the steady growth of industrial refrigeration and the trade of refrigerated produce from the last quarter of the nineteenth century, domestic mechanical refrigeration became widespread only several decades later. Even in the United States, where refrigeration with ice boxes has been extensively used since the 1890s, there was adverse criticism about this new technique of food preservation at least until the 1920s (Anderson 1953: 130). By 1941 just over half of the American households

[1] The secondary literature on food and foodways is vast and is associated mostly with the fields of anthropology, geography, sociology, history, cultural studies, medicine, public health and business. For a comprehensive bibliographical essay up to 1982 see Camp (1982). For a recent collection of readings on a social and cultural analysis of food see (Watson and Caldwell 2005). For a history of commercial farming and processed food in the US see Vileisis (2007). Similarly on a comparative analysis of attitudes towards food in four European countries and the US see Fischler and Masson (2008). Note also the existence of the specialised international and interdisciplinary journal *Food and Foodways* that publishes a wide variety of articles on the history and culture of nourishment.

[2] For the use of the term 'fresh' for non-processed products that may have been harvested, stored and travelled months before they reach the hands of the consumer see also Golfrank (2005).

had an electrical refrigerator, whereas the following decade the percentage rose to 80 % (Cowan 1985: 196). In Europe, and in particular in Greece, domestic electrical refrigeration started entering the household only after the Second World War.

Between the 1920s and the 2000s there is a long story, still untold, about consumers' attitudes towards the development of the Cold Chain in Greece and the radical changes it brought to the habits of eating.[3] This section will attempt to give a first glimpse of this story, by focusing on the period between 1950 and 1970, a period characterised by the development of the Cold Chain and the gradual diffusion of domestic refrigerators in the Greek urban household.

The Cold Chain refers to a large technological system that extends from producers to consumers and includes refrigeration machinery, transportation, distribution and storing facilities, which maintain a given temperature. As such, the development of the Cold Chain is closely related to industrialisation, the development of critical infrastructures, such as networks of electricity, railway, sea transport and national roads, and it is closely linked to urbanisation. The Cold Chain is also inseparable from more general social and cultural issues that have to do with the habits of eating and house economy.

The history of the Cold Chain in Greece cannot be told without taking into consideration the overall political, economic and social context of the period. Greece, at the beginning of World War II, was under occupation by the forces of the Axis, during which the country experienced an extended food crisis and famine that killed approximately 5 % of the population (Hionidou 2006). The 3-year Civil War between 1946 and 1949, had left the country in great economic distress. Despite the existence of cold stores from 1922 and their 60-fold increase in volume capacity within the next two decades, it was only in the 1950s that the Cold Chain in Greece begun to develop in view of the country's integration to the European Economic Community in the late 1960s (Vamvakas 1975: 8).[4]

14.1.1 Frozen Fish

The fishing industry was probably one of the few sectors that was not affected considerably by the wars. Studies in the mid 1950s had shown that the Mediterranean Sea, although rich in variety, was poor in quantities of fish and could not cover the country's needs for protein-rich foods (Giannopoulos 1955;

[3] An interesting study on the emergence of 'cultural diversity' through food in Greece during the 1990s is cultural anthropologist's Yiakoumaki, V. 2006. 'Local', 'Ethnic', and 'Rural' Food: On the Emergence of 'Cultural Diversity' in Post-EU-Accession Greece. *Journal of Modern Greek Studies*, 24: 415–45.

[4] Greece became an associate member of the EEC in 1962. We are still in demand of a more detailed story of the development of the Cold Chain in Greece, the contribution of technical laboratories and experimental stations working on the preservation of perishable produce in the low temperatures, the role of foreign expertise and so on.

Kalogirou 1944; Kouvelis 1964). A viable alternative was to expand fishing activities past the Strait of Gibraltar in the Atlantic Ocean, where other European countries were already active.

The first fishing boat that sailed for the Atlantic in 1952 was *Eurydice*, whose owners came from powerful shipping families. She was the first Greek vessel equipped with a unit capable of freezing fish aboard at below 40 °C (Giannopoulos 1955). *Eurydice* and later other freezers were praised by the financial press for the 'scientific organisation' of the work taking place aboard and the 'perfect freezing of the fish, which entered the port of destination totally fresh and frozen' (Kyriazis 1957). The same company had established around 35 selling points for frozen fish, mostly in Athens, Piraeus and other urban centres, and advertised its product as 'fresher than fresh'.[5] In the eyes of people involved in the Atlantic fishing industry, business analysts, and advertisers there was no contradiction between the notion of 'fresh' and that of 'frozen'. Within a decade the Greek freezer fleet increased to around 20 ships, owned by different ship owners. Production rose from a mere 1,000 t in 1955 to 18,000 t in 1964 (Vayanos 1964: 301).[6] Although the consumption of frozen fish increased, most articles and studies on the fishing industry mentioned a certain reluctance from the side of Greek consumers towards frozen fish. This reservation was explained due to negative experiences from the importation of bad quality of frozen fish in the past, frequent cases of food poisoning,[7] but also lack of knowledge for this 'novel product' (Kyriazis 1957). Financial articles between 1955 and 1968 underlined the importance of Atlantic fishing for Greek economy and criticized the lack of interest and support from the State (Anon. 1956: 10). Complaints were expressed on behalf of Greek ship owners for the duty-free importation of frozen fish from other countries, especially Scandinavia, Turkey, and Japan, which compromised Greek activities. The trade was also slowed down by the lack of fish landing stations in the major ports of the country.[8] Complaints targeted also Greek consumers, who were still not convinced of the qualities of frozen fish. Producers were trying to persuade consumers that frozen fish was abundant, cheap and as tasty as fresh fish. An important aspect of the rhetoric of legitimation was that frozen fish was bought and consumed in the 'developed countries of Europe' and the US, and most significantly it was argued that it was an indispensable addi-

---

[5] The distribution of frozen fish in the countryside was very limited due to the difficulties of transportation and storage in freezers, and due to the fact that the network of electricity did not cover the entire country.

[6] Of course, the quantity of fish harvested by other European countries far superseded that of Greece. Denmark had reached an annual production of 450,000 t, Spain 770,000 t, the USSR around 2,500,000 t etc. (Kyriazis 1957: 6).

[7] Individual experience seems to be a very powerful factor that undermines statistical data. For an analysis of risk perception concerning food see Fischler, Claude, 'Food Selection and Risk Perception': http://www.researchgate.net/publication/228793839_Food_selection_and_risk_perception. Accessed 20 May 2013.

[8] See for example Anon. 1956. The economy of Piraeus and issues of the port. *Oikonomikos Tachidromos* (1956, December 20/1957, April 11/1958, October 23/1959, June 25/1960, February 11/1960): 10.

tion to the nutrition of the Greek population, whose diet was very poor in protein compared to the 'standards set by science' (Anon. 1960).

The cartoon is indicative of consumers' preference towards fresh fish (and sellers' attempts to conceal the fact that they sold frozen produce). The legend reads: 'He is lying to you that I was caught last night. He has me in the fridge for a month now' (Pavlidis, P. 1968. *Romantzo,* issue 1146: 28).

14.1.2 Agricultural Products

On the agricultural sector, the development of the Cold Chain was intensified in view of the country's integration in the EEC. This is expressed from the numerous articles in the daily and periodical press on the Cold Chain, and the various reports from local and foreign experts on the rationalisation of agricultural production that were published from the 1960s.[9] Because of the short distances that agricultural products had to travel within Greece, the existence of a fully operational Cold Chain for fulfilling local needs was not a priority (perhaps not even an issue) for the Greek State. More than half of the cold stores in pre- and post-war period were private. Moreover, Greece exported mostly agricultural products that did not need refrigeration, such as tobacco, olive oil, olives, fermented grapes and raisins (Mazower 1991: 396).

After the Second World War, the increase of fruit production in the Greek region of Macedonia led to the creation of 300,000 m^3 of cold stores, a great part of which were used for exports.[10] Of course, problems of coordination due to lack of experience were all but absent. Producers complained that the cold stores did not suffice for storing surplus production or that they were not close enough to the production areas. In some instances, producers from Northern Greece had to send their products in cold stores located in the Peloponnese before sending them off to Europe through the northern boarders of the country (Sklias 1958). Moreover, there were several complaints about the absence of pre-cooling, sorting and packaging stations, and about the need for extra refrigerator wagons rented from Interfrigo.[11] On the other hand, state authorities were convinced that fruit and vegetable exports were successful (Anon. 1958), and the government encouraged trading activities related to perishable products in the European Markets (Kypriadis 1963: 12–14).

Despite the relatively successful venture in the commercialisation of frozen fish, Greek farmers were more reluctant in employing freezing techniques for their harvest. The first exportation of frozen agricultural products took place in 1964, on an experimental basis and under the supervision of experts sent by the Organisation for Economic Co-operation and Development (OECD) (Vamvakas 1975: 29). The quantities exported were negligible and a significant increase was observed only

[9] See for example Chevalier, C. 1961, The creation of the Cold Chain in Greece. *Oikonomikos Tachidromos*, June 22: 8; Vayanos, J. 1963. Greece in the Common Market: The role of the developing Cold Chain in the promotion of exports in the EEC. *Oikonomikos Tachidromos*, February 7: 8; Maniatis, G., 1964. On the smooth diffusion of perishable produce in the Greek and International Market. The Cold Chain. *Oikonomikos Tachidromos,* November 26: 8; G.L. 1967. New perspectives on fruits and vegetables. Frozen. Strawberries and Vegetables. *Oikonomikos Tachidromos*, March 9: 12; Zoumpoulis, A. 1969. For the good preservation of agricultural products during transport. The Cold Chain is completed with the automatic containers. The latest international destinations, and the Greek transport policy. *Oikonomikos Tachidromos,* July 31: 23 etc.

[10] Most exports were directed to the Fruit and Vegetable market in Munich. See Anon. (1969).

[11] Interfrigo was a Belgium-based company created in 1949, which was later transformed into a consortium of various European and neighbouring countries. See also Sklias (1957).

after the mid-1970s. In 1969, the first Greek brand of frozen vegetables, *Barba Stathis*, was created, but it was only at the end of the following decade that the company actually took off. According to the official story of this company, it started its activities in adverse conditions due to consumption habits and widespread belief that frozen food was a cheap and inferior alternative to fresh produce.[12] It may also be the case that trust in frozen produce could not precede the widespread use of domestic electric refrigerators.

14.1.3 Gaining Consumer's Trust: Domestic Refrigeration

Almost every summer, several articles in the daily and periodical press referred to the 'eternal suffering of the summer months' because of the insufficient production of ice (Anon. 1951). Athenians, a journalist noticed, could not preserve any food from May to October because the 20 ice producers of the capital could not meet demand for ice. Most of the ice produced (around 45,000 ice blocks per day) was consumed for the preservation of fish, creating shortage for individual consumers (Tsibidaros 1950: 3).[13]

In the early 1950s, electric refrigerators were not an affordable solution because they were luxury products, encumbered with heavy import tariffs.[14] In 1935 with a population of about 6 million people, there were only 600 electric refrigerators throughout Greece (K., A. 1935),[15] while immediately after the war the cost of an electric refrigerator was forbidden for consumers. According to contemporary sources, in the US, an electric refrigerator cost around 200$, whereas in Greece the same model was sold at 400$ (Tsibidaros 1950). Its price was equivalent to about three monthly salaries of a University Professor. Moreover, legislation concerning imports of 'luxury' goods was so strict that even if someone received from the US an electric refrigerator as a present from a relative, s/he would have to pay in taxes 100 % of its value in addition to the tariffs paid at the port. The importers of the American refrigerators *Philco* had addressed the problem with a number of letters to Greek newspapers. In 1950, they underlined the importance of electric refrigeration for public health, and used as an example the Americans living in Greece 'who would not accept to live in our climate without a refrigerator' (Lozos 1950a, b).

---

[12] It was only in the 1980s that the same brand offered the first frozen pre-made meals in Greece. See http://www.arathra.eu/~superbrn/cms/images/stories/pdf_files/barba_stathis.pdf. Accessed 20 May 2013.

[13] Almost the same amount of ice was produced in the pre-war period from the 20 ice factories of Athens (K., A. 1935).

[14] Cartoons in general periodicals of the early 1950s associate, in a satirical way, electrical refrigerators with the 'new rich'. See for example Christodoulou, V. 1951. New rich. *Thesauros*, November 19: 15.

[15] We should of course take into consideration that these 600 refrigerators were probably installed in the urban centres since the network of electricity did not extend to the countryside.

However, in the late 1950s, there was a significant rise in the use of domestic electrical refrigeration. This rise was due to a new phase of economic growth the country had entered, and also because of the emergence of Greek manufacturers, who founded companies assembling electrical refrigerators and made them available to the Greek public at relatively reasonable prices.[16] In 1952 the Greek manufacturer of domestic appliances, *Izola*, produced the first 'Greek' electric refrigerator. Despite *Izola's* effective advertisement campaign for all electric appliances,[17] there were very few advertisements for electrical refrigerators before 1954.[18] Up to 1954 we can find several ads of iceboxes that were marketed as being as efficient as the new electric ones, inferring that the use of electric refrigerators was still not widespread (Anon. 1955).

The financial difficulty to buy electric domestic appliances is expressed in numerous articles, satirical short stories and cartoons published in the press (Epikairos 1953). The strategy followed by almost all companies was that of selling their product in numerous instalments. Instalment credit was first launched by *Izola* in 1953, and was presented as an American conception thanks to which 'all American houses' were able to buy an electric kitchen and an electric refrigerator.[19] In many advertisements, *Izola* stressed the fact that their fridges were much cheaper than the imported ones, while other companies, such as *General Electric*, tried to convince the public that their products were worth more than their actual price.[20] Particularly during Christmas holidays, most brands urged consumers to spend their 13th salary, i.e. their extra Christmas salary bonus, on an electric refrigerator.[21]

A careful study of advertisements for electric refrigerators may provide useful insights not only of marketing strategies but also of consumers' beliefs concerning the necessity of this modern but overpriced household machinery. Apart from being tools of persuasion, advertisements also reflect ideologies and lifestyles and may give us useful information about consumers' responses to new products. A telling example is *Izola's* advertising campaign on the need and usefulness of refrigerators

[16] The diffusion of electrical refrigeration in the Greek (urban) household is made evident from the rapid decrease in ice production during the 1960s. Within a period of 6 years, from 1960 to 1966, the total ice production of the country for domestic needs decreased by half. See Vamvakas (1975: 44).

[17] Some of the first advertisements of *Izola* kitchens underlined the fact that the company's products were produced by Greek workmen, and therefore consumers contributed to the well being of Greek workforce. See *Izola* Advertisement. 1953. *Embros*, November 4: 3; November 12: 5 etc.

[18] The only reference to an *Izola* refrigerator I came across was a 1952 piece of news that announced *Izola*'s offering an electric refrigerator to the King, who had visited the factory in order to see 'the progress of the Greek industry of electrical appliances' see Anon. (1952: 5).

[19] Advertisement: Electric kitchen and electric refrigerator. *Izola*. 1953. *Romantzo,* issue 518, February 3: 39, or Why be sad for not being able to buy an electric refrigerator in cash?. 1953 *Romantzo*, issue 520, February 17: 39.

[20] Advertisement: Worth more than what you pay. 1960. *Eikones*, issue 263, November 4: 2.

[21] Advertisement: During the holidays add happiness in your house with an electric fridge Kelvinator. 1960. *Eikones,* issue 269, December 16: 6.

at all seasons of the year. Trying to convince people that refrigerators are useful throughout the year wouldn't make sense, if, at the time, people were already convinced that the possession of an electrical refrigerator during wintertime was essential.[22] In a satirical column at the daily *Embros*, a journalist suggested that one should be able to buy a fridge at a better price in winter because it was an off-season product (Giannoukakis 1952). Long before having to think about the implications of refrigeration concerning the seasonality of foodstuff, Greek consumers had to come in terms with the non-seasonality of the machinery itself.[23]

Even by the late 1950s, family magazines kept publishing frequent articles that advised the Greek housewife to shop daily in order to prepare the meals of the family and not to store meat or fish for more than 1 day (Anon. 1959). Having to overcome such habits, *Izola*'s advertisers employed some basic principles of Home Economics in their advertising strategies that had to do with health, parsimony, and convenience. An *Izola* fridge ensured the family's health, spared the housewife from everyday shopping, contributed to the economy of the household (since excess food would not be thrown away), and offered the convenience of eating whatever one wanted at any time.[24] It was about making visible the invisible, based on a better understanding of the science of food which revealed that during winter food decayed without visible signs.[25] On the other hand, the benefits of an electric refrigerator for health and hygiene during the hot days of the summer were so obvious that were hardly mentioned. Emphasis was placed on comfort and convenience. An electric refrigerator promised 'a cool breath from paradise',[26] or made people wish that 'August came twice'.[27]

If one was to trust food stored for more than 1 day in a domestic refrigerator, one had to trust the machinery designed to serve this purpose. And if there was general mistrust concerning the quality of products manufactured in Greece, 'trust' had to be imported from abroad. In most advertisements, emphasis was placed on refrigerators manufactured according to American standards or with American mechanical components. For the American brands, such as *General Electric*, *American Electric*, *Frigidaire*, *Kelvinator*, *Philco* etc., a reference to their many years of experience and tradition in the refrigeration industry was just about enough.[28] Superior

[22] Advertisement: Winter? It is now that your fridge is indispensable!. 1961. *Embros*, December 17: 7.

[23] The 'seasonality' of the machine in the mind of Greek consumers of the early 1950s is also shown in cartoons depicting costumers rushing to buy a refrigerator over the hot summer days. See Heat. 1951. *Romantzo,* issue 440, August 7: 14.

[24] Advertisement: Every time you drink a glass of water, think of *Izola*. 1958. *Eikones*, issue 153, September 29: 7.

[25] Advertisement: During winter the electric refrigerator is necessary for you. 1959. *Eikones,* issue 203, September 14: 37; Advertisement: Winter! Now you need an *Izola* fridge. 1960. *Eikones,* issue 268, December 9: 15.

[26] Advertisement: So that I can enjoy in the mid of the summer the wonderful coolness of an *Izola* fridge. 1960. *Eikones,* issue 253, August 26: 65.

[27] Advertisement: August, my good month, I wished you came twice a year. 1960. *Eikones*, issue 253, August 26: 65.

[28] Advertisements: The brand that symbolises many years of superiority!. 1960. *Eikones*, issue 233, April 8: 6; 80 years of experience. 1961. *Eikones,* issue 285, April 7: 12; Defrosting? Which one

quality was almost taken for granted: *American Electric* was a 'magnificent fridge with the quality of American materials'.[29] *Kelvinator* refrigerators, although manufactured in Greece, were closely supervised by American experts.[30] Foreign brands were mostly concerned about how to present their product as suitable for the Greek household.[31] The situation was not that different for the Greek brands since the vital parts of a refrigerator, such as the compressor, the heat-exchanging pipes, the expansion valves and the thermostats, were all imported from abroad and assembled locally.[32] *Pitsos*, a Greek newcomer in the refrigeration industry, claimed service to the Greek household from 1865,[33] quality of American standards and prices adjusted to Greek reality.[34] *Izola*, on the other hand, turned its 'Greek origins' into an advantage since maintenance services could be provided at the very factory where the product was manufactured.[35]

One would assume that advertisements of refrigerators and articles about how they are used were exclusively addressed to women. Indeed articles offering practical advice concerning the preservation of food and cooking were explicitly addressed to the Greek housewife. However, a closer look at advertisements indicates that there was a differentiation in roles according to the message conveyed. When it came to using the appliance, the advertisement usually depicted a woman with an apron near the product. When the ad focused on convincing the costumer to buy the appliance, it was men who appeared near the product because they were the ones to make the 'smart buy'.[36] Similarly, male figures appeared when persuasion came in the guise of scientific advice. For instance, in an ad of *American Electric* a male doctor advised readers that 'an electric fridge is neces-

do you prefer? 1967. *Eikones,* issue 597, March 31: 11.

[29] Advertisement: A magnificent fridge with the quality of American materials. 1964. *Eikones*, issue 451, June 12: 8.

[30] Advertisement: 15.000.000 families have Kelvinator fridges. 1960. *Eikones,* issue 249, July 29: 13.

[31] Advertisements: Leonard. The most suitable fridge for every Greek house. 1959. *Embros*, July 13: 46; Exactly as you want them. The new Kelvinator fridges. 1960. *Eikones*, issue 240, May 27: iii; Welcome!...metallic compartment, cooling motor for tropical climates, automatic defrosting. 1965. *Eikones,* issue 498, May 7: 57 etc.

[32] Kerassotis. *The story of a fake legend and a big fraud. Izola. The profit-making industrial carcinoma and Drakos. What is hidden behind the glamour and advertisements* (n.p./n.d) [probably published in Athens in 1959]: 26, and oral interview with Greek manufacturer Apostolos Pitsos (4.10.2008).

[33] *Pitsos* manufactured its first fridge in 1959. However, the company dated back in 1865 with the production of small household equipment. Advertisement: New Line. 1963. *Eikones,* issue 404, July 19: 9.

[34] Advertisement: Once for ever. 1960. *Eikones,* issue 270, December 23: 73.

[35] Advertisement: Electric fridges *Izola*. 1958. *Eikones,* issue 159. November 10: 57

[36] Advertisement: Buy a Frigidaire. The best in the world. A product of General Motors Corp. 1960. *Eikones,* issue 248, July 22: 7; Advertisement: Being smart…is to buy an electric fridge Frigidaire. The best in the world. 1960. *Eikones,* issue 250, August 5: 6.

sary for the health of your family'.[37] Men also appeared in front of open, empty fridges, usually praised for their technical characteristics, while women appeared in front of open fridges full of foodstuff so as to indicate their role in preparing the meals of the family.[38]

Advertisements are not the sole indicators of the emergence of a new domestic appliance that promised to transform the traditional Greek household into a modern one. During the period under examination, the press hosted a number of popular science articles on artificial cold. Most of them referred to the origins of mechanical refrigeration and most specifically to the 'father of refrigeration', Charles Tellier, whose biography embodied the typical characteristics of an attractive science story (an ingenious inventor who died in poverty).[39] Other articles referred to the actual use of the artefact and offered specific scientific content in relation to food preservation. Two were the issues at stake: the decrease of the rate of food decay, and the preservation of the nutritional value of refrigerated or frozen food.[40] The danger of spoiled food for public health was usually attributed to lack of strict market regulations but also to individual negligence and lack of knowledge concerning the use of an electric refrigerator (D. Th. 1964). A third related category of articles that does not belong to the popular science genre is that of articles, which gave practical advice of how to use an electric refrigerator and how to handle refrigerated and frozen food. Even in these articles there were references to scientific research on nutrition.[41]

---

[37] Advertisement: Every doctor will tell you that…an electric fridge is necessary for your health. 1961. *Eikones*, issue 281, March 10: 4.

[38] Advertisement: Fridges in a new panoramic line. 1961. *Eikones*, issue 299, July 14: 56

[39] Tellier was presented as the archetype of an inventive genius, who died in poverty after his inability to secure the commercial exploitation of his inventions. Anon. 1932, Daily encyclopedia: The "father" of cold. *Athinaika Nea*, October 19: 3; Anon. 1939. Daily encyclopedia: The refrigerator. *Athinaika Nea*, November 10: 3; Anon. 1939. Daily encyclopedia: Artificial cold. *Athinaika Nea*, November 11: 3; Anon. 1954. The miracles of refrigeration. *Romantzo*, issue 592, July 5: 35; Anon. 1954. The inventor of the refrigerator. *Thesauros*, issue 821, August 22: 57; Anon. 1955. The inventor of frozen food. *Thesauros*, issue 849, March 6: 62; Anon. 1955. Frozen meat. *Thesauros*, issue 889, December 11: 30; Anon. 1962. How was the electric refrigerator invented. *Romantzo*, issue 1011, July 17: 33; Anon. 1963. How was ice discovered. The history of the invention of refrigerators. *Romantzo*, issue 1051, April 23: 24.

[40] Anon. 1952. Science for all: Vitamins of frozen [meat]. *Thesauros*, issue 705, June 1: 32; Zervos, K., 1955. Frozen meat. *Thesauros*, issue 889, December 11: 30; Anon. 1958. Fresh vegetables and fruits. How to preserve their vitamins. *Thesauros*, issue 1015, May 8: 37; Scientific news: Anon. 1960. Food in the refrigerator. *Thesauros*, issue 1143, October 20: 32; Anon. 1960. Preventions about foodstuff. *Thesauros*, issue 1144, October 27: 59.

[41] Anon. 1952. The electric fridge: how to store foodstuff and how to take care of them. *Romantzo*, issue 482, May 30: 55; Anon. 1953. Small secrets for women: Frozen meat. *Thesauros*, issue 753, May 3: 58; Anon. 1956. How to use your electric refrigerator. *Thesauros*, issue 917, June 24: 30; Anon. 1959. What do you have to know about your electric refrigerator. *Romantzo*, issue 848, June 2: 14; Anon. 1962. Frozen fish. *Thesauros*, issue 1223, May 3: 57; Anon. 1963. Do you know how to use your refrigerator? Everything housewives should know about. *Thesauros*, issue 1290,

14.2 Concluding Remarks

Let me close with a quote from a well-known journalist, writer and academician, Grigorios Xenopoulos, published in 1933:

> Here is a satanic invention that banned almost all fresh products: the refrigerator. Because of it we eat only stale food… preserved (Xenopoulos 1933).

Xenopoulos referred mostly to the products sold at the Athenian market, and complained about the loss of fresh local products. By 'local' he meant products that were cultivated in the fields around Athens that were gradually turning into roads and houses. Xenopoulos was writing at a time when refrigeration was only starting to make its appearance in Greece, and when domestic refrigeration was certainly a concept alien to large segments of the population.

Based on a wide range of printed primary sources, I have attempted to provide some preliminary and fragmented remarks concerning consumers' attitudes towards refrigeration. Despite the recovery and gradual growth of the country's economy during the 1950s, Greek society was hardly a society of affluence or a society of consumerism. Refrigeration entered the urban household through the installment credit and it took almost two decades before it was considered to be a standard household appliance of the modern Greek kitchen. Moreover, due to the short distances agricultural products had to travel within the country, the necessity to have an organized Cold Chain did not appear before the country's prospects to enter the EEC.

Especially on what concerns food production and daily patterns of food shopping, preparation and consumption, one should be particularly aware of the importance of the specific economic, social and cultural conditions of each locality but also of geographical factors which create a set of constrains as to what is produced, distributed and consumed. A certain delay in the establishment of the Cold Chain in Greece in respect to Scandinavian countries for instance, cannot be explained only in terms of economics, politics, and available technologies but should be also seen in cultural terms. In countries where temperature is often below 0 °C, refrigeration and freezing technologies seem almost a natural extension of people's ways of life. In countries of the European South, where temperatures often climb over 35 °C, artificial cold can be perceived as an unnatural intruder into the household.

News-items, popular science and practical advice articles published in the 1950s and 1960s, all reveal strategies for gaining social legitimacy for this new house appliance. Moreover, advertising messages show that strategies of legitimisation and strategies of persuasion were sometimes part of marketing strategies as well. Apart from commentaries in the Greek daily press, funny short stories and cartoons which all rendered attitudes towards refrigeration a literary *topos* of the period under examination, I chose to recover perceptions towards refrigeration through

---

August 15: 68–9; Anon. 1964. Guide for a young housewife: When she works outside the house. *Thesauros*, issue 1348, September 24: 60–1 etc.

perceptions of the machinery itself. If fruits and vegetables were still considered seasonable products, and iceboxes were used with or without ice according to the weather, what could the status of an electric refrigerator be? Advertisers had to present it as a year-round necessity, which once bought would not consume more electricity than that of an incandescent lamp. Two issues remained at stake concerning the use of artificial cold for the preservation of food: the danger of food poisoning, and the danger of malnourishment. The former was a reality because of slack market controls and food adulteration; the later was a remembrance, a residual fear from the famine during the German occupation, which is still referred to nowadays as 'the syndrome of the Occupation'. Popularization of a number of scientific and technological 'facts' came to dissipate those fears and explained how artificial cold prevented food spoilage and how the nutritional values were kept intact.

Acknowledgements This research is part of a larger project on the emergence of refrigeration technologies in Greece funded by a Marie Curie European Reintegration Grant. Earlier versions of this chapter were presented in workshops in Trondheim, Norway ('Food, Technology and Trust', March 2010) and Lisbon, Portugal (5th HoST annual workshop, September 2010). I would like to thank the participants of both workshops, as well as Kostas Gavroglu for their comments and suggestions.

References

Anderson, O.E. 1953. *Refrigeration in America. A history of a new technology and its impact.* Princeton: Princeton University Press.

Anon. 1951. Ice. *Romantzo*, issue 428, May 15: 15.

Anon. 1952. The King at *Izola*. *Embros*, November 14: 5.

Anon. 1955. The history and development of nature. Where there refrigerators in antiquity? Today the need for an electric refrigerator is even greater. *Embros*, May 28: 9.

Anon. 1956. Atlantic fishing. *Oikonomikos Tachidromos,* April 26: 10.

Anon. 1958. Sufficient refrigerator wagons for the increased fruit exports. *To Vima*, August 27: 6.

Anon. 1959. Buy fresh and healthy food. *Romantzo,* issue 844, May 5: 59.

Anon. 1960. Greek shipping: More proteins from the sea for the Greek people. *Oikonomikos Tachidromos*, April 21: 22.

Anon. 1969. The fruit and vegetable market of Munich. The biggest in Europe. *Oikonomikos Tachidromos*, October 16: 25.

Camp, C. 1982. Foodways in everyday life. *American Quarterly* 34(3): 278–89.

Cowan, R.S. 1985. *More work for mother. The ironies of household technology from the open hearth to the microwave*. New York: Basic Books.

D., Th. 1964. The terrible danger of food poisoning threatens the Athenians. *Thesauros*, issue 1345, September 3: 9.

Epikairos. 1953. Spicy news of the week: Importation of refrigerators. *Romantzo,* issue 524, March 24: 23.

Fischler, C., and E. Masson (eds.). 2008. *Manger. Français, Européens et Américains face à l'Alimentation*. Paris: Odile Jacob.

Freidberg, S. 2009. *Fresh. A perishable history*. Cambridge, MA: Belknap.

Giannopoulos, C. 1955. Addressing the problem of nutrition for the Greek population: Greek fishing and its potential. A productive sector whose importance is not fully understood. *Oikonomikos Tachidromos,* August 25: 8.

Giannoukakis. 1952. Serious and funny incidents of the day. Thinking ahead. *Embros*, November 14: 2.

Golfrank, W.L. 2005. Fresh demand: The consumption of Chilean produce in the United States. In *The cultural politics of food and eating*, ed. J. Watson and M. Caldwell, 42–53. Oxford: Blackwell Publishing.

Hionidou, V. 2006. *Famine and death in occupied Greece, 1941–1944*. Cambridge: Cambridge University Press.

K., A. 1935. We consume 40.000 columns of ice per day. Crisis in the ice factories. *Athinaika Nea*, September 3: 2.

Kalogirou, P. 1944. *After-war problems of nutrition of the Greek people*. Athens: A. Karavia.

Kouvelis, P. 1964. *Nutritional level of urban households. A survey carried out in the urban areas of Greece during 1957/8*. Athens: National Statistical Service of Greece.

Kypriadis, M. 1963. Refrigeration transportation and exports. Athens.

Kyriazis, P. 1957. A large research on our commercial shipping: Greek Atlantic fishing. *Oikonomikos Tachidromos*, September 26: 6.

Lozos, A. 1950a. Electric refrigerators. *To Vima*, May 7: 5.

Lozos, A. 1950b. About electric refrigerators. *To Vima*, July 20: 4.

Mazower, M. 1991. *Greece and the Inter-war economic crisis*. Oxford: Oxford University Press. (Greek Edition: Mazower, M. 2002. *Η Ελλάδα και η οικονομική κρίση του Μεσοπολέμου* (μτφ. Μαρκέτος, Σ.). Αθήνα: Μορφωτικό Ίδρυμα Εθνικής Τραπέζης).

Petrick, G. 2006. *The arbiters of taste: Producers, consumers and the industrialization of taste in America, 1900–1960*. PhD diss., University of Delaware.

Sklias. 1957. The first steps for the export of fruits have been taken. Encouraging potential, but also problems that need immediate solutions. *To Vima*, January 6: 5.

Sklias. 1958. The wonderful miracle of fruit production is in danger because of State negligence. *To Vima*, September 18: 1, 6.

Tsibidaros, V. 1950. The eternal suffering of the summer months. *Embros*, June 8: 3.

Vamvakas, C. 1975. *The Cold Chain in Greece*. Athens: Ministry of Agriculture.

Vayanos, J. 1964. *L'Evoluzione dell' industria dei prodotti congelati in Grecia (estratto dagli Atti del XIII Congresso Nazionale del Freddo)*. Padova.

Vileisis, A. 2007. *Kitchen literacy: How we lost knowledge of where food comes from and why we need to get it back*. Washington, DC: Island Press.

Watson, J., and M. Caldwell (eds.). 2005. *The cultural politics of food and eating*. Oxford: Blackwell Publishing.

Xenopoulos, Gr. 1933. Athenian evenings: Foodstuff. *Athinaika Nea*, May 9: 1.

Index

K. Gavroglu (ed.), *History of Artificial Cold, Scientific, Technological and Cultural Issues*, Boston Studies in the Philosophy and History of Science 299, DOI 10.1007/978-94-007-7199-4, © Springer Science+Business Media Dordrecht 2014

R

S

T

MIX
Papier aus verantwortungsvollen Quellen
Paper from responsible sources
FSC® C105338

If you have any concerns about our products,
you can contact us on
ProductSafety@springernature.com

In case Publisher is established outside the EU,
the EU authorized representative is:
Springer Nature Customer Service Center GmbH
Europaplatz 3, 69115 Heidelberg, Germany

Printed by Libri Plureos GmbH
in Hamburg, Germany